Compressible Flow with Applications to Engines, Shocks and Nozzles

Mathematics and Physics for Science and Technology

Series Editor: L.M.B.C. Campos
former Director of the Center for Aeronautical
and Space Science and Technology
Lisbon University

Volumes in the series:

TOPIC A – THEORY OF FUNCTIONS AND POTENTIAL PROBLEMS
Volume I (Book 1) – Complex Analysis with Applications to Flows and Fields
L.M.B.C. Campos
Volume II (Book 2) – Elementary Transcendentals with Applications to Solids and Fluids
L.M.B.C. Campos
Volume III (Book 3) – Generalized Calculus with Applications to Matter and Forces
L.M.B.C. Campos

TOPIC B – BOUNDARY AND INITIAL-VALUE PROBLEMS
Volume IV – Ordinary Differential Equations with Applications to Trajectories and Oscillations
L.M.B.C. Campos
Book 4 – Linear Differential Equations and Oscillators
L.M.B.C. Campos
Book 5 – Non-Linear Differential Equations and Dynamical Systems
L.M.B.C. Campos
Book 6 – Higher-Order Differential Equations and Elasticity
L.M.B.C. Campos
Book 7 – Simultaneous Differential Equations and Multi-Dimensional Vibrations
L.M.B.C. Campos
Book 8 – Singular Differential Equations and Special Functions
L.M.B.C. Campos
Book 9 – Classification and Examples of Differential Equations and their Applications
L.M.B.C. Campos
Volume V – Partial Differential Equations with Applications to Waves and Diffusion
L.M.B.C. Campos & L.A.R. Vilela
Book 10 – Vector Fields with Applications to Thermodynamics and Irreversibility
L.M.B.C. Campos & L.A.R. Vilela
Book 11 – Compressible Flow with Applications to Engines, Shocks and Nozzles
L.M.B.C. Campos & L.A.R. Vilela

For more information about this series, please visit: http://www.crcpress.com/
Mathematics-and-Physics-for-Science-and-Technology/book-series/CRCMATPHYSCI

Compressible Flow with Applications to Engines, Shocks and Nozzles

Mathematics and Physics for Science and Technology

Luis Manuel Braga da Costa Campos

and

Luís António Raio Vilela

CRC Press
Taylor & Francis Group
Boca Raton London New York

CRC Press is an imprint of the
Taylor & Francis Group, an **informa** business

First edition published 2023
by CRC Press
6000 Broken Sound Parkway NW, Suite 300, Boca Raton, FL 33487-2742

and by CRC Press
2 Park Square, Milton Park, Abingdon, Oxon, OX14 4RN

Library of Congress Cataloging-in-Publication Data
Names: Campos, Luis Manuel Braga da Costa, author. | Vilela, Luís António
Raio, author.
Title: Partial differentials with applications to thermodynamics and compressible flow / Luis Manuel Braga da Costa Campos and Luís António Raio Vilela.
Description: First edition. | Boca Raton, FL : CRC Press, 2021. | Series:
Mathematics and physics for science and technology | Includes
bibliographical references and index. | Summary: "This book is part of the series "Mathematics and Physics Applied to Science and Technology." It combines rigorous mathematics with general physical principles to model practical engineering systems with a detailed derivation and interpretation of results"-- Provided by publisher.
Identifiers: LCCN 2021005670 (print) | LCCN 2021005671 (ebook) | ISBN 9781032029870 (v. 1 ; hardback) | ISBN 9781032030838 (v. 1 ; paperback) | ISBN 9781032030838 (v. 2 ; hardback) | ISBN 9781032030814 (v. 2 ; paperback) | ISBN 9781003186595 (v. 1 ; ebook) | ISBN 9781003186588 (v. 2 ; ebook)
Subjects: LCSH: Thermodynamics. | Fluid dynamics. | Gases,
Compressed--Mathematical models. | Differential equations, Partial.
Classification: LCC QC168 .C29 2021 (print) | LCC QC168 (ebook) | DDC
515/.353--dc23
LC record available at https://lccn.loc.gov/2021005670
LC ebook record available at https://lccn.loc.gov/2021005671

ISBN: 978-1-032-02989-4 (hbk)
ISBN: 978-1-032-03081-4 (pbk)
ISBN: 978-1-003-18658-8 (ebk)

DOI: 10.1201/9781003186588

Typeset in Times
by SPi Technologies India Pvt Ltd (Straive)

to Lev Davidovich Landau

Contents

Notes, Tables, and Diagrams

NOTES

TABLES

DIAGRAMS

Preface

The preface to the series "Mathematics and Physics Applied to Science and Technology" can be found in book 10 and is not repeated in this book 11 of the same series. Likewise the preface to volume V of the series on "Partial Differential Equations with Applications to Waves and Diffusion" is found in book 10, which is the first book of volume V, and is concerned with "Vector Fields with Application to Thermodynamics and Irreversibility". Among the major applications of reversible and irreversible thermodynamics is "Compressible Flow with Applications to Engines, Shocks and Nozzles", which is book 11 of the series and is the second book of volume V. Book 10 consists of chapter 1 of volume V plus sections 2.1 to 2.4 of chapter 2, covering reversible (irreversible) thermodynamics in section 2.1 (2.3) with application to the constitutive (diffusive) properties of matter in section 2.2 (2.4). Book 11 of the series and second book of volume V completes chapter 2, starting with the basic thermodynamics with section 2.5 on the equation of state and applications to the Carnot, Atkinson and Stirling cycles of engines, refrigerators and heat pumps. In the case of jet propulsion, including the Barber-Brayton cycle, compressible flow must be considered. The simplest compressible flow is adiabatic (section 2.6) but it is limited to moderate supersonic speeds beyond which heat exchanges are inevitable. One important case of irreversible compressible flow is shock waves, either normal (section 2.7) or oblique (section 2.8). Adiabatic flow (section 2.6) and shock waves (sections 2.7-2.8) can coexist in exhaust nozzles of jet and rocket engines (section 2.9) operating in the atmosphere of the earth or vacuum of space. The final notes consider the general form of the equations of compressible viscous two-phase flows allowing for triple dissipation by viscous stresses, heat conduction and mass diffusion.

ORGANIZATION AND PRESENTATION OF THE SUBJECT MATTER

Volume V (Partial Differential Equation with Applications to Waves and Diffusion) is organized like the preceding four volumes of the series Mathematics and Physics applied to Science and Technology: (volume IV) *Ordinary Differential Equations with Applications to Trajectories and Oscillations*; (volume III) *Generalized Calculus with Applications to Matter and Forces*; (volume II) *Elementary Transcendentals with Applications to Solids and Fluids*; and (volume I) *Complex Analysis with Applications to Flows and Fields*. Volume V consists of ten chapters: (i) the odd-numbered chapters present mathematical developments; (ii) the even-numbered chapters contain physical and engineering applications; (iii) the last chapter is a set of 20 detailed examples of (i) and (ii). The chapters are divided into sections and subsections, for example, chapter 1, section 1.1, and subsection 1.1.1. The formulas are numbered by chapters in curved brackets, for example (1.2) is equation 2 of chapter 1. When referring to volume I the symbol I is inserted at the beginning, for example: (i) chapter I.36, section I.36.1, subsection I.36.1.2; (ii) equation (I.36.33a). The final part of each chapter includes: (i) a conclusion referring to the figures as a kind of visual summary; (ii) the notes, lists, tables, diagrams, and classifications as additional support. The latter (ii) appear at the end of each chapter, and are numbered within the chapter (for example, diagram-D2.1, note-N1.1, table-T2.1); if there is more than one diagram, note or table, they are numbered sequentially (for example, notes-N2.1 to N2.20). The chapter starts with an introductory preview, and related topics may be mentioned in the notes at the end. The lists of mathematical symbols and physical quantities appear before the main text, and the index of subjects and bibliography are found at the end of the book.

About the Authors

L.M.B.C. Campos was born on March 28, 1950, in Lisbon, Portugal. He graduated in 1972 as a mechanical engineer from the Instituto Superior Técnico (IST) of Lisbon Technical University. The tutorials as a student (1970) were followed by a career at the same institution (IST) through all levels: assistant (1972), assistant with tenure (1974), assistant professor (1978), associate professor (1982), chair of Applied Mathematics and Mechanics (1985). He has served as the coordinator of undergraduate and postgraduate degrees in Aerospace Engineering since the creation of the programs in 1991. He was the coordinator of the Scientific Area of Applied and Aerospace Mechanics in the Department of Mechanical Engineering and also the director (and founder) of the Center for Aeronautical and Space Science and Technology until retirement in 2020.

In 1977, Campos received his doctorate on "waves in fluids" from the Engineering Department of Cambridge University, England. Afterwards, he received a Senior Rouse Ball Scholarship for research at Trinity College, while on leave from IST. In 1984, his first sabbatical was as a Senior Visitor at the Department of Applied Mathematics and Theoretical Physics of Cambridge University, England. In 1991, he spent a second sabbatical as an Alexander von Humboldt scholar at the Max-Planck Institüt fur Aeronomie in Katlenburg-Lindau, Germany. Further sabbaticals abroad were excluded by major commitments at the home institution. The latter were always compatible with extensive professional travel related to participation in scientific meetings, individual or national representation in international institutions, and collaborative research projects.

Campos received the von Karman medal from the Advisory Group for Aerospace Research and Development (AGARD) and Research and Technology Organization (RTO). Participation in AGARD/RTO included serving as a vice-chairman of the System Concepts and Integration Panel, and chairman of the Flight Mechanics Panel and of the Flight Vehicle Integration Panel. He was also a member of the Flight Test Techniques Working Group. With AGARD support, he was involved in the creation of an independent flight test capability, active in Portugal during the last 30 years, which has been used in national and international projects, including Eurocontrol and the European Space Agency. The participation in the European Space Agency (ESA) has afforded Campos the opportunity to serve on various program boards at the levels of national representative and Council of Ministers.

His participation in activities sponsored by the European Union (EU) has included: (i) 27 research projects with industry, research, and academic institutions; (ii) membership of various Committees, including Vice-Chairman of the Aeronautical Science and Technology Advisory Committee; (iii) participation on the Space Advisory Panel on the future role of EU in space. Campos has been a member of the Space Science Committee of the European Science Foundation, which liaises with the Space Science Board of the National Science Foundation of the United States. He has been a member of the Committee for Peaceful Uses of Outer Space (COPUOS) of the United Nations. He has served as a consultant and advisor on behalf of these organizations and other institution. His participation in professional societies includes member and vice-chairman of the Portuguese Academy of Engineering, fellow of the Royal Aeronautical Society, Royal Astronomical Society and Cambridge Philosophical Society, associate fellow of the American Institute of Aeronautics and Astronautics, and founding and life member of the European Astronomical Society.

Campos has published and worked on numerous books and articles. His publications include 15 books as a single author, one as an editor, and one as a co-editor. He has published 166 papers (82 as the single author, including 12 reviews) in 60 journals, and 262 communications to symposia. He has served as reviewer for 40 different journals, in addition to 28 reviews published in *Mathematics Reviews*. He is or has been member of the editorial boards of several journals, including *Progress in Aerospace Sciences*, *International Journal of Aeroacoustics*, *International Journal of Sound and Vibration*, and *Air & Space Europe*.

Campos' areas of research focus on four topics: acoustics, magnetohydrodynamics, special functions, and flight dynamics. His work on acoustics has concerned the generation, propagation, and refraction of sound in flows with mostly aeronautical applications. His work on magnetohydrodynamics has concerned magneto-acoustic-gravity-inertial waves in solar-terrestrial and stellar physics. His developments on special functions have used differintegration operators, generalizing the ordinary derivative and primitive to complex order; they have led to the introduction of new special functions. His work on flight dynamics has concerned aircraft and rockets, including trajectory optimization, performance, stability, control and atmospheric disturbances.

Campos' professional activities on the technical side are balanced by other cultural and humanistic interests. Complementary non-technical interests include classical music (mostly orchestral and choral), plastic arts (painting, sculpture, architecture), social sciences (psychology and biography), history (classical, renaissance and overseas expansion) and technology (automotive, photo, audio). Campos is listed in various biographical publications, including *Who's Who in the World* since 1986, *Who's Who in Science and Technology* since 1994, and *Who's Who* in America since 2011.

L.A.R. Vilela was born on March 11, 1994, in Vila Real, Portugal. He was in his school's honours board for best academic results every year from the 5th to the 10th grade. In 2012, he started taking the first year of the program of study for the Informatics Engineering bachelor's degree at Universidade de Trás-os-Montes e Alto-Douro (UTAD), but changed, in 2013, to the Integrated Master's degree in Aerospace Engineering at Instituto Superior Técnico (IST) of Lisbon University, and is currently completing it.

Besides physical sciences, Vilela has a special interest in philosophy, psychology, anthropology, world politics, and, in general, an advanced understanding of the human mind.

Acknowledgements

The fifth volume of the series justifies renewing some of the acknowledgments made in the first four volumes. Thanks are due to those who contributed more directly to the final form of this book: L. Sousa for help with manuscripts; Mr. J. Coelho for all the drawings; and at last, but not least, to my wife, my companion in preparing this work.

Physical Quantities

The mathematical symbols used in book 11 are the same as in book 10 and most are not repeated. The physical symbols below refer to the present book 11. The location of first appearance is indicated, for example, "2.7" means *section 2.7*; "6.8.4" means *subsection 6.8.4*; "N.8.8" means *note 8.8*; and "E10.13.1" means *example 10.13.1*.

1 – SUFFIXES

c	Carnot cycle
e	engine
n	normal component
p	heat pump
r	refrigerator
s	Stirling cycle
t	transversal, tangential
v	viscous
w	water
0	stagnation (at rest)
$*$	critical (velocity equal to sound speed)

2 – SMALL ARABIC LETTERS

$\vec{a}$	acceleration: 2.6.2.
c_s	adiabatic sound speed: 2.6.3
c_t	isothermal sound speed: N2.8
c_V	specific heat at constant volume per unit mass: 2.5.6
c_p	specific heat at constant pressure per unit mass: 2.5.6
f	number of degrees of a molecule: 2.5.4
$\vec{g}$	acceleration of gravity: 2.6.29
h	enthalpy per unit mass: 2.6.4
h_0	stagnation enthalpy: 2.6.4
h^-/h^+	enthalpy upstream/downstream of a shock wave: 2.7.3
j	mass flux: 2.7.2
k	thermal conductivity: N2.4
	Boltzmann's constant: 2.5.2
m	mass: 2.5.4
$\dot{m}$	mass flow rate: 2.6.30
n	polytropic exponent: 2.6.13;
p	pressure: 2.5.1
p_0/p_*	stagnation/critical pressure: 2.6.13
p^-/p^+	pressure upstream/downstream of a shock wave: 2.7.2
s	entropy per unit mass: 2.6.3

s^-/s^+ entropy upstream/downstream of a shock wave: 2.7.6.
t time: 2.6.1
u internal energy per unit mass: 2.6.3
u_0 stagnation internal energy: 2.6.4
u^-/u^+ internal energy upstream/downstream of a shock wave: 2.7.5
$\vec{v}$ velocity vector: 2.6.1
v_n velocity normal to a surface: 2.7.1
v_t velocity tangent to a surface: N2.16
x Cartesian coordinate: N2.16
y Cartesian coordinate: N2.7
z altitude: 2.9.16

3 – CAPITAL ARABIC LETTERS

A Avogadro number: 2.5.4
C_V specific heat at constant volume: 2.5.2
C_p specific heat at constant pressure: 2.5.2
E energy density: 2.6.4
F free energy: 2.2.3
           thrust: 2.6.32
G free enthalpy: 2.5.3
$\vec{G}$ energy flux: 2.6.4
$\vec{G}^c$ convective energy flux: 2.6.4
$\vec{G}^d$ dissipative energy flux: N2.6
$\vec{G}^q$ heat flux: N1.3
$\vec{G}^m$ energy flux due to mass diffusion: N2.13
$\vec{G}^v$ energy flux due to viscosity: N2.12
$\vec{G}^{qm}$ energy flux due to combined heat and mass fluxes: N2.12
H enthalpy: 2.2.3
$\vec{I}$ diffusive mass flux: 2.4.18
K_p coefficient of thermal expansion: 2.5.6
K_T coefficient of thermal compression: 2.5.6
M Mach number: 2.6.13
M_0 stagnation Mach number: 2.6.13
M_* critical Mach number: 2.6.13
M^-/M^+ Mach number upstream/downstream of a shock wave: 2.7.8
N mole number: 2.5.4
R ideal gas constant: 2.5.2
$\bar{R}$ ideal gas constant per mole: 2.5.4
$\bar{\bar{R}}$ ideal gas constant per unit mass: 2.5.4
S entropy: 2.5.2
S_V non-adiabatic volume coefficient: 2.5.6
S_p non-adiabatic pressure coefficient: 2.5.6
$\dot{S}_{qm}$ rate of entropy production by heat conduction and mass diffusion: N2.14
T temperature: 2.5.2

T_0/T_* stagnation/critical temperature: 2.6.13
T^-/T^+ temperature upstream/downstream of a shock wave: 2.7.7
U internal energy: 2.5.2
V volume: 2.5.1
W work: 2.5.1
$\dot{W}$ power or activity: 2.6.20

4 – SMALL GREEK LETTERS

α thermal-mass diffusion coupling coefficient: N2.13
β thermal-mass diffusion coupling coefficient: N.2.13
δ_j^i identity matrix: N2.4
γ adiabatic exponent: 2.5.3
η shear viscosity: N2.4
η_e efficiency of engine: 2.5.20
η_p efficiency of heat pump: 2.5.21
η_r efficiency of refrigerator: 2.5.22
φ angle of deflection of the velocity across a shock wave: 2.7.4
ψ mass diffusion factor: N2.17
$\bar{v}$ relative affinity: N2.12
ρ mass density per unit volume: 2.5.4
ρ_0/ρ_* stagnation/critical mass density: 2.6.13
ρ^-/ρ^+ mass density upstream/downstream of a shock wave: 2.7.2
τ_{ij} viscous stress tensor: N2.4
ϑ specific volume: 2.5.4
ϑ^-/ϑ^+ specific volume upstream/downstream of a shock wave: 2.7.5.
$\bar{\varpi}$ vorticity: 2.6.6
ξ convected coordinate: N1.7
     mass fraction: N2.10
ζ bulk viscosity: N2.9
$\bar{\zeta}$ total viscosity: N2.16
$\bar{\bar{\zeta}}$ modified total viscosity: N2.16
χ_m mass diffusivity N2.13

2 Thermodynamics and Irreversibility

2.5 EQUATION OF STATE AND THERMODYNAMIC CYCLES

The work and heat are not functions of state, for example, the work performed between the same initial and final state depends on the thermodynamic process (subsection 2.5.1). The thermodynamic properties of a basic thermodynamic system are completely determined by specifying an equation of state relating the pressure, volume and temperature, for example, for: (i) for an ideal gas, whose specific heats depend only on temperature (subsection 2.5.2); (ii) in the particular case of a perfect gas, for which the specific heats are constant (subsection 2.5.4); (iii) in the less simple case of condensed matter, like a liquid (subsection 2.5.8). The equation of state allows in principle explicit calculation of all 12 thermodynamic coefficients (derivatives) in Table 2.4 (2.5), for example, for: (i) an ideal gas (subsection 2.5.3); (ii) a perfect gas (subsections 2.5.5–2.5.7); (iii) condensed matter (subsection 2.5.9). The four most important thermodynamic processes can be detailed for a particular equation of state (for example, a perfect gas), namely: (i) adiabatic [subsection 2.3.4 (2.5.10)]; (ii) isochoric [subsection 2.3.5 (2.5.11)]; (iii) isobaric [subsection 2.3.6 (2.5.12)]; (iv) isothermal [subsection 2.3.7 (2.5.13)]. All four are particular or limiting cases of a polytropic process (subsection 2.5.14). Several thermodynamic processes may be combined in a thermodynamic cycle which brings the thermodynamic system back to the initial condition, for example, the Carnot and Atkinson (Stirling) cycles [subsections 2.5.15–2.5.25 (2.5.26–2.5.29)] that is open (closed) in the sense it uses (does not use) atmospheric air. The thermodynamic cycles have applications in: (i) engines that use a combustion process to generate heat that is converted into work to propel a vehicle, like a car, airplane, train, surface ship or submarine (subsections 2.5.15–2.5.20); (ii/iii) refrigerators (heat pumps) that consume work [subsection 2.5.21–2.5.22)] to extract heat from a cold body (to transfer heat to a hot body).

2.5.1 WORK FOR DISTINCT PRESSURE – VOLUME RELATIONS

Both the heat exchanged (2.6b) and the work performed, for example, by the pressure in a volume change (2.70c): (i) are specified by inexact differentials; (ii) are not functions of state; (iii) they depend not only on the initial and final state but also on the thermodynamic process between them. This is illustrated (Figure 2.12) by calculating the work between the same initial and final states along different paths in the thermodynamic space of pressure and volume. The pressure generally depends on

DOI: 10.1201/9781003186588-1

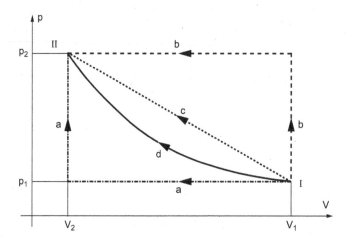

FIGURE 2.12 The work of the pressure in a volume change is not a function of state because it depends on the path between the same initial (p_1, V_1) and final (p_2, V_2) states. For example, if there is compression $V_2 < V_1$ and increase in pressure $p_2 > p_1$ between initial I and final II states, the work: (i,ii) is maximum (minimum) along the path "b" ("a") with pressure increase all at maximum V_1 (minimum V_2) volume; (iii) is intermediate along the path "c" of pressure a linear function of volume $p \sim V$; (iv) lower along the path "d" of a polytropic relation $p \sim V^{-n}$ with $n > 1$ between pressure and volume.

the volume, the work performed in changing the volume of a thermodynamic system from V_1 to V_2 is given by (2.70c) $\equiv$ (2.444):

$$W_{12} = -\int_{V_1}^{V_2} p(V)dV, \tag{2.444}$$

and corresponds to the area under the curve $p(V)$ in Figure 2.12 in the interval $V_1 \le V \le V_2$. For the same initial $(\dot{p}_1, V_1)$ and final (p_2, V_2) state, the mechanical work done depends (Figure 2.12) on the path, for example: (a) if (i) the volume is first reduced from the initial V_1 to the final V_2 value at constant initial pressure $=p_1$, followed by (ii) an increase of pressure from initial p_1 to final p_2 value at constant final volume $V = V_2$, then work is done only in the first process (i):

$$W_a = -p_1 \int_{V_1}^{V_2} dV = p_1(V_1 - V_2); \tag{2.445a, b}$$

(b) inversely if (i) the pressure is increased from p_1 to p_2 at constant volume V_1, and then (ii) the volume is reduced from V_1 to V_2 at constant pressure, then work is done only in the second process (ii):

$$W_b = -p_2 \int_{V_1}^{V_2} dV = p_2(V_1 - V_2); \tag{2.446a, b}$$

(c) if the pressure is a linear function of the volume (2.447a):

$$p - p_1 = \frac{p_2 - p_1}{V_2 - V_1}(V - V_1),$$

(2.447a)

the corresponding work (2.447b–g):

$$W_c = -\int_{V_1}^{V_2}\left[p_1 + \frac{p_2 - p_1}{V_2 - V_1}(V - V_1)\right]dV = \left[-p_1 V - \frac{p_2 - p_1}{V_2 - V_1}\left(\frac{V^2}{2} - V_1 V\right)\right]_{V_1}^{V_2}$$

$$= p_1(V_1 - V_2) - \frac{p_2 - p_1}{V_2 - V_1}\frac{V_2^2 + V_1^2 - 2V_1 V_2}{2} = p_1(V_1 - V_2) + \frac{p_2 - p_1}{V_1 - V_2}\frac{(V_1 - V_2)^2}{2} \quad (2.447\text{b–h})$$

$$= (V_1 - V_2)\left(p_1 + \frac{p_2 - p_1}{2}\right) = \frac{p_1 + p_2}{2}(V_1 - V_2) = \frac{W_a + W_b}{2},$$

is the arithmetic mean (2.447h) of (2.445b) and (2.446b) because the line c in Figure 2.10 is the diagonal of the rectangle formed by the lines a and b; (d) if the pressure is related to the specific volume by the **polytropic law** (2.448c) with positive **coefficient** (2.448a) and **exponent** (2.448b) the work is given by (2.448d–g):

$$K > 0, n > 1, p(V) = KV^{-n}$$

$$W_d = -K\int_{V_1}^{V_2} V^{-n}\, dV = \frac{K}{n-1}\left[V^{1-n}\right]_{V_1}^{V_2}$$

$$= \frac{K}{n-1}\left(V_2^{1-n} - V_1^{1-n}\right) = \frac{p_2 V_2 - p_1 V_1}{n-1}.$$

(2.448a–g)

Assuming that the initial state has lower pressure (2.449a) and larger volume (2.449b) than the final state, the successive processes a, b, c, d lead to decreasing work (2.449c):

$$p_1 < p_2, \quad V_1 > V_2: \quad W_b > W_c > W_d > W_a > 0$$

(2.449a–c)

because work equals the area under the curves a, b, c, d in the range $V_1 \geq V \geq V_2$ in Figure 2.12, that is progressively smaller. The relation between any three among the four variables pressure, volume, temperature and entropy is the **equation of state**, that completes the description of a basic thermodynamic system, beyond the three principles of thermodynamics (Sections 2.1–2.4), distinguishing different types of matter, for example: (i/ii) an ideal (perfect) gas [subsections 2.5.2–2.5.4 (2.5.5–2.5.7)] whose specific heats at constant volume and pressure depend only on temperature (are constant); (iii) condensed matter like a liquid (subsections 2.5.8–2.5.9).

2.5.2   EQUATION OF STATE (CLAPEYRON 1834) FOR AN IDEAL GAS

An **ideal gas** consists of molecules with density so low that all interactions, such as mechanical collisions and electric, magnetic and gravity attractions or repulsions, can be neglected or are **elastic**, that is do not change the total kinetic energy during the interaction. It follows that for an ideal gas: (i) the internal energy depends only on the kinetic energy of molecules; (ii) the **absolute temperature** is defined as twice the average kinetic energy of the microscopic molecular motion (2.450a) with the **Boltzmann constant** (2.450c) as coefficient with the dimensions of energy (Joule) per degree of absolute temperature (Kelvin):

$$<mv^2> = kT \geq 0, \quad k = (1.380622 \pm 0.000059) \times 10^{-23} \, J\,K^{-1}; \quad (2.450a, b)$$

(iii) the absolute temperature is always positive (2.450b) ≡ (2.281d), and increases from zero with all molecules at rest to larger values for stronger thermal agitation; (iv) the internal energy (enthalpy) depends only on temperature (2.451c) [(2.452c)], vanishes at zero temperature (2.451a) [(2.452a)] and its derivative with regard to temperature is the specific heat at constant volume (2.252e) ≡ (2.451a) [pressure (2.252f) ≡ (2.452a)] that generally depends on temperature:

$$U(0) = 0, \quad \frac{dU}{dT} = C_V(T), \quad U(T) = \int_0^T C_V(\theta)\,d\theta, \qquad (2.451a\text{--}c)$$

$$H(0) = 0, \quad \frac{dH}{dT} = C_p(T), \quad H(T) = \int_0^T C_p(\theta)\,d\theta; \qquad (2.452a\text{--}c)$$

(v) thus the internal energy (2.451c) [enthalpy (2.452c)] is the integral of specific heat at constant volume (pressure) with regard to temperature with zero as lower limit of integration.

The difference (2.223a) between the enthalpy (2.452c) and internal energy (2.451c) specifies (2.453a) the equation of state (2.453b) for an ideal gas:

$$pV = H(T) - U(T) = \int_0^T \left[C_p(\theta) - C_V(\theta)\right] d\theta, \qquad (2.453a, b)$$

that relates pressure p and volume V to temperature T. In the limit of zero temperature (2.353c, 2.354c) the r.h.s. of (2.453b) vanishes (2.454a–c):

$$\lim_{T\to 0} \int_0^T \left[C_p(\theta) - C_V(\theta)\right] d\theta = \lim_{T\to 0} \int_0^T O(\theta)\,d\theta = \lim_{T\to 0} O(T^2) = 0, \quad (2.454a\text{--}c)$$

$$\lim_{T\to 0} pV = \lim_{T\to 0}\left[H(T) - U(T)\right] = \lim_{T\to 0} O(T^2) = 0, \qquad (2.455a\text{--}c)$$

and so does (2.332d, 2.333d) the l.h.s. (2.455a–c). Thus there is no added constant in (2.453b) in agreement with (2.451a, 2.452a). Since the volume is not generally zero, the pressure vanishes at zero temperature. The pressure on a wall is the linear momentum $m\bar{v}$ imparted by molecules colliding with it. At zero temperature the molecules are at rest, and thus exert no pressure on any wall. Since the r.h.s. of (2.453b) vanishes at zero temperature (2.454c) *the equation of state of an ideal gas (2.453b) can be written in the form (2.456a) and its shown next that the relation is linear on temperature, because the function R(T) is a constant (2.456b), namely the* **ideal gas constant:**

$$pV = TR(T): \quad R(T) \equiv const \Leftrightarrow \frac{dR}{dT} = 0, \qquad (2.456a\text{–}c)$$

otherwise the entropy would not be a function of state.

To prove (2.456b) the internal energy (2.451b) [enthalpy (2.452b)] and equation of state (2.456a) of an ideal gas are substituted in the entropy (2.225a) ≡ (2.457a) [(2.225b) ≡ (2.458a)] leading to (2.457b) [(2.458b)]:

$$dS = \frac{dU}{T} + \frac{P}{T}dV = C_V(T)\frac{dT}{T} + R(T)\frac{dV}{V}, \qquad (2.457a, b)$$

$$dS = \frac{dH}{T} - \frac{V}{T}dp = C_p(T)\frac{dT}{T} - R(T)\frac{dp}{p}. \qquad (2.458a, b)$$

In order for the entropy to be a function of state (2.457b) [(2.458b)] it must be an exact differential, that is the cross-derivatives must be equal (2.459a–c) [(2.459d–f)]

$$0 = \frac{\partial}{\partial V}\left[\frac{C_V(T)}{T}\right] = \frac{\partial}{\partial T}\left[\frac{R(T)}{V}\right] = \frac{1}{V}\frac{dR}{dT},$$

$$0 = -\frac{\partial}{\partial p}\left[\frac{C_p(T)}{T}\right] = \frac{\partial}{\partial T}\left[\frac{R(T)}{p}\right] = \frac{1}{p}\frac{dR}{dT}. \qquad (2.459a\text{–}f)$$

Both cases (2.459c) [(2.459f)] lead to (2.456c) proving the existence of a gas constant for an ideal gas.

Thus *an ideal gas is defined by an internal energy (enthalpy) that depend only on temperature (2.451c) [(2.452c)] and vanish at zero temperature (2.451a) [(2.452a)]. Their derivatives with regard to temperature specify the specific heat at constant volume (2.451b) [(2.452b)] that generally depend on temperature. The adiabatic exponent that is their ratio generally depends on temperature (2.460a) but their difference does not and specifies the ideal gas constant (2.460b) that appears in the* **equation of state of an ideal gas** *(2.460c) ≡ (2.456a, b):*

$$\gamma(T) \equiv \frac{C_p(T)}{C_V(T)}, \quad C_p(T) - C_V(T) = R, \quad pV = RT. \qquad (2.460a\text{–}c)$$

The proof of (2.460b) follows (2.461a–d) from (2.451b; 2.452b), (2.453a) and (2.457a,2.445a, b):

$$C_p(T)-C_V(T)=\frac{dH}{dT}-\frac{dU}{dT}=\frac{d}{dT}(pV)=\frac{d}{dT}(RT)=R. \qquad (2.461\text{a–d})$$

The equation of state of an ideal gas (2.460c) together with the internal energy (2.451c) and enthalpy (2.452c) specifies all other functions of state (subsection 2.5.3), namely the entropy, free energy and free enthalpy.

2.5.3 FUNCTIONS OF STATE FOR AN IDEAL GAS

Taking logarithms (2.462a) in (2.460c) and differentiating leads to (2.462b):

$$\log p + \log V = \log R + \log T \Rightarrow \frac{dp}{p}+\frac{dV}{V}=\frac{dT}{T} \Leftarrow p\,dV +$$

$$V\,dp = PV\frac{dT}{T}=RdT, \qquad (2.462\text{a–c})$$

that can also be obtained differentiating (2.460c) leading to (2.462c) and dividing by (2.460c) leading to (2.462b). The *equation of state of an ideal gas in differential form* (2.462b) ≡ (2.463a) leads to the differential of entropy in terms of pressure and volume (2.463b):

$$\frac{dp}{p}+\frac{dV}{V}-\frac{dT}{T}=0 \quad dS=C_V(T)\frac{dp}{p}+C_p(T)\frac{dV}{V}, \qquad (2.463\text{a, b})$$

that can be obtained (2.464a; 2.463a) [(2.465a; 2.463a)] from the differentials of entropy in terms of temperature and volume (2.457b) [pressure (2.458b)]:

$$dS=C_V(T)\left(\frac{dp}{p}+\frac{dV}{V}\right)+R\frac{dV}{V}=C_V(T)\frac{dp}{p}+\left[R+C_V(T)\right]\frac{dV}{V} \qquad (2.464\text{a, b})$$

$$dS=C_p(T)\left(\frac{dp}{p}+\frac{dV}{V}\right)-R\frac{dp}{p}=C_p(T)\frac{dV}{V}+\left[C_p(T)-R\right]\frac{dp}{p} \qquad (2.465\text{a, b})$$

since (2.465b) ≡ (2.464b) ≡ (2.463b) by (2.460b).

Thus *the entropy of an ideal gas is given: (i/ii) by (2.457b) ≡ (2.466a) [(2.458b) ≡ (2.466b)] in terms of temperature and volume (pressure); (iii) by (2.463b) ≡ (2.466c) in terms of pressure and volume:*

$$S(T,V)=S_0+\int_{T_0}^{T}C_V(\theta)\frac{d\theta}{\theta}+R\,\log\left(\frac{V}{V_0}\right), \qquad (2.466\text{a})$$

$$S(T,p) = S_0 + \int_{T_0}^{T} C_p(\theta)\frac{d\theta}{\theta} - R\, log\left(\frac{p}{p_0}\right), \tag{2.466b}$$

$$S(p,V) = S_0 + \int_{p_0}^{p} C_V\left(\frac{\xi V}{R}\right)\frac{d\xi}{\xi} + \int_{V_0}^{V} C_p\left(\frac{p\xi}{R}\right)\frac{d\xi}{\xi}. \tag{2.466c}$$

In an adiabatic process (2.467a): (i/ii) the temperature is related to volume (pressure) by (2.457a, 2.460a, b) ≡ (2.467b–e) [(2.458a, 2.460a, b) ≡ (2.467f–i)]; (iii) the pressure to volume by (2.463b) ≡ (2.467j–l):

$$dS = 0: \quad \frac{dT}{dV} = -\frac{RT}{C_V V} = -\frac{1}{\gamma-1}\frac{T}{V} = -\frac{p}{C_V} = \frac{1-\gamma}{R}p, \tag{2.467a–e}$$

$$\frac{dT}{dp} = \frac{RT}{C_p p} = \frac{\gamma-1}{\gamma}\frac{T}{p} = \frac{V}{C_p} = \frac{\gamma}{\gamma-1}\frac{V}{R}, \tag{2.467f–i}$$

$$\frac{dp}{dV} = -\frac{C_p}{C_V}\frac{p}{V} = -\gamma\frac{p}{V} = -\gamma\, RT. \tag{2.467j–l}$$

The free energy (2.222a) [free enthalpy (2.224c)] are specified by the internal energy (2.451c) [enthalpy (2.452c)] and entropy in the form (2.466a) [(2.466b)] leading to (2.468a) [(2.468b)]:

$$F(T,V) = U(T) - T\,S(T,V), \quad G(T,p) = H(T) - T\,S(T,p). \tag{2.468a, b}$$

A particular case of the ideal gas (subsections 2.5.2–2.5.3) for which the specific heats at constant volume and pressure do not depend on temperature is the perfect gas (subsections 2.5.4–2.5.7).

2.5.4 EQUATION OF STATE FOR A PERFECT GAS

An ideal gas (subsections 2.5.2–2.5.3) consists of molecules that have only kinetic energy, that may be of two types: (i) translational and rotational for motion without links; (ii) vibrational if the atoms linked within molecules and can oscillate. The oscillations of molecules are neglected for a **perfect gas** whose energy is entirely kinetic due to translations (rotations), each with up to 3 degrees-of-freedom specified by coordinates (angles). The average kinetic energy for each degree of freedom (2.450a) is proportional to one-half of the absolute temperature T through the Boltzmann constant (2.450b), implying that absolute temperature zero corresponds to molecules at rest; thus the difference of internal energy from zero at rest is (2.469b) proportional to $\frac{1}{2}kT$ multiplied by the number of degrees-of-freedom f and number n of particles:

$$U(T) = \frac{1}{2}f\,kT\,n: \quad n = N\,A, \quad A = (6.022169 \pm 0.000040)\times 10^{23}\, mole^{-1}, \tag{2.469a–c}$$

the number of particles n is the **mole number** N multiplied (2.469b) by the number of particles per mole, that is the **Avogadro number** (2.86b) $\equiv$ (2.469c). Thus the internal energy of a perfect gas (2.469a, b) $\equiv$ (2.470a) is given by (2.470b) [(2.470c)]:

$$U(T) = \frac{1}{2} f k T A N = \frac{1}{2} f R T = \frac{1}{2} f \bar{R} T N \qquad (2.470a-c)$$

$$\bar{R} = \frac{R}{N} = k A = 8.31434 \pm 0.00035 \, J \, K^{-1} \, mole^{-1} \qquad (2.470d-f)$$

where $R(\bar{R})$ in (2.470b, e) [(2.470c, d)] is the gas constant (**gas constant per mole**). Thus *the internal energy of a perfect gas (2.470b)* $\equiv$ *(2.471a) is proportional to the temperature through the specific heat at constant volume (2.471b, c) that equals (2.471d) the gas constant multiplied by one-half the number of degrees-of-freedom of a molecule:*

$$U(T) = C_V \, T = \frac{1}{2} f R T : \quad C_V = \frac{dU}{dT} = \frac{1}{2} f R. \qquad (2.471a-d)$$

The gas constant appears in the equation of state appears with constant coefficients in (2.460c) $\equiv$ *(2.472a, b) for an ideal gas and (2.472c–e) also hold for an ideal (perfect) gas with coefficients dependent on temperature (constant coefficients):*

$$pV = RT = N \bar{R} T = (C_p - C_V) T = (\gamma - 1) C_V \, T = \left(1 - \frac{1}{\gamma}\right) C_p T. \quad (2.472a-e)$$

*The **specific volume** (2.473a) is the volume per unit mass or inverse of the **mass density** (2.473b); the mass equals (2.473c) the mole number multiplied by the **molecular mass** $\bar{m}$, that is the mass of one mole, or the mass of the Avogadro number (2.470c) of molecules:*

$$\vartheta = \frac{V}{m} = \frac{1}{\rho}, \quad m = N \bar{m} : \quad \frac{p}{\rho} = p \vartheta = \frac{pV}{N\bar{m}} = \frac{\bar{R}T}{\bar{m}} = \bar{\bar{R}} T, \quad \bar{\bar{R}} = \frac{\bar{R}}{\bar{m}} = \frac{R}{N\bar{m}}; \quad (2.473a-i)$$

introducing the specific volume (2.473a) as the volume per unit mass m, and its inverse the mass density (2.473b) as the mass per unit volume, the equation of state (2.472a) becomes (2.473d–g) where (2.473h, i) is the gas constant (2.470d–f) divided by the molecular mass. Thus the equation of state for a perfect gas can be written in two equivalent forms: (i) in terms of the total volume (2.472a–e) involving the gas constant per mole (2.470d–f); (ii) in terms of the specific volume (2.473d–g) $\equiv$ *(IV.5.111a, b) involving the gas constant divided by the molecular mass (2.473h, i)* $\equiv$ *(IV.5.111c, d). It follows that the enthalpy (2.223a)* $\equiv$ *(2.474a) is also proportional to the temperature (2.474b):*

$$H(T) = U(T) + pV = C_p T = \frac{1}{2} (f + 2) R T : \quad C_p = \frac{dH}{dT} = \frac{f+2}{2} R, \quad (2.474a-e)$$

through the specific heat at constant pressure (2.474c, d) that is similar to constant volume (2.471c) adding two degrees-of-freedom.

*The specific heat at constant volume (2.471c) [pressure (2.474d)] of a perfect gas are both constant and proportional to the gas constant (2.471d) [2.474e)]. Their ratio specifies the **adiabatic exponent** of an ideal gas, that depends on temperature (2.460a) for an ideal gas; for a perfect gas the adiabatic exponent is a constant, specified by the number of degrees-of-freedom of a molecule:*

$$\gamma \equiv \frac{C_p}{C_V} = 1 + \frac{2}{f} = \begin{cases} \dfrac{5}{3} = 1.667 \text{ if } f = 3\,(monoatomic), \\[2mm] \dfrac{7}{5} = 1.400 \text{ if } f = 5\,(diatomic), \\[2mm] \dfrac{4}{3} = 1.333 \text{ if } f = 6\,(poliatomic), \end{cases} \qquad (2.475a\text{--}c)$$

namely: (i) a monoatomic molecule has 3 translational degrees-of-freedom (2.475a); (ii) a diatomic molecule (or polyatomic molecule with all atoms in a row) adds two rotational degrees-of-freedom (2.475b); (iii) a polyatomic molecule (with the atoms not all in a row) adds another rotational degree-of-freedom for the six like a rigid body (2.475c). Table 2.8 compares for ideal and perfect gases and Table 2.9 gives more detail on perfect gases. Vibrational degrees-of-freedom are excluded (allowed) for a perfect (ideal) gas, hence the constant specific heats (2.471d; 2.474e) [specific heats depending on temperature (2.451b) [(2.452b)]. The same equation of state applies both to ideal (2.460c) and perfect (2.472a) gases, and in the case of perfect gases, the constancy of specific heats at constant pressure and volume simplifies: (i) all functions of state, including the entropy and thus the adiabatic conditions (subsection 2.5.5); (ii/iii) all 12 thermodynamic [subsection 2.5.6 (2.5.7)] coefficients (derivatives).

2.5.5   FUNCTIONS OF STATE AND ADIABATIC CONDITIONS

Using the equation of state of a perfect gas (2.472a) the entropy is given by (2.457b) $\equiv$ (2.476a) [(2.458b) $\equiv$ (2.476b)] as a function of temperature and volume (pressure) and by (2.463b) $\equiv$ (2.476c) as a function of the last two:

$$dS = C_V \frac{dT}{T} + R\frac{dV}{V} = C_p \frac{dT}{T} - R\frac{dp}{p} = C_V \frac{dp}{p} + C_p \frac{dV}{V}. \qquad (2.476a\text{--}c)$$

Since all coefficients in (2.476a–c) are constant *for a perfect gas, the entropy is given by (2.477a) [(2.477b)] as a function of the temperature and volume (pressure) and by (2.477c) as a function of the last two:*

$$\begin{aligned} S - S_0 &= C_V \log\left(\frac{T}{T_0}\right) + R\log\left(\frac{V}{V_0}\right) = C_p \log\left(\frac{T}{T_0}\right) - R\log\left(\frac{p}{p_0}\right) \\[2mm] &= C_V \log\left(\frac{p}{p_0}\right) + C_p \log\left(\frac{V}{V_0}\right) = C_V \log\left(\frac{p}{p_0}\right) - C_p \log\left(\frac{\rho}{\rho_0}\right) \\[2mm] &= C_V \log\left(\frac{T}{T_0}\right) - R\log\left(\frac{\rho}{\rho_0}\right) \end{aligned} \qquad (2.477a\text{--}e)$$

TABLE 2.8

Comparison of Ideal and Perfect Gases

Quantity	Symbol	Ideal	Perfect
		Gas	

Quantity	Symbol	Ideal	Perfect
Specific heat at constant volume	C_v	$C_v(T)$	(2.471c,d)
Specific heat at constant pressure	C_p	$C_p(T)$	(2.474d,e)
Gas constant	$R = C_p - C_v$	$R = \text{const}$	(2.470d-f)
Adiabatic exponent	$\gamma = \dfrac{C_p}{C_v}$	$\gamma(T)$	(2.475a-c)
Internal energy	$U(T)$	(2.451c)	(2.471a,b)
Enthalpy	$H(T)$	(2.452c)	(2.474a-c)
Entropy	$S(T,V)$	(2.466a)	(2.477a)
	$S(T,p)$	(2.466b)	(2.477b)
	$S(p,V)$	(2.466c)	(2.477c)
Adiabatic Relation	T,V	(2.467b-e)	(2.478g)
	T,p	(2.467f-i)	(2.478f)
	P,V	(2.467j-l)	(2.478b)
Free energy	$F(T,V)$	(2.468a)	(2.485a)
Free enthalpy	$G(T,p)$	(2.468b)	(2.485b)

Note: The difference between ideal (perfect) gases (Diagram 2.2) is that the specific heats at constant pressure and volume depend only on temperature (are constant), and this has implications for the: (i–ii) gas constant and adiabatic exponent; (iii–iv) internal energy and enthalpy; (v–vii) entropy as a function of any two variables among pressure, volume and temperature; (viii–x) the corresponding adiabatic relations for constant entropy; (xi–xii) the free energy and free enthalpy.

and the volume in (2.477c) [(2.477a)] may be replaced (2.473b) by the mass density in (2.477d) [(2.477e)]. It follows that in an adiabatic process (2.467a) ≡ (2.478a) for a perfect gas the pressure and volume are related by (2.478b) and are related to the temperature, respectively, by (2.478e) and (2.478g) that all involve (2.478c, f, h) the adiabatic exponent (2.475a–c) and the mass density replaces the volume in (2.478d, i):

$$S = S_0 : \quad \frac{p}{p_0} = \left(\frac{V}{V_0}\right)^{-C_p/C_V} = \left(\frac{V}{V_0}\right)^{-\gamma} = \left(\frac{\rho}{\rho_0}\right)^{\gamma},$$

$$\frac{p}{p_0} = \left(\frac{T}{T_0}\right)^{C_p/R} = \left(\frac{T}{T_0}\right)^{\gamma/(\gamma-1)}, \quad \frac{T}{T_0} = \left(\frac{V}{V_0}\right)^{-R/C_V} = \left(\frac{V}{V_0}\right)^{1-\gamma} = \left(\frac{\rho}{\rho_0}\right)^{\gamma-1}. \quad (2.478a\text{–}i)$$

The adiabatic relations for perfect gases are indicated in Table 2.10, distinguishing monoatomic, diatomic and polyatomic in Table 2.11.

TABLE 2.9
Properties of Perfect Gases

Quantity	Value	Gas Molecules		
		Monatomic	Diatomic*	Polyatomic**
Degrees-of-freedom	f	3	5	6
Specific heat at constant volume	$C_v = \dfrac{f}{2} R$	$\dfrac{3}{2} R$	$\dfrac{5}{2} R$	3R
Specific heat at constant pressure	$C_p = \dfrac{f+2}{2} R$	$\dfrac{5}{2} R$	$\dfrac{7}{2} R$	4R
Adiabatic exponent	$\gamma = 1 + \dfrac{2}{f}$	$\dfrac{5}{3}$	$\dfrac{7}{5}$	$\dfrac{4}{3}$
Internal energy	$U = \dfrac{f}{2} RT$	$\dfrac{3}{2} RT$	$\dfrac{5}{2} RT$	3RT
Enthaldy	$H = \dfrac{f+2}{2} RT$	$\dfrac{5}{2} RT$	$\dfrac{7}{2} RT$	4RT

Note: Properties of monoatomic/diatomic/polyatomic perfect gases based on the respective number 3/5/6 of degrees of freedom of a molecule, namely: (i–ii) specific heats at constant volume and pressure; (iii) adiabatic exponent; (iv–v) internal energy and enthalpy.

TABLE 2.10
Adiabatic Relations for Perfect Gases

Variables	p/p_0	T/T_0	$V/V_0 = \vartheta/\vartheta_0$	$\rho/\rho_0 = \vartheta_0/\vartheta$
p/p_0	p/p_0	$(T/T_0)^{\gamma/(\gamma-1)}$	$(V/V_0)^{-\gamma}$	$(\rho/\rho_0)^{\gamma}$
T/T_0	$(p/p_0)^{1-1/\gamma}$	T/T_0	$(V/V_0)^{1-\gamma}$	$(\rho/\rho_0)^{\gamma-1}$
$\dfrac{V}{V_0} = \dfrac{\vartheta}{\vartheta_0}$	$(p/p_0)^{-1/\gamma}$	$(T/T_0)^{1/(1-\gamma)}$	V/V_0	ρ_0/ρ
ρ/ρ_0	$(p/p_0)^{1/\gamma}$	$(T/T_0)^{1/(\gamma-1)}$	V_0/V	ρ/ρ_0

Note: Adiabatic relations for perfect gases between any two of three thermodynamic variables: temperature, pressure and mass density (or its inverse, the specific volume). The exponents involve the adiabatic exponent (Table 2.9).

The adiabatic (2.258c) ≡ (2.479b) [isothermal (2.258c) ≡ (2.480b)] sound speed is specified for an ideal gas by the relation (2.467k) ≡ (2.479a) [equation of state (2.473g) ≡ (2.480a)]:

$$\frac{dp}{d\rho} = \gamma \frac{p}{\rho} : \left(c_s\right)^2 \equiv \left(\frac{\partial p}{\partial \rho}\right)_s = \gamma \frac{p}{\rho} = \gamma \overline{\overline{R}} T, \qquad (2.479a\text{–}d)$$

TABLE 2.11
Monatomic, Diatomic and Polyatomic Perfect Gases: Adiabatic Relations

Relation	Gas Molecules		
	Monatomic	Diatomic*	Polyatomic
Adiabatic exponent	$\gamma = \dfrac{5}{3}$	$\gamma = \dfrac{7}{5}$	$\gamma = \dfrac{4}{3}$
$p \sim V^{-\gamma}$	$p \sim V^{-5/3}$	$p \sim V^{-7/5}$	$p \sim V^{-4/3}$
$p \sim \rho^{\gamma}$	$p \sim \rho^{5/3}$	$p \sim \rho^{7/5}$	$p \sim \rho^{4/3}$
$p \sim T^{\gamma/(\gamma-1)}$	$p \sim T^{5/2}$	$p \sim T^{7/2}$	$p \sim T^4$
$V \sim p^{-1/\gamma}$	$V \sim p^{-3/5}$	$V \sim p^{-5/7}$	$V \sim p^{-3/4}$
$V \sim T^{1/(1-\gamma)}$	$V \sim T^{-3/2}$	$V \sim T^{-5/2}$	$V \sim T^{-3}$
$\rho \sim p^{1/\gamma}$	$\rho \sim p^{3/5}$	$\rho \sim p^{5/7}$	$\rho \sim p^{3/4}$
$\rho \sim T^{1/(\gamma-1)}$	$\rho \sim T^{3/2}$	$\rho \sim T^{5/2}$	$\rho \sim T^3$
$T \sim V^{1-\gamma}$	$T \sim V^{-2/3}$	$T \sim V^{-2/7}$	$T \sim V^{-1/3}$
$T \sim \rho^{\gamma-1}$	$T \sim \rho^{2/3}$	$T \sim \rho^{2/5}$	$T \sim \rho^{1/3}$
$T \sim p^{1-\frac{1}{\gamma}}$	$T \sim p^{2/5}$	$T \sim p^{2/7}$	$T \sim p^{1/4}$

* or polyatomic with atoms in a row

Note: Combining the adiabatic relations (Table 2.10) with the values of the adiabatic exponent for mono-atomic/diatomic/polyatomic perfect gases (Table 2.9) specifies (Table 2.11) the adiabatic relations between any two of: temperature, pressure and mass density (or specific volume) for monoatomic/diatomic/polyatomic perfect gases.

$$p = \rho \overline{\overline{R}} T: \; (c_t)^2 \equiv \left(\frac{\partial p}{\partial \rho}\right)_T = \overline{\overline{R}} T = \frac{p}{\rho} = \frac{(c_s)^2}{\gamma}, \qquad (2.480\text{a–e})$$

in terms of the temperature (2.479d) [(2.480c)] or pressure and mass density (2.479c) [(2.480d)] and the ratio (2.480e) ≡ (2.264b) agrees with the expressions for general matter (2.262d; 2.263d). The adiabatic sound speed (2.479d) ≡ (2.481a) is also a thermodynamic variable related to the temperature and hence to the pressure (2.481b)/volume (2.481c)/mass density (2.481d):

$$\left(\frac{c}{c_0}\right)^2 = \frac{T}{T_0} = \left(\frac{p}{p_0}\right)^{1-1/\gamma} = \left(\frac{V}{V_0}\right)^{1-\gamma} = \left(\frac{\rho}{\rho_0}\right)^{\gamma-1}, \qquad (2.481\text{a–d})$$

using (2.478f, h, i). The adiabatic relation (2.479a) for an ideal gas follows from (2.473b) ≡ (2.482a) leading (2.482b–d) to (2.482e) ≡ (2.480a):

$$\frac{dp}{d\rho} = \frac{dp}{d(m/V)} = \frac{1}{m}\frac{dp/dV}{d(1/V)/dV} = -\frac{V^2}{m}\times\left(-\frac{\gamma p}{V}\right) = \frac{\gamma pV}{m} = \gamma \frac{p}{\rho}. \qquad (2.482\text{a–e})$$

The free energy (2.222a) ≡ (2.483a) [free enthalpy (2.223a) ≡ (2.484a)] is given for a perfect gas (2.471b) [(2.474b)] by (2.483b) [(2.484b)]:

$$F = U - T S_0 - T(S - S_0) = \frac{1}{2} f RT - T S_0 - T(S - S_0), \qquad (2.483a, b)$$

$$G = H - T S_0 - T(S - S_0) = \frac{f+2}{2} RT - T S_0 - T(S - S_0), \qquad (2.484a, b)$$

and substitution of (2.477a) [(2.477b)] leads to (2.485a) [(2.485b)]:

$$F(T,V) = \left(\frac{fR}{2} - S_0 \right) T - T \left[C_V \log \left(\frac{T}{T_0} \right) + R \log \left(\frac{V}{V_0} \right) \right] \qquad (2.485a)$$

$$G(T,p) = \left(\frac{f+2}{2} R - S_0 \right) T - T \left[C_p \log \left(\frac{T}{T_0} \right) - R \log \left(\frac{p}{p_0} \right) \right] \qquad (2.485b)$$

From (2.477a–c) also follows that for a perfect gas: (i,ii) at constant volume the entropy increases with temperature and pressure; (iii,iv) at constant pressure the entropy increases with temperature and volume; (v,vi) at constant temperature the entropy increases with volume and decreases with pressure. The cases (i–v) [(vi)] of increase (decrease) in entropy correspond to a larger (smaller) number of accessible states. The equation of state of an ideal gas is used next [subsection 2.5.6 (2.5.7)] to specify all 12 thermodynamic coefficients (derivatives) in Table 2.4 (2.5), including the particular case of a perfect gas.

2.5.6 THERMODYNAMIC COEFFICIENTS FOR AN IDEAL GAS

The 12 non-inverse thermodynamic derivatives (2.267a–f) in Table 2.4 are considered both for general matter (subsections 2.3.10–2.3.15 and 2.3.19) and for ideal (perfect) gases [subsections 2.5.1–2.5.2 (2.5.3–2.5.4) using the sign $\underset{=}{IG}$ $\left(\underset{=}{PG} \right)$ in the equations (2.486a) [(2.486b)]:

$$\text{ideal gas} \quad \underset{=}{IG}; \quad \text{perfect gas} \quad \underset{=}{PG}, \qquad (2.486a, b)$$

For example, the specific heats at constant volume (pressure) are given by (2.451b) ≡ (2.487a) [(2.452b) ≡ (2.487c)] for ideal gases, and by (2.471d) ≡ (2.487b) [(2.474e) ≡ (2.487d)] for perfect gases:

$$C_V \underset{=}{IG} \frac{dU}{dT} \underset{=}{PG} \frac{f}{2} R, \quad C_p \underset{=}{IG} \frac{dH}{dT} \underset{=}{PG} \frac{f+2}{2} R. \qquad (2.487a\text{–}d)$$

The adiabatic (isothermal) sound speed is given by (2.262a–d) [(2.263a–d)] for general matter, and by (2.479b–d) [(2.480b–d)] for ideal gases that include as a particular case perfect gases. Of the 12 thermodynamic coefficients in Table 2.4 for an ideal gas have already been considered 4, namely: (i/ii) the specific heats at constant volume (2.487b) $\equiv$ (2.488d–f) [pressure (2.487d) $\equiv$ (2.488g–i)] for a perfect gas, respectively (2.488a–c) for monoatomic (2.475a)/diatomic (2.475b)/polyatomic (2.475c) molecules (2.488a–c):

$$f \underset{=}{PG} \{3,5,6\}; \quad C_p \underset{=}{PG} \left\{ \frac{3}{2}, \frac{5}{2}, 3 \right\} R, \quad C_p \underset{=}{PG} \left\{ \frac{5}{2}, \frac{7}{2}, 4 \right\} R; \qquad (2.488a\text{–}i)$$

(iii/iv) the gas constant (2.460d–f) and the adiabatic exponent (2.475a–c) for a perfect gas; (v) the coefficient of thermal expansion for an ideal gas is calculated alternatively by (2.253a; 2.472a) $\equiv$ (2.489a–c) [(2.253b; 2.458b) $\equiv$ (2.489d–f)]:

$$K_p \equiv \frac{1}{V} \left(\frac{\partial V}{\partial T} \right)_p \underset{=}{IG} \frac{R}{Vp} = \frac{1}{T}, \quad K_p = -\frac{1}{V} \left(\frac{\partial S}{\partial p} \right)_T \underset{=}{IG} \frac{R}{pV} = \frac{1}{T}; \qquad (2.489a\text{–}f)$$

(vi–vii) the coefficient of isothermal compression for an ideal gas (2.253c; 2.472a) $\equiv$ (2.490a–c) agrees (2.490d) with the coefficient of isothermal expansion (2.253d; 2.472a) $\equiv$ (2.490e–g):

$$K_T \equiv -\frac{1}{V} \left(\frac{\partial V}{\partial p} \right)_T \underset{=}{IG} \frac{RT}{Vp^2} = \frac{1}{p} = \frac{1}{\beta V}, \quad \beta = -\left(\frac{\partial p}{\partial V} \right)_T \underset{=}{IG} \frac{RT}{V^2} = \frac{p}{V}; \qquad (2.490a\text{–}g)$$

(viii) the thermal pressure coefficient for an ideal gas is given equivalently by (2.254a; 2.472a) $\equiv$ (2.491a–c) and (2.254b; 2.457b) $\equiv$ (2.491d, e):

$$\alpha \equiv \left(\frac{\partial p}{\partial T} \right)_V \underset{=}{IG} \frac{R}{V} = \frac{p}{T}, \quad \alpha \equiv \left(\frac{\partial S}{\partial V} \right)_T \underset{=}{IG} \frac{R}{V}; \qquad (2.491a\text{–}e)$$

(ix–x) the adiabatic (2.262b; 2.478b) $\equiv$ (2.492a–e) [isothermal (2.263b; 2.472a) $\equiv$ (2.493a–e)] sound speed for an ideal gas in agreement with (2.479a–d) [(2.480a–e)]:

$$(c_s)^2 = -\vartheta V \left(\frac{\partial p}{\partial V} \right)_S \underset{=}{IG} \vartheta V \frac{C_p}{C_V} \frac{p}{V} = \gamma \, p \vartheta = \frac{\gamma \, p}{\rho} \underset{=}{PG} \gamma \, \overline{\overline{R}} T, \qquad (2.492a\text{–}e)$$

$$(c_t)^2 = -\vartheta V \left(\frac{\partial p}{\partial V} \right)_T \underset{=}{IG} \vartheta V \frac{p}{V} = p \vartheta = \frac{p}{\rho} \underset{=}{PG} \overline{\overline{R}} T; \qquad (2.493a\text{–}e)$$

(xi–xii) the non-adiabatic pressure (volume) coefficient for an ideal gas (2.265a; 2.463b) ≡ (2.494a–c) [(2.266a; 2.463b) ≡ (2.494d–f)]:

$$S_p \equiv \left(\frac{\partial S}{\partial p}\right)_V \stackrel{IG}{=} \frac{C_V}{p} = \frac{R}{\gamma-1}p, \quad S_V \equiv \left(\frac{\partial S}{\partial V}\right)_p \stackrel{IG}{=} \frac{C_p}{V} = \frac{\gamma}{\gamma-1}\frac{R}{V}. \quad (2.494a–f)$$

The equation of state of an ideal gas can also be used to calculate all 12 thermodynamic derivatives (subsection 2.5.7) in Table 2.5 and check the general relations among thermodynamic coefficients in Table 2.4.

2.5.7 THERMODYNAMIC DERIVATIVES FOR PERFECT GASES

The 12 thermodynamic derivatives (2.267a–l) listed in Table 2.5 simplify for an ideal IG (perfect gas PG): (i/ii) the derivative of the entropy with regard to the temperature at constant volume (2.457b) ≡ (2.495a) [pressure (2.458b) ≡ (2.495c)] relates to the specific heat at constant volume (2.471d) ≡ (2.495b) [pressure (2.474e) ≡ (2.495d)]:

$$\left(\frac{\partial S}{\partial T}\right)_V \stackrel{IG}{=} \frac{C_V}{T} \stackrel{PG}{=} \frac{fR}{2T}, \quad \left(\frac{\partial S}{\partial T}\right)_p \stackrel{IG}{=} \frac{C_p}{T} \stackrel{PG}{=} \frac{(f+2)R}{2T}; \quad (2.495a–d)$$

(iii–iv) the derivative (2.496a, b) ≡ (2.458b) agrees (2.243d) with (2.496c, d) ≡ (2.472a):

$$\left(\frac{\partial S}{\partial p}\right)_T \stackrel{IG}{=} -\frac{R}{p} = -\frac{V}{T}, \quad \left(\frac{\partial V}{\partial T}\right)_p \stackrel{IG}{=} \frac{R}{p} = \frac{V}{T}; \quad (2.496a–d)$$

(v–vi) the derivative (2.497a, b) ≡ (2.463b) agrees (2.243a) with (2.497c,d) ≡ (2.467b,c):

$$\left(\frac{\partial S}{\partial p}\right)_V \stackrel{IG}{=} \frac{C_V}{p} = \frac{R}{\gamma-1}\frac{1}{p}, \quad \left(\frac{\partial V}{\partial T}\right)_S \stackrel{IG}{=} -\frac{C_V}{p} = \frac{R}{1-\gamma}\frac{1}{p}; \quad (2.497a–d)$$

(vii–viii) the derivative (2.498a, b) ≡ (2.457b) agrees (2.243b) with (2.498c, d) ≡ (2.472a):

$$\left(\frac{\partial S}{\partial V}\right)_T \stackrel{IG}{=} \frac{R}{V} = \frac{p}{T}, \quad \left(\frac{\partial p}{\partial T}\right)_V \stackrel{IG}{=} \frac{R}{V} = \frac{p}{T}; \quad (2.498a–d)$$

(ix–x) the derivative (2.499a, b) ≡ (2.463b) agrees (2.243c) with (2.499c,d) ≡ (2.467f,g):

$$\left(\frac{\partial S}{\partial V}\right)_p \stackrel{IG}{=} \frac{C_p}{V} = \frac{\gamma}{\gamma-1}\frac{R}{V}, \quad \left(\frac{\partial p}{\partial T}\right)_S \stackrel{IG}{=} \frac{C_p}{V} = \frac{\gamma}{\gamma-1}\frac{R}{V}; \quad (2.499a–d)$$

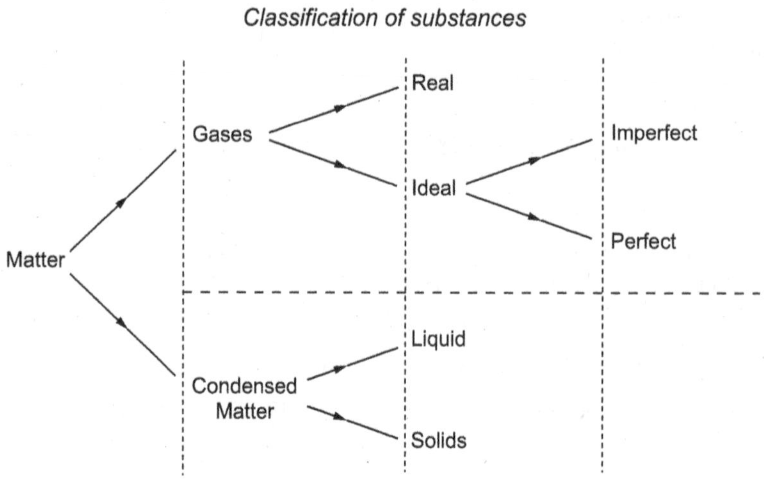

Classification of substances

DIAGRAM 2.2 Matter may be classified into gases and condensed matter, the latter consisting of liquids and solids. Among the gases, the ideal (perfect) gases are particularly simple since the internal energy and enthalpy depend only on temperature (are linear functions of temperature), and thus the specific heats at constant volume and pressure depend only on temperature (are constant).

(xi–xii) the derivatives (2.500a) [(2.500e)] are evaluated by (2.467h–j) [(2.472a)] leading to (2.500b–d) [(2.500e)]:

$$\left(\frac{\partial p}{\partial V}\right)_S \overset{IG}{=} -\frac{C_p}{C_V}\frac{p}{V} = -\gamma\frac{p}{V} = -\gamma\,RT, \quad \left(\frac{\partial p}{\partial V}\right)_T \overset{IG}{=} -\frac{RT}{V^2} = -\frac{p}{V}. \qquad (2.500a\text{–}e)$$

Substances may be classified (Diagram 2.2) as: (i) **condensed matter**, that is **liquids** or **solids**, and gases; (ii) the gases may be ideal if the specific heats at constant pressure and volume depend only on temperature, and as **real gases** otherwise; (iii) the ideal gases are classified as perfect gases if the specific heats at constant volume and pressure are constant and as **imperfect gases** otherwise. An imperfect ideal gas with the specific heats at constant volume and pressure quadratic functions of temperature is considered in example 10.5. A different approach from that used to arrive at equations of state of ideal (perfect) gases [subsections 2.5.2–2.5.3 (2.5.4–2.5.7)] is to consider the lowest order terms of a Taylor series (I.23.32a, b) expansion of the pressure as a function of the entropy and mass density (subsections 2.5.8–2.5.9).

2.5.8   EQUATION OF STATE OF FIRST (SECOND) ORDER IN ENTROPY (MASS DENSITY)

The equation of state is a relation among three of the following four independent thermodynamic variables: (i) pressure p; (ii) temperature T; (iii) entropy S; (iv) volume V or specific volume ϑ or mass density ρ in (2.473a, b). The equation of state was initially derived in the form $p(V,T)$ for an ideal gas (2.460c). An alternative form of equation of state is $p(\rho,S)$ replacing volume V by mass density ρ and the temperature T by entropy S. The approximation of the equation of state (2.501a) to first (second) order in the entropy (mass density) is (2.501b):

$$p = p(\rho,S)$$

$$p = p_0 + \left(\frac{\partial p_0}{\partial \rho_0}\right)_{S_0} (\rho - \rho_0) + \frac{1}{2}\left(\frac{\partial^2 p_0}{\partial \rho_0^2}\right)_{S_0} (\rho - \rho_0)^2 + \left(\frac{\partial p_0}{\partial S_0}\right)_{\rho_0} (S - S_0) + \dots \quad (2.501a, b)$$

where the derivatives are calculated in equilibrium conditions: (i) the adiabatic derivative of pressure with regard to mass density specifies the adiabatic sound speed (2.262a–d) for an ideal gas by (2.479b; 2.473d) leading to (2.502a–e):

$$\left(c_s\right)^2 = \left(\frac{\partial p}{\partial \rho}\right)_S = -\vartheta^2 \left(\frac{\partial p}{\partial \vartheta}\right)_S = \vartheta^2 \gamma \frac{p}{\vartheta} = \gamma\, p\vartheta = \gamma \frac{p}{\rho} = \gamma \overline{\overline{R}} T; \quad (2.502a\text{–}e)$$

(ii) this specifies the second-order adiabatic derivative of the pressure with regard to the mass density (2.503b–e):

$$\gamma = const \quad \left(\frac{\partial^2 p}{\partial \rho^2}\right)_S = \left[\frac{\partial\left(c_s^2\right)}{\partial \rho}\right]_S = \left[\frac{\partial\left(\dfrac{\gamma\, p}{\rho}\right)}{\partial \rho}\right]_S$$

$$= \frac{\gamma}{\rho}\left(\frac{\partial p}{\partial \rho}\right)_S - \frac{\gamma\, p}{\rho^2} = \frac{c_s^2}{\rho}(\gamma - 1), \quad\quad (2.503a\text{–}e)$$

where was assumed a constant adiabatic exponent (2.503a); (iii) the remaining thermodynamic derivative in (2.501b), of pressure with regard to entropy at constant mass density is given by the inverse non-adiabatic pressure coefficient (2.504a–e):

$$K_s \equiv \left(\frac{\partial p}{\partial S}\right)_\rho = \left(\frac{\partial p}{\partial S}\right)_V = \frac{1}{S_p} = \frac{C_p - C_V}{C_V K_p V} = \frac{\gamma - 1}{m}\frac{\rho}{K_p}, \quad (2.504a\text{–}e)$$

that is the inverse of (2.265a–f; 2.473b). This completes the coefficients (2.502a–e; 2.503a–e; 2.504a–e) in the equation of state (2.501a, b) that is compared next (subsection 2.5.9) with other forms.

2.5.9 TAIT (1888) EQUATION OF STATE FOR WATER

Substitution of (2.502a; 2.503e; 2.504e) in (2.501b) shows that *the equation of state (2.501a) for the pressure to second order in the mass density and first order in the entropy is:*

$$p - p_0 = c_0^2 (\rho - \rho_0) + \left(\frac{c_0^2}{2 \rho_0} \right) (\gamma_0 - 1)(\rho - \rho_0)^2 + \frac{\gamma_0 - 1}{m} \left(\frac{p_0}{K_{p0}} \right) (S - S_0), \quad (2.505)$$

with all coefficients calculated for the mean state with index zero and involving: (i) the sound speed (2.502a–e) as for an ideal gas with constant adiabatic exponent (2.503a); (ii) the inverse non-adiabatic pressure coefficient (2.504a–e). This is compared with: (i) the exact equation of state for a perfect gas in the form (2.477c) ≡ (2.506a) relating pressure, entropy and volume (mass density):

$$\frac{p}{p_0} \overset{PG}{=} \left(\frac{V}{V_0} \right)^{-C_p / C_V} \exp\left(\frac{S - S_0}{C_v} \right) = \left(\frac{\rho}{\rho_0} \right)^{\gamma} \exp\left[(\gamma - 1) \frac{S - S_0}{R} \right]; \quad (2.506a, b)$$

(ii) the **Tait (1888) equation of state for water** (2.507c) where (2.507a, b) are constants:

$$\gamma, A = const \quad p - p_0 = A \left[\left(\frac{\rho}{\rho_0} \right)^{\gamma} - 1 \right]. \quad (2.507a\text{–}c)$$

For the latter: (i) the adiabatic sound speed is given by:

$$(c_s)^2 = \left(\frac{\partial p}{\partial \rho} \right)_S = \frac{A \gamma}{\rho} \left(\frac{\rho}{\rho_0} \right)^{\gamma} = \frac{\gamma}{\rho} (p - p_0 + A); \quad (2.508a\text{–}c)$$

(ii) its derivative with regard to the mass density:

$$\left(\frac{\partial^2 p}{\partial \rho^2} \right)_S = \left[\frac{\partial (c_s^2)}{\partial \rho} \right]_S = \frac{\gamma}{\rho} \left[\left(\frac{\partial p}{\partial \rho} \right)_S - \frac{p - p_0 - A}{\rho} \right]$$

$$= (p - p_0 - A) \left(\frac{\gamma^2}{\rho^2} - \frac{\gamma}{\rho^2} \right) = c_s^2 \frac{\gamma - 1}{\rho}, \quad (2.509a\text{–}d)$$

coincides with (2.503e) ≡ (2.509d). The Tait equation of state (2.507c) ≡ (2.510a):

$$\frac{p}{p_0} = \frac{A}{p_0} \left(\frac{\rho}{\rho_0} \right)^{\gamma} + 1 - \frac{A}{p_0}; \quad A = p_0 : \quad \frac{p}{p_0} = \left(\frac{\rho}{\rho_0} \right)^{\gamma} = \left(\frac{V}{V_0} \right)^{-\gamma}, \quad (2.510a\text{–}d)$$

leads for (2.510b) to the adiabatic equation (2.510c) ≡ (2.510d) ≡ (2.478c, d) for a perfect gas. The four main thermodynamic processes are adiabatic/isochoric/isobaric/isothermal (respectively, subsections 2.3.4/2.3.5/2.3.6/2.3.7) that are reconsidered for ideal and perfect gases (subsections 2.5.10/2.5.11/2.5.12/2.5.13) as a precursor to the analysis of thermodynamic cycles (subsections 2.5.14–2.5.27).

2.5.10 ADIABATIC PROCESS FOR A PERFECT GAS

Since there are two extensive (intensive) parameters, namely, the entropy S and volume V (temperature T and pressure p), there are four particular thermodynamic processes for which one of these parameters is constant. In an adiabatic process the entropy is constant (2.511a):

$$dS = 0; \quad dQ = T\,dS = 0, \quad dW = -p\,dV = dU, \quad dH = V\,dp \qquad (2.511a\text{–}f)$$

and hence no heat (2.6b) $\equiv$ (2.511b) is exchanged (2.511c), and the work (2.70c) $\equiv$ (2.511d) is a function of state, because it coincides, in this particular case, with the internal energy (2.221d) $\equiv$ (2.511e); the enthalpy (2.223b) is given by (2.511f) in adiabatic conditions. For an ideal gas the specific heats at constant volume (2.451b) [pressure (2.452b)] depend only on temperature leading to the adiabatic relations (2.467a–l). In the case of a perfect gas the specific heats at constant volume (2.471d) and pressure (2.474e) are constant, and also the adiabatic exponent (2.475a–c), and thus (2.467c/g/k) may be integrated, respectively (2.512c, d/e /f, g):

$$\rho \equiv \frac{dm}{dV} \equiv \frac{1}{\vartheta}; \quad T V^{\gamma-1} = const = T\rho^{1-\gamma}, \quad T^{\gamma} p^{1-\gamma} = const, \qquad (2.512a\text{–}e)$$

$$pV^{\gamma} = const = p\rho^{-\gamma}, \qquad (2.512f, g)$$

using also as alternative to the volume V the mass density (2.512a) that is the inverse (2.512b) of the specific volume. The relations (2.512c, e, f) are equivalent as follows from the equation of state (2.472a) applied to (2.512f) $\equiv$ (2.513a) and leading to (2.513b, c) $\equiv$ (2.512e) and (2.513d, e) $\equiv$ (2.512c):

$$const \approx pV^{\gamma} = p\left(\frac{RT}{p}\right)^{\gamma} = R^{\gamma} T^{\gamma} p^{1-\gamma} = \left(\frac{RT}{V}\right)V^{\gamma} = RTV^{\gamma-1}. \qquad (2.513a\text{–}e)$$

The adiabatic relations for a perfect gas (2.512c–g) are equivalent to (2.478b–i), more precisely (2.512c/ d/e/f/g) $\equiv$ (2.478h/ i/f/c/d). *The adiabatic relations (2.512c–g) [$\equiv$ (2.478b–i)] for a perfect gas are mutually consistent (2.513a–e) and can be obtained alternatively integrating (2.467c/ g/k) [setting (2.478a) in (2.477a–e)]. In an adiabatic process (2.511a) there is no heat (2.511c) and the work equals the internal energy (2.511e) that for a perfect gas is given by (2.471a; 2.472a) $\equiv$ (2.514a–d):*

$$W_{12} = U_2 - U_1 = C_V\left(T_2 - T_1\right) = \frac{R}{\gamma - 1}\left(T_2 - T_1\right) = \frac{p_2 V_2 - p_1 V_1}{\gamma - 1}. \qquad (2.514a\text{–}d)$$

The adiabatic (isochoric) processes [subsection 2.5.10 (2.5.11)] contrast in having work (heat) and no heat (work).

2.5.11   ISOCHORIC PROCESS FOR AN IDEAL GAS

In an isochoric process (2.515a), that is with constant volume, there is (2.515c) no work (2.70c) ≡ (2.515b) and thus the heat (2.6b) ≡ (2.515d) equals (2.221d) the internal energy (2.515e):

$$dV = 0: \quad dW = -p\,dV = 0, \quad dQ = T\,dS = dU. \qquad (2.515a\text{–}e)$$

For an ideal gas (2.460e) in an isochoric process (2.516a) the pressure is proportional to the temperature (2.516b) and for a perfect gas the entropy (2.477a, c) is given by (2.516c, d):

$$V = const: \quad \frac{p}{T} = const, \quad S - S_0 = C_V \log\left(\frac{T}{T_0}\right) = C_V \log\left(\frac{p}{p_0}\right). \qquad (2.516a\text{–}d)$$

The heat in the isochoric process (2.517a) of an ideal gas is specified by the internal energy (2.515e) ≡ (2.517b), and is given by (2.514a–d) ≡ (2.517c–e):

$$V = const: Q_{12} = U_2 - U_1 = C_V(T_2 - T_1) = \frac{R}{\gamma - 1}(T_2 - T_1) = \frac{RV}{\gamma - 1}(p_2 - p_1). \qquad (2.517a\text{–}e)$$

The isochoric (isobaric) processes [subsection 2.5.11 (2.5.12)] contrast in that the heat is specified by the internal energy (enthalpy) and there is no (there is) work.

2.5.12   ISOBARIC PROCESS (CHARLES 1787, GAY-LUSSAC 1802, DALTON 1802)

In an isobaric process (2.518a), that is at constant pressure, the enthalpy (2.223b) ≡ (2.518b) reduces to the heat (2.6b) ≡ (2.518c) that is a function of state like the work (2.70c) ≡ (2.518d, e):

$$dp = 0: \quad dH = T\,dS = dQ, \quad dW = -p\,dV = -d(pV). \qquad (2.518a\text{–}e)$$

For any substance the work is (2.518e) ≡ (2.519a):

$$W_{12} = p(V_1 - V_2), \quad Q_{12} = H_2 - H_1 = C_p(T_2 - T_1)$$
$$= \frac{\gamma R}{\gamma - 1}(T_2 - T_1) = \frac{\gamma R p}{\gamma - 1}(V_2 - V_1), \qquad (2.519a\text{–}e)$$

and the heat is specified by the enthalpy, leading for a perfect gas to (2.474b; 2.472a) ≡ (2.519b–e). For an isobaric process (2.520a) of an ideal gas (2.472a) the *Charles (1787) – Gay-Lussac (1802) – Dalton (1802) law* states that volume is proportional to the temperature (2.520b) and for a perfect gas the entropy (2.477b,c) is given by (2.520c–e):

$$p = const; \quad \frac{V}{T} = const; \quad S - S_0 = C_p \log\left(\frac{T}{T_0}\right) = C_p \log\left(\frac{V}{V_0}\right) = -C_p \log\left(\frac{\rho}{\rho_0}\right). \quad (2.520\text{a--e})$$

The isobaric (isothermal) process [subsection 2.5.12 (2.5.13)] both have heat and work, but given by different expressions, in contrast with the adiabatic (isochoric) process that has only work (heat).

2.5.13 ISOTHERMAL PROCESS (BOYLE 1662, MARIOTTE 1876)

In an isothermal process, that is at constant temperature (2.521a) the heat (2.6b) ≡ (2.521b, c) is a function of state, and the work (2.70c) also because it coincides (2.222b) with the free energy (2.521d, e):

$$dT = 0: \quad dQ = T\, dS = d(T\,S), \quad dW = -p\, dV = dF. \quad (2.521\text{a--e})$$

The heat in an isothermal process (2.522a) is given by (2.522b) for any substance, and for a perfect gas the entropy is given by (2.477a, b) ≡ (2.522c–e):

$$T = const: \quad Q_{12} = T\left(S_2 - S_1\right), \quad (2.522\text{a--b})$$

$$S - S_0 = R \log\left(\frac{V}{V_0}\right) = -R \log\left(\frac{p}{p_0}\right) = -R \log\left(\frac{\rho}{\rho_0}\right). \quad (2.522\text{c--e})$$

In an isothermal process (2.523a) of an ideal gas the pressure varies inversely with the volume (2.523b) according to the **Boyle (1662) – Mariotte (1876) law**:

$$T = const: \quad pV = const';$$
$$W_{12} = -\int_{V_1}^{V_2} p\, dV = -RT \int_{V_1}^{V_2} \frac{dV}{V} = RT \log\left(\frac{V_1}{V_2}\right)$$
$$= RT \log\left(\frac{p_2}{p_1}\right) = RT \log\left(\frac{\rho_2}{\rho_1}\right), \quad (2.523\text{a--g})$$

and the work is given by (2.70c; 2.472a; 2.523b) ≡ (2.523c–g). The polytropic relation between the pressure and the volume (subsection 2.5.14) includes all four preceding cases (subsections 2.5.10–2.5.13).

2.5.14 POLYTROPIC RELATION AND THERMODYNAMIC CYCLES

The **polytropic relation** *between the pressure and the volume:*

$$pV^n = const \quad \begin{cases} n = \gamma : adiabatic : dS = 0, \\ n = 1 : isothermal : dT = 0, \\ n = 0 : isobaric : dp = 0, \\ n = \infty : isochoric : dV = 0, \end{cases} \quad (2.524\text{a--d})$$

includes for specific values of the exponent all four thermodynamic processes, namely: (i) adiabatic (2.524a) for the adiabatic exponent (2.512f); (ii) isothermal (2.524b) for exponent unity (2.523b); (iii) isobaric (2.524c) for exponent zero (2.520a); (iv) isochoric (2.524d) in the limit of infinite exponent (2.516a). The polytropic process is usually considered with exponent in the range $1 \le n \le \gamma$ from isothermal (2.524b) to adiabatic (2.524a).

The Boyle (1662) – Mariotte (1876) law (2.523a, b) $\equiv$ (2.525a, b) and the Gay-Lussac (1802) – Dalton (1802) law (2.516a, b) $\equiv$ (2.525c, d) attributed by the former to Charles (1787):

$$T = const \Rightarrow pV = const; \quad p = const \Rightarrow \frac{V}{T} = const,$$

$$\frac{pV}{T} = const = R \tag{2.525a–e}$$

when combined lead to the **Clapeyron (1834) law** or the ideal gas law (2.525e) $\equiv$ (2.460c). The ideal gas law can be used to quantify the theory of heat (**Clausius 1850, 1854**) and machines (**Carnot 1878**) that is the basis of: (i) engines that supply work; (ii) refrigerators that extract heat from a cold body; (iii) heat pumps that add heat to a hot body.

The thermodynamic processes can be combined to construct a **thermodynamic cycle**, where matter is returned to the initial state. Thus a thermodynamic cycle (subsections 2.5.15–2.5.20) describes the heat/work output/input and efficiency of a **thermodynamic machine**, such as [subsections 2.5.17–2.5.19 (2.5.20–2.5.21)] an internal combustion engine (a refrigerator or heat pump). The thermodynamic machines include the internal combustion engines, such as the piston and the jet or turbine engines, which are used in the propulsion of various vehicles (cars, trains, ships, aircraft) and as standard or emergency power units.

Besides the Carnot (1878) and **Atkinson (1882)** cycles used in piston engines using atmospheric air, there are others like the **Stirling (1816)** cycle for air independent propulsion, and the **Barber (1891) – Brayton (1930)** cycles for jet engines that are considered in Sections 2.5 and 2.6. There are many variations, and one of the most widely used is the four-stroke piston engine in petrol (**Otto 1878**) or **Diesel (1913)** forms. The detailed analysis of engines, refrigerators and pumps is preceded by the general principles of heat and work (efficiency) in a thermodynamic cycle [subsection 2.5.15 (2.5.16)].

2.5.15   HEAT AND WORK IN A THERMODYNAMIC CYCLE

A thermodynamic cycle is a sequence of thermodynamic processes in which the final state coincides with the initial state; thus *the integral of the internal energy over a thermodynamic cycle is zero (2.526a):*

$$0 = \oint dU = \oint dW + \oint dQ \equiv W + Q, \tag{2.526a–c}$$

implying (2.6a) ≡ (2.526b) that in order (2.526c) to obtain work (2.527a, c) heat must be supplied (2.527b, d, e):

$$W \equiv \oint dW, \quad Q = \oint dQ; \quad W < 0 \Rightarrow Q = -W > 0. \tag{2.527a–e}$$

The integral of the entropy in a reversible (2.273a) [irreversible (2.273b)] thermodynamic cycle is zero (2.528a) [positive (2.528b)]:

$$\oint dS \begin{cases} = 0 \text{ reversible cycle: } Q = -W, \\ > 0 \text{ irreversible cycle: } Q > -W, \end{cases} \tag{2.528a–d}$$

implying that the heat equals (2.528b) [exceeds (2.528d)] the work. The second irreversible case is inevitable in practice because: (i) heat losses, for example, by thermal conduction, reduce the heat available to perform the work; (ii) mechanical friction between moving components wastes work into heat.

In a reversible isothermal (2.529a) thermodynamic cycle there is no heat (2.529b–d) and hence no work (2.529e):

$$T = const: \quad Q = \oint \frac{dS}{T} = \frac{1}{T} \oint dS = 0 \Rightarrow W = 0. \tag{2.529a–e}$$

Thus a thermodynamic engine requires temperature differences to provide work. The simplest engine cycle is the **Carnot (1878) cycle** consisting (Figure 2.13a) of two adiabatics and two isothermals, that is a rectangle in a temperature versus entropy diagram, specified by high (low) temperatures T_+ (T_-) and entropies S_+ (S_-). Heat is supplied at high temperature (2.530a) and withdrawn at low temperature (2.530b):

$$Q_{34} = T_+ \left(S_+ - S_- \right) > 0 \quad Q_{12} = T_- \left(S_- - S_+ \right) < 0, \tag{2.530a, b}$$

so that the total heat (2.530b, c) is positive (2.530a), with the + sign implying that the thermodynamic system receives heat:

$$0 < Q = Q_{34} + Q_{12} = \left(T_+ - T_- \right)\left(S_+ - S_- \right) \geq -W, \tag{2.531a–d}$$

and equals (exceeds) the work (2.531d) provided in a reversible (irreversible) process.

2.5.16 EFFECIENCY OF THE CARNOT (1878) CYCLE FOR ANY SUBSTANCE

The **efficiency** is defined as the ratio (2.532a) of the work output to the heat input:

$$\eta_C \equiv -\frac{W}{Q_{34}} \leq \frac{Q}{Q_{34}} = \frac{Q_{12} + Q_{34}}{Q_{34}} = 1 + \frac{Q_{12}}{Q_{34}} = \frac{T_+ - T_-}{T_+} = 1 - \frac{T_-}{T_+} \leq 1, \tag{2.532a–g}$$

and cannot exceed: (i) the total heat divided by the heat input (2.532b); (ii) unity plus the ratio of heat lost to heat gained (2.532c, d), that have opposite signs; (iii) unity minus the

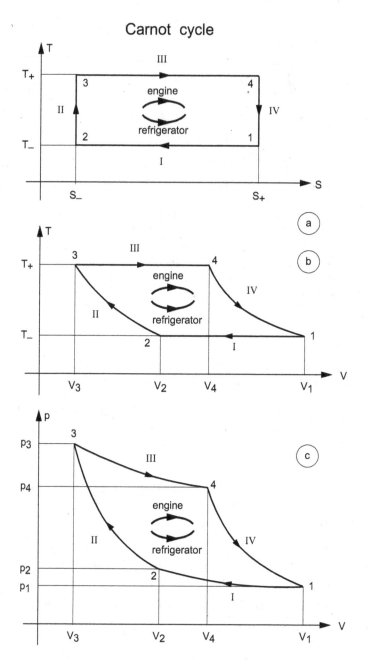

FIGURE 2.13   The Carnot thermodynamic cycle consists of two adiabatics 23 (41) at low S_- (high S_+) entropy linked by two isothermals 12 (34) at low T_- (high T_+) temperature and thus: (a) has the simplest representation as a rectangle in a temperature T versus entropy S diagram; (b) the adiabatics II and IV become curves in a temperature T versus volume V diagram; (c) the isothermals I and III also become curves in a pressure p versus volume V diagram.

ratio of cold to hot temperature (2.532d, e), that cannot exceed unity. This proves the **Carnot (1878) principle**: *an engine using a thermodynamic cycle consisting of two isothermals and two adiabatics has a maximum efficiency (2.532f) that depends only on the hot and cold temperatures, and is independent of the material substance used; the maximum efficiency is possible only in a reversible cycle, and is reduced by irreversibility at any stage.* The Carnot principle can be used as an alternative to the second principle of thermodynamics (subsection 2.3.16); this alternative is quite indirect and convoluted, like the Caratheodory principle (subsection 2.3.17), and this is why the stability principle of minimum internal energy or maximum entropy has been preferred (subsection 2.3.16) as a statement of the second principle of thermodynamics.

The maximum efficiency of the Carnot engine (2.532a–g) was calculated using only heat input (2.530a) and output (2.530b), for simplicity, avoiding the calculation of the work. For completeness the work in the Carnot cycle is calculated directly next in four stages with − sign for work supplied by the system to the outside: (i) in the adiabatic processes 23 and 41 the work (2.511e) is specified by the internal energy (2.533a, b); (ii) in the isothermal processes 34 and 12 the work (2.521e) is specified by the free energy (2.533c, d):

$$W_{23} = U_3 - U_2, \quad W_{41} = U_1 - U_4; \quad W_{34} = F_4 - F_3, \quad W_{12} = F_2 - F_1, \quad (2.533\text{a–d})$$

(ii) the total work (2.534a) is the sum over all four processes; comprising two adiabatic (2.533a, b) and two isothermal (2.533c, d) leading to (2.534b):

$$-W = -W_{12} - W_{23} - W_{34} - W_{41} = F_1 - U_1 + U_2 - F_2 + F_3 - U_3 + U_4 - F_4; \quad (2.534\text{a, b})$$

(iii) the difference (2.222a) between free and internal energy at the four corners of the Carnot cycle is (2.531a–d):

$$F_1 - U_1 = T_- S_+, \quad F_2 - U_2 = T_- S_-, \quad F_3 - U_3 = T_+ S_-, \quad F_4 - U_4 = T_+ S_+; \quad (2.535\text{a–d})$$

(iv) substituting (2.535a–d) in (2.534b) leads to the total work (2.536a, b):

$$W = T_- S_+ - T_- S_- + T_+ S_- - T_+ S_+ = -(T_+ - T_-)(S_+ - S_-) = -Q, \quad (2.536\text{a–c})$$

that equals (2.536c) minus the total heat (2.531c) in agreement with (2.526c). *The calculation of the heat (2.530a, 2.501a, b; 2.531a–d), work (2.536a–c) and efficiency (2.532a–g) of the Carnot cycle in terms of temperature and entropy (Figure 2.10a) is valid for any substance,* since no equation of state was used. The analysis of the Carnot cycle in terms of temperature and volume (pressure) requires [Figure 2.13] an equation of state (subsections 2.5.18–2.5.22), for example, for a perfect gas. Before is described an implementation of the Carnot cycle as a four-stroke piston engine (subsection 2.5.17).

2.5.17   THE FOUR-STROKE PISTON ENGINE (OTTO 1878; DIESEL 1913)

The piston engine consists of (Figure 2.14): (i) one (or more) cylinder(s) moving up and down in a chamber, with o-rings to seal the space above, and not let gases escape; (ii) the space in the cylinder above the piston is the combustion chamber that has valves to let inlet air in and exhaust gases out; (iii) the reciprocating motion of the piston is transmitted by a rod articulated at a crank, leading to a rotation of the crank shaft. Thus the piston engine produces a rotatory motion associated with a torque, for example, to drive the wheels of a car through a gearbox. The cycle of a four-stroke piston engine (Figure 2.14) consists of: (I) admission of air through the open inlet valve, by the suction due to the piston moving down; (II) when after half-a-rotation the piston moves up, the valves are closed and the air inside the combustion chamber is compressed; (III) at or near the maximum compression fuel is injected and burned either by compression (diesel engine) or ignited by a spark (petrol engine), driving the piston down; (IV) the piston is driven only in the combustion phase, and by inertia causes the crank to rotate, so that when the piston comes up again it expels the burned gases through the open exhaust valve. The four-stroke cycle starts again when the piston moves down, the exhaust valve is closed and the inlet valve is opened. The piston engine does not self-start, and needs external power such as an electric motor to crank-up to idle speed. The piston engine is limited by the maximum pressure, temperature and rotation speed that the various components can withstand.

The opening and closing of the inlet and exhaust valves have to be well synchronized with the motion of the piston to ensure safe, correct and efficient functioning of the engine. This synchronization may be insured by a timing belt (chain) with usually limited (almost unlimited) duration that connects the crankshaft below the engine to camshafts above the valves; the eccentric shape of the cams ensures opening and closing of the valves at the right moment in the engine cycle. There are three main catastrophic failures of a piston engine: (i) broken timing belt or chain causing piston and valves to collide; (ii) lack of cooling causing engine over heating; (iii) lack of lubrication causing the pistons to seize in the cylinder. In both cases of lack of cooling (ii) [lubrification (iii)] the cause may be a failed pump or a burst pipe. The thermodynamic cycle of an engine can be idealized with the help of: (i) gas sources/sinks which provide whatever quantity of gases is admitted at the inlet and (ii) heat reservoirs supply or withdraw any amount of heat without changing temperature. The (i) gas source/sinks and (ii) heat reservoirs allow a reversible thermodynamic process consisting of equilibrium states; this neglects the heat losses by entropy production in a real process, which inevitably is not at equilibrium at all stages. Thus the relevant thermodynamic cycle determines the maximum power and efficiency which can be expected from a real thermodynamic machine. The latter is the limit of what any technology can do, and is an indicator of how close or far from the optimum a particular implementation is. The Carnot cycle, relevant for the piston engine, is described alternatively in three thermodynamic diagrams namely in terms of temperature and entropy (pressure/volume) for [Figure 2.13] an arbitrary substance (perfect gas) in subsections 2.5.15–2.5.16 (2.5.18–2.5.22).

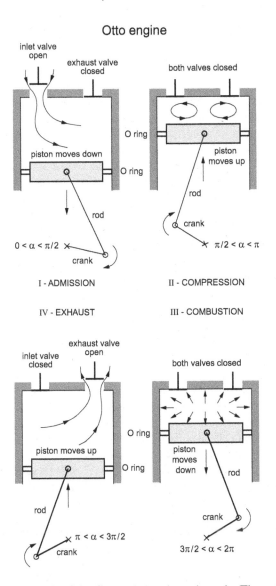

FIGURE 2.14 An example of the Carnot thermodynamic cycle (Figure 2.13) is the four-stroke piston engine: (I) the admission of air, when the piston moves down, with open inlet valve and closed outlet valve, is isothermal at low ambient temperature and increases the volume, thus decreases the entropy; (II) the compression, when the piston moves up, with inlet and exhaust valves both closed is adiabatic and increases the temperature from low to high; (III) the combustion, with inlet and exhaust valves still closed, and piston near the top of the travel, is isothermal at the high temperature, and the heat release increases entropy from low to high; (IV) when the piston is pushed down during the combustion phase (III) and moves up again (IV), with the inlet valve closed and the exhaust valve open, pushes the burned gases outside, in an adiabatic process, that reduces the temperature from the high value of combustion to the low atmospheric value. Thus the Carnot cycle of the piston engine can restart after two rotations of the crank and the rod moving the piston up and down twice.

2.5.18   THREE THERMODYNAMIC DIAGRAMS FOR THE CARNOT CYCLE

Of the three thermodynamic diagrams (Figure 2.13) for the Carnot cycle, the simplest uses as coordinates the entropy and temperature leading to a rectangle (Figure 2.13) because the four strokes or half-turns of crankshaft correspond to: (I) the admission $1 \rightarrow 2$ at constant low temperature T_-, as air is sucked in through the inlet valve, by the downward motion of the piston, decreasing the entropy from S_+ to S_-; (II) the compression $2 \rightarrow 3$ as the piston moves up with closed valves trapping the air, in an adiabatic process at low entropy S_-, that due to the compression increases the temperature from T_- to T_+; (III) combustion $3 \rightarrow 4$ near the top of the course of the piston, with valves closed, when fuel is injected and burned at constant high temperature T_+, increasing the entropy from S_- to S_+, as the piston is pushed down; (IV) exhaust $4 \rightarrow 1$ as the piston moves up again, and the exhaust valve opens, letting the burned gas escape to the atmosphere, in an adiabatic process at constant high entropy S_+, with the burning temperature T_+ reduced to the atmospheric temperature T_- of inlet, when the cycle repeats itself.

The Carnot cycle can also be represented in a thermodynamic diagram with temperature and volume as coordinates (Figure 2.13), for the four strokes: (I) admission $1 \rightarrow 2$ at constant temperature T_- decreasing the volume from V_1 corresponding to S_+ to V_2 corresponding to S_-; (II) adiabatic compression $2 \rightarrow 3$ at constant low entropy S_- decreasing the volume further from V_2 to V_3; (III) combustion $3 \rightarrow 4$ at constant high temperature T_+, increasing the entropy from S_- to S_+ and the volume from V_3 to V_4; (IV) adiabatic exhaust $4 \rightarrow 1$ at high entropy S_+, reducing the temperature from T_+ to T_- and the volume from V_4 back to V_1 as at the inlet. The Carnot cycle in the (T,S) diagram [(T,V) diagram] in Figure 2.13, has: (i) two horizontal lines, for admission (combustion) at constant low T_- (high T_+) temperature; (ii) two vertical lines $S = const$ [parabolas $p\,V_\gamma = const$ in (2.513a)] for the adiabatic compression and expansion, respectively, at low S_- and high S_+ entropy.

The Carnot cycle can also be represented in a thermodynamic diagram with pressure and volume as coordinates (Figure 2.13) for the four strokes: (I) admission $1 \rightarrow 2$ at constant low temperature T_- is a hyperbola (2.460c) $\equiv$ (2.537a) with decreasing volume $V_2 < V_1$ and increasing pressure $p_2 > p_1$; (II) the adiabatic compression $2 \rightarrow 3$ increases further the pressure $p_3 > p_2$ and (2.513a) $\equiv$ (2.537b) decreases further the volume $V_3 < V_2$; (III) the combustion $3 \rightarrow 4$ at constant high temperature (2.460c) $\equiv$ (2.537c) and increasing entropy increases the volume $V_4 > V_3$ and decreases the pressure $p_4 < p_3$; (IV) the adiabatic exhaust $4 \rightarrow 1$ reduces further the pressure $p_1 < p_4$ and (2.513a) $\equiv$ (2.537d) at constant low temperature T_- increases the volume $V_1 > V_4$:

$$p_1 V_1 = R T_- = p_2 V_2, \quad p_1 V_1^\gamma = p_2 V_2^\gamma, \quad p_2 V_2 = R T_+ = p_3 V_3, \quad p_4 V_4^\gamma = p_1 V_1^\gamma. \qquad (2.537\text{a--d})$$

Any of the three thermodynamic diagrams in Figure 2.13 can be used to calculate the heat and work in the Carnot cycle (subsection 2.5.17 and 2.5.19), although the simplest is the rectangle in the (T,S) diagram of Figure 2.13.

2.5.19   CARNOT CYCLE FOR A PERFECT GAS

The **Carnot cycle** described next consists of four thermodynamic processes modelling those in an engine. The positive (negative) sign applies to work or heat that the thermodynamic system receives from (supplies to) the exterior, so that its internal energy increases (decreases), heat is gained (lost) or work is received (performed). The Carnot cycle (Figure 2.14) for a perfect gas applies to a **four-stroke engine** described next. The (I) admission stroke is an isothermal process (subsection 2.5.13) at the "cold" temperature (2.538a) during which: (i) the volume and entropy are reduced, corresponding to a heat loss (2.538b, c):

$$I - \text{admission}: T = T_-, \quad Q_{12} = T_-\left(S_- - S_+\right) < 0, \qquad (2.538a\text{-}c)$$

$$W_{12} = F_2 - F_1 = RT_- \log\left(\frac{V_1}{V_2}\right) > 0; \qquad (2.538d\text{-}f)$$

(ii) the received work (2.538d) coincides with the variation of free energy (2.538e) and is given for a perfect gas by $(2.523e) \equiv (2.538e)$.

The (II) compression stroke is an adiabatic process (subsection 2.5.10) at constant low entropy (2.539a) and thus (i) there is no heat exchanged (2.539b):

$$II - \text{compression}: \quad S = S_-, \quad Q_{23} = 0, \qquad (2.539a, b)$$

$$W_{23} = U_3 - U_2 = \frac{R}{\gamma - 1}\left(T_+ - T_-\right) > 0; \qquad (2.539c, d)$$

(ii) the work received equals the variation of internal energy (2.539c) specified by $(2.514c) \equiv (2.539d)$.

The (III) combustion stroke is an isothermal process at the hot temperature (2.540a) and thus (2.538a–f) apply with changes of variable $(T_-, 1 \to 2) \to (T_+, 3 \to 4)$ in (2.540b–f):

$$III - \text{combustion}: T = T_+, \quad Q_{34} = T_+\left(S_+ - S_-\right) > 0, \qquad (2.540a\text{-}c)$$

$$W_{34} = F_4 - F_3 = RT_+ \log\left(\frac{V_3}{V_4}\right) < 0, \qquad (2.540d\text{-}f)$$

so that: (i) heat is received (2.540b, c); (ii) work is performed (2.540d–f).

The (IV) exhaust stroke is adiabatic thus similar to (2.539a–d) with the changes $(S_-, 2 \to 3) \to (S_+, 4 \to 1)$, and thus:

$$IV - \text{exhaust}: S = S_+, \quad Q_{41} = 0, \qquad (2.541a, b)$$

$$W_{41} = U_1 - U_4 = \frac{R}{\gamma - 1}\left(T_- - T_+\right) < 0, \qquad (2.541c\text{-}e)$$

(i) the high entropy is constant (2.541a) so there is no heat exchange (2.541b); (ii) the work $(2.514c) \equiv (2.541e)$ is supplied to the exterior.

The work and energy balance of the Carnot cycle specify the power and efficiency (subsection 2.5.20).

2.5.20 POWER OF THE CARNOT CYCLE FOR A PERFECT GAS

Since the Carnot cycle returns to the original state, the internal energy is conserved, and the work performed follows from the balance of received heat (2.538b; 2.540b) ≡ (2.542a–d):

$$W = -Q = -Q_{12} - Q_{34} = -\left(T_+ - T_-\right)\left(S_+ - S_-\right) < 0, \qquad (2.542a\text{–}d)$$

because there is heat exchange only during combustion (gain) and admission (loss), and the net heat received or work performed is the area within the cycle in the T-S diagram; in this case a rectangle (Figure 2.13) with $T_- \le T \le T_+$ and $S_- \le S \le S_+$, leading to (2.542d) that applies to any substance (2.536a–c) ≡ (2.542a–c). In the case of a perfect gas, the work in the compression (2.539c) and exhaust (2.541c) phases cancel (2.543a) and only the work in admission (2.538e) and combustion (2.540e) phases contributes to the total work:

$$W_{23} + W_{41} = 0;$$
$$W^e = W_{12} + W_{34} = RT_- \log\left(\frac{V_1}{V_2}\right) + RT_+ \log\left(\frac{V_3}{V_4}\right) \qquad (2.543a\text{–}f)$$
$$= T_-\left(S_+ - S_-\right) + T_+\left(S_- - S_+\right) = -\left(T_+ - T_-\right)\left(S_+ - S_-\right) < 0,$$

that can be expressed in terms of the entropy (2.543d, e), confirming (2.542d) ≡ (2.543f).

The Carnot cycle driven clockwise (Figure 2.15) represents an engine which: (i) receives heat by combustion (2.544a, b); (ii) looses heat in the admission (2.544c, d)

$$Q_+^e = T_+\left(S_+ - S_-\right) > 0 > T_-\left(S_- - S_+\right) = Q_-^e; \qquad (2.544a\text{–}d)$$

(iii) provides work in the process (2.545a–c):

$$-W^e = Q_+^e + Q_-^e = \left(T_+ - T_-\right)\left(S_+ - S_-\right) > 0; \quad \dot{W}^e = \frac{1}{2}\Omega\left(T_+ - T_-\right)\left(S_+ - S_-\right); \quad (2.545a\text{–}d)$$

*(iv) the **power** is the work per unit time, and involves (2.545d) the product by half the angular velocity, since a cycle takes two rotations, and increases for larger difference of temperatures and entropies. The aims of engine design are: (i) to provide the maximum work, that is – W^e maximum (2.545c; 2.546a); (ii) to use the least possible heat in combustion, that is – Q_+ minimum in (2.544b; 2.546b) to minimize fuel consumption. Thus the **maximum efficiency of an engine**, defined as the ratio of work delivered to heat consumed (2.546c):*

$$\left(-W^e\right)_{max},\left(Q_+^e\right)_{min}: \quad \eta_e \equiv \frac{-W^e}{Q_+^e} = \frac{T_+ - T_-}{T_+} = 1 - \frac{T_-}{T_+}, \qquad (2.546a\text{--}e)$$

increases (2.546d, e) for a larger ratio of hot to cold temperatures of, respectively, combustion and admission. The efficiency is zero for equal temperatures; it is unity

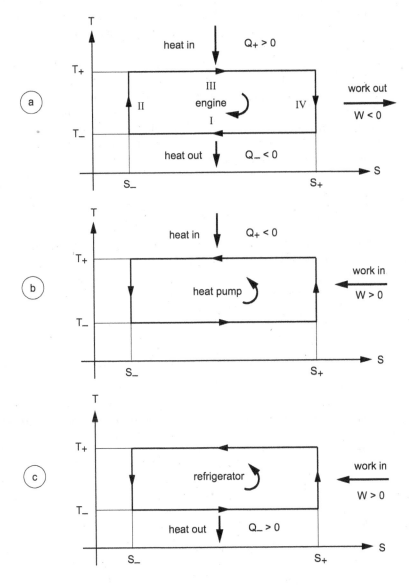

FIGURE 2.15 The Carnot thermodynamic cycle (Figure 2.13) can be used in the piston engine (Figure 2.14) to extract work by supplying heat (Figure 2.15) by combustion of fuel with air. (a) The Carnot cycle in reverse can be used supplying work from an external engine: (b) in a heat pump to transfer heat to a hot body; (c) in a refrigerator to extract heat from a cold body.

for hot body at infinite temperature, regardless of cold body temperature; also for any hot body the efficiency is unity for a cold body at zero absolute temperature.

In practice engine design aims to improve efficiency by: (i) achieving the highest hot temperature which the engine tolerates without failure or excessive wear; (ii) the lowest inlet/exhaust temperature to minimize waste heat. The exhaust-driven turbo achieves (ii) by: (i) reducing exhaust temperature; (ii) providing work to compress inlet air. Compressing inlet air with a turbo increases inlet temperature; this undesirable effect may be countered with an intercooler. The turbo increases engine power by increasing the fluid mass for the same size of cylinder; the turbo does not consume engine energy, because it is driven by exhaust gases, and only becomes effective at sufficiently high exhaust velocity, and hence high gas temperature and engine r.p.m. (revolutions per minute). An engine driven compressor is effective at low r.p.m. but consumes engine power instead of using waste energy. The combination of an engine-driven compressor at low r.p.m. and exhaust-driven turbo at high r.p.m. gives a wide power band from low to high r.p.m. As an example of Carnot cycle, for inlet air at ambient temperature (2.547a) and exhaust gas at temperature (2.547b) has an efficiency (2.490h):

$$T_- = 15°C = 288\,K, \quad T_+ = 200°C = 473\,K: \quad \eta_e = 0.391. \qquad (2.547a\text{–}c)$$

The thermodynamic efficiency is an upper limit since it does not take into account: (i) work losses by friction that reduce the numerator in (2.546c); (ii) heat losses by conduction that imply that the heat supply in the denominator of (2.546c) must be increased. The Carnot cycle driven in reverse consumes power and can transfer heat from a cold to a hot body, for example, [subsection 2.5.21 (2.5.22)] in a heat pump (refrigerator).

2.5.21   WORK TO TRANSFER HEAT FROM A COLD TO A HOT BODY

The Carnot cycle driven in reverse, that is counterclockwise in Figure 2.15, can act as **heat pump**, adding heat to the hot body, which is possible only if it receives work. *The aims of the design of a heat pump are: (i) to deliver the maximum (2.549a) heat to the hot body (2.548a, b); (ii) to do so with the least minimum (2.549b) work, which has to come from an engine or power supplies (2.548c):*

$$-Q_+^p = T_+\left(S_+ - S_-\right) > 0, \quad -W^p = \left(T_+ - T_-\right)\left(S_+ - S_-\right). \qquad (2.548a\text{–}c)$$

*The **coefficient of heat dump performance** is defined as the ratio of heat delivered to the hot body to work received from the outside (2.549c):*

$$\left(Q_+^p\right)_{max}, \quad \left(W^p\right)_{min}; \quad \eta_p \equiv \frac{-Q_+^p}{W_p} = \frac{T_+}{T_+ - T_-} = \left(1 - \frac{T_-}{T_+}\right)^{-1}, \qquad (2.549a\text{–}e)$$

and increases (2.549d, e) with the temperature of the hot body and a small temperature difference from the cold body. The coefficient of heat pump performance would be infinite if the temperatures of the hot and cold body were the same, because then there is no "thermal barrier" between the two. If the cold body is at zero temperature the performance coefficient takes the lowest possible value unity. A heat pump needs a large amount of work if the cold body is at very low temperature. As an example for a cold body at ambient temperature (2.550a) to transfer heat to a hot body at temperature (2.550b) the heat pump coefficient must be (2.550c):

$$T_- = 15°C = 288\,K, \quad T_+ = 90°C = 363\,K: \quad \eta_p = 4.84. \qquad (2.550a\text{--}c)$$

The thermal efficiencies for the engine (2.546e) and heat pump (2.549e) assume that there are no heat losses. In practice there are heat losses due to thermal conduction (subsections 2.4.1–2.4.2), viscosity (subsections 2.4.9–2.4.12), friction, imperfect combustion and other processes reducing the heat input. Work losses through friction, exhaust back pressure above the atmospheric pressure and other processes have a similar effect of reducing efficiency. Transferring heat to a hot body using a heat pump (subsection 2.5.21) is similar to extracting heat from a cold body using a refrigerator (subsection 2.5.22), and in both cases work must be supplied.

2.5.22   REVERSED ENGINE CYCLES: HEAT PUMP AND REFRIGERATOR

The Carnot cycle can also operate as a **refrigerator** (Figure 2.15) taking heat away from the cold body, which requires a work input. Like the efficiency of an engine (2.546a–e) or the performance of a heat pump (2.549a–e) the performance of a refrigerator depends only on the temperatures of the hot and cold bodies. The heat exchanged (2.544a–d) and work performed (2.545a–c) are the same in all cases, apart from sign, but the objectives are different. For the refrigerator to follow a reversible thermodynamic process variation of the total entropy must be zero (2.551a–c):

$$0 = dS = dS_+ + dS_- = \frac{dQ_+}{T_+} + \frac{dQ_-}{T_-}; \quad dW = -dQ_- - dQ_+ \qquad (2.551a\text{--}d)$$

since the internal energy is conserved the work is minus the total heat (2.551d). From (2.551c, d) follow (2.552a, b)

$$\frac{dQ_-}{dW} = -1 - \frac{dQ_+}{dW} = -1 + \frac{T_+}{T_-}\frac{dQ_-}{dW}: \quad 1 = \left(\frac{T_+}{T_-} - 1\right)\frac{dQ_-}{dW}, \qquad (2.552a\text{--}c)$$

implying (2.552c). *The objectives of refrigerator design are: (i) to extract the maximum heat from the cold body (2.553a); (ii) to do so consuming as little work as possible (2.553b). The* **coefficient of refrigerator performance** *is defined as the ratio of heat extracted from the cold body to the work performed to do so (2.553c) leading to (2.553c)* ≡ *(2.553d, e):*

$$\left(Q^r\right)_{max}, \quad \left(W^r\right)_{min}: \quad \eta_r = \frac{dQ^r}{dW^r} = \left(\frac{T_+}{T_-}-1\right)^{-1} = \frac{T_-}{T_+-T_-}, \qquad (2.553\text{a--e})$$

and increases with the temperature of the cold body and the small difference of temperature from the hot body. The performance coefficient of the refrigerator is infinite if the two bodies are at the same temperature, that is when there is no thermal barrier, either for refrigerator (2.553a–d) or for heat pump (2.549a–c). The performance coefficient of the refrigerator is zero if: (i) the hot body is at infinite temperature; (ii) the cold body is at zero temperature. Thus a large amount of work is needed to refrigerate a very cold body passing heat to a very hot body. This suggests that absolute zero temperature is difficult to attain, since an increasing amount of work is needed to extract the same amount of heat as the temperature reduces. For example, cooling water at ice temperature (2.554a) at ambient temperature (2.554b) leads to a refrigerator efficiency (2.554c):

$$T_- = 0°C = 273\,K, \quad T_+ = 15°C = 288\,K; \quad \eta_r = 18.2. \qquad (2.554\text{a--c})$$

From the general values of the efficiency of the Carnot cycle for engines (2.546a–e), heat pumps (2.549a–e) and refrigerators (2.553a–e) follow the relations (2.555a–f):

$$\eta_p = \frac{T_+}{T_+-T_-} = \frac{1}{\eta_e} = 1+\eta_r, \quad \eta_r = \frac{T_-}{T_+-T_-} = \eta_p-1 = \frac{1}{\eta_e}-1. \qquad (2.555\text{a--f})$$

The efficiency objectives are different depending on how the Carnot cycle is used: (i) maximum work output and minimum heat input for an engine (2.546a, b); (ii) minimum work input and maximum heat output to hot body for a heat pump (2.549a, b); (iii) minimum work input and maximum heat output from cold body for a refrigerator (2.553a, b). These distinct objectives are compared in Table 2.12 for all combinations of five cold (hot) body temperatures 0, 4, 100, 273, 288 K (30, 100, 300, 1727, infinite K), indicating for the Carnot cycle: (i) the efficiency (2.564e) of an engine (Figure 2.15); (ii/iii) the performance coefficient (2.549e) [(2.553c)] of [Figure 2.15] a heat pump (refrigerator). The same engine, or heat pump or refrigerator, can be operated with different thermodynamic cycles by means of mechanical changes; for example, the same four-stroke piston engine (Figure 2.14) with different valve openings for equal compression and expansion phases (expansion longer than compression) leads [subsections 2.5.16–2.5.22 (2.5.23–2.5.24)] to the Carnot (1878) [Atkinson (1882)] cycle; the Atkinson cycle (subsection 2.5.23) has higher efficiency (subsection 2.5.24) than the Carnot cycle.

2.5.23 ATKINSON (1882) CYCLE AND VARIABLE VALVE TIMING

The Carnot cycle (Figure 2.13) applies to a piston engine (Figure 2.14) with symmetric motion, so that the compression lasts as long as the expansion. The **Atkinson**

TABLE 2.12
Efficiency of Carnot Cycles for Engines, Heat Pumps and Refrigerators

Hot T_+	30 C	100 C	300 C	1727 C	∞ C
Cold T_-	303 K	373 K	673 K	2000 K	∞ K
0 K −273 C	$\eta_e = 1.000$ $\eta_p = 1.000$ $\eta_r = 0.000$	$\eta_e = 1.000$ $\eta_p = 1.000$ $\eta_r = 0.000$	$\eta_e = 1.000$ $\eta_p = 1.000$ $\eta_r = 0.000$	$\eta_e = 1.000$ $\eta_p = 1.000$ $\eta_r = 0.000$	$\eta_e = 1.000$ $\eta_p = 1.000$ $\eta_r = 0.000$
4 K −269 C	$\eta_e = 0.987$ $\eta_p = 1.013$ $\eta_r = 0.013$	$\eta_e = 0.989$ $\eta_p = 1.011$ $\eta_r = 0.011$	$\eta_e = 0.994$ $\eta_p = 1.006$ $\eta_r = 0.006$	$\eta_e = 0.998$ $\eta_p = 1.002$ $\eta_r = 0.002$	$\eta_e = 1.000$ $\eta_p = 1.000$ $\eta_r = 0.000$
100 K −173 C	$\eta_e = 0.670$ $\eta_p = 1.494$ $\eta_r = 0.493$	$\eta_e = 0.732$ $\eta_p = 1.366$ $\eta_r = 0.366$	$\eta_e = 0.851$ $\eta_p = 1.175$ $\eta_r = 0.175$	$\eta_e = 0.950$ $\eta_p = 1.053$ $\eta_r = 0.053$	$\eta_e = 1.000$ $\eta_p = 1.000$ $\eta_r = 0.000$
273 K +0 C	$\eta_e = 0.099$ $\eta_p = 10.10$ $\eta_r = 9.100$	$\eta_e = 0.268$ $\eta_p = 3.730$ $\eta_r = 2.730$	$\eta_e = 0.594$ $\eta_p = 1.683$ $\eta_r = 0.683$	$\eta_e = 0.863$ $\eta_p = 1.158$ $\eta_r = 0.158$	$\eta_e = 1.000$ $\eta_p = 1.000$ $\eta_r = 0.000$
288 K +15 C	$\eta_e = 0.050$ $\eta_p = 20.20$ $\eta_r = 19.20$	$\eta_e = 0.228$ $\eta_p = 4.388$ $\eta_r = 3.388$	$\eta_e = 0.572$ $\eta_p = 1.743$ $\eta_r = 0.748$	$\eta_e = 0.856$ $\eta_p = 1.168$ $\eta_r = 0.168$	$\eta_e = 1.000$ $\eta_p = 1.000$ $\eta_r = 0.000$

η_e - efficiency of engine;
η_p - coefficient of heat pump performance;
η_r - coefficient of refrigerator performance;

$$\eta_e = \frac{T_+ - T_-}{T_+}, \eta_p = \frac{T_+}{T_+ - T_-}, \eta_r = \frac{T_-}{T_+ - T_-},$$

Note: The efficiency of the Carnot cycle for engines η_e, heat pumps η_p and refrigerators η_r depends only on the temperature of hot T_+ and cold T_- bodies. The three efficiencies are given for all 25 combinations of: (i) five cold temperatures $T_- = 0, 4, 100, 273, 288\ K$; (ii) five hot temperatures $T_+ = 303, 373, 673, 2000, \infty\ K$.

(1882) **cycle** corresponds to a shorter compression and longer expansion, to increase the efficiency. The Atkinson engine (Figure 2.16) replaces the crank-and-rod mechanism of the Otto-Diesel engine (Figure 2.14) by a double crank-and-rod mechanism: the first crank rotates and is linked by a second rod to a junction of the second crank that oscillates and a first rod that drives the alternating motion of the piston in the cylinder. The same effect of asymmetry in the compression and expansion phases can be achieved with the single crank-and-rod mechanism (Figure 2.14) by changing the opening and closing times of the inlet and outlet valves, leading to the engine with **variable valve timing** (VVT), that may be adjusted to the rotation speed. The Carnot (Atkinson) cycles [Figure 2.13 (2.17)] seen in the pressure – volume diagram are: (i) similar in the adiabatic compression 23 and expansion 41 phases; (ii) different in the combustion 34 and cooling 12 phases that is isothermal (initially isochoric 33' and 11' because the valves are closed, and after isobaric 3'4 and 1'2 because the valves are open).

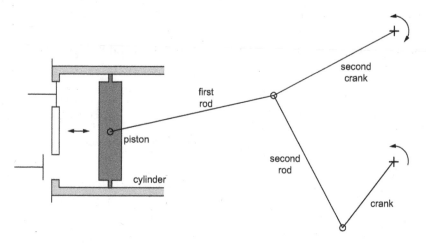

FIGURE 2.16 The piston engine with a single crank and rod (Figure 2.14) leads to the Carnot cycle (Figure 2.13) implying that the compression takes as long as the expansion. By using two cranks and rods (Figure 2.16) the expansion is longer and the compression shorter, increasing the efficiency in the Atkinson cycle (Figure 2.17).

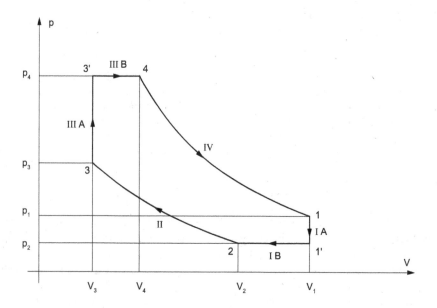

FIGURE 2.17 The Atkinson cycle with higher efficiency due to shorter compression and longer expansion (Figure 2.17) also applies to the piston engine with a single crank and rod (Figure 2.14) by changing the times of opening and closing of inlet and exhaust valves, as a function of rotation speed; this variable valve timing (VVT) increases efficiency over a range of rotation speeds. The difference between the Carnot (Figure 2.13) and Atkinson (Figure 2.17) cycles is that the admission (I) and combustion (III) is no longer isothermal but consists of isochoric and isobaric sub-phases.

In the Atkinson cycle (Figure 2.17) the highest temperature is T_4 at the end of the combustion phase: (i) that is higher (2.556a) than T_3' at the same pressure and lower volume; (ii) and in turn higher (2.556b) than T_3 at the same volume and lower pressure. Likewise the lowest temperature in the Atkinson cycle is T_2 at the end of the cooling phase: (iii) that is less (2.556c) than T_1' at the same pressure and larger volume; (iv) and in turn less (2.556d) than T_1 at the same volume and higher pressure:

$$T_4 > T_3' > T_3, \quad T_2 < T_1' < T_1 : \quad \frac{T_3}{T_4} < 1 < \frac{T_1}{T_2}, \quad \frac{T_3'}{T_4} < 1 < \frac{T_1'}{T_2}, \qquad (2.556\text{a–h})$$

The inequalities (2.556a–d) imply (2.556e–h). The heat input in the Atkinson cycle consists of: (i) an isochoric sub-phase 33' when it is specified by the internal energy (2.514a, c) ≡ (2.557a) for a perfect gas (2.517c) ≡ (2.557b); (ii) an isobaric sub-phase 3'4 when it is specified by the enthalpy (2.519a, c) ≡ (2.557c) for a perfect gas (2.519c) ≡ (2.557d):

$$Q_{33'} = U_3' - U_3 = C_V\left(T_3' - T_3\right), \quad Q_{3'4} = H_4 - H_3' = C_p\left(T_4 - T_3'\right). \quad (2.557\text{a–d})$$

Thus the total heat input (2.558d) in the combustion phase of the Atkinson cycle for a perfect gas is given (2.558a–c) by the sum of (2.557b, d):

$$Q_{34} = Q_{33'} + Q_{3'4} = C_V\left(T_3' - T_3\right) + C_p\left(T_4 - T_3'\right) \qquad (2.558\text{a, b})$$

$$= C_p T_4 - C_V T_3 - R T_3' > 0, \qquad (2.558\text{c, d})$$

where was used the gas constant (2.460b). Likewise the heat loss (2.559d) in the cooling phase 12 consists of isochoric 11' and isobaric 1'2 sub-phases leading to (2.559a–c):

$$Q_{12} = Q_{11'} + Q_{1'2} = C_V\left(T_1' - T_1\right) + C_p\left(T_2 - T_1'\right) \qquad (2.559\text{a, b})$$

$$= C_p T_2 - C_V T_1 - R T_1' < 0. \qquad (2.559\text{c, d})$$

The efficiency of the Atkinson cycle is obtained substituting (2.558c; 2.559c) in (2.532d) leading to (2.560a) and using (2.264d, e) to (2.560b, c):

$$\eta_a = 1 + \frac{Q_{12}}{Q_{34}} = 1 + \frac{C_p T_2 - C_V T_1 - R T_1'}{C_p T_4 - C_V T_3 - R T_3'} = 1 + \frac{T_2 - \dfrac{1}{\gamma}T_1 - \dfrac{\gamma-1}{\gamma}T_1'}{T_4 - \dfrac{1}{\gamma}T_3 - \dfrac{\gamma-1}{\gamma}T_3'}. \qquad (2.560\text{a–c})$$

The efficiency of the Atkinson cycle (2.560b) is compared next (subsection 2.5.24) with that of the Carnot cycle (2.532f) for the same highest and lowest temperatures.

2.5.24 COMPARISON OF EFFICIENCIES OF ATKINSON AND CARNOT CYCLES

The efficiency of the Atkinson cycle (2.560b) is compared with the efficiency of the Carnot cycle (2.532f) ≡ (2.561c, d) for the same highest (2.561a) and lowest (2.561b) temperatures:

$$T_+ = T_4 > T_- = T_2: \quad \eta_c = 1 - \frac{T_-}{T_+} = 1 - \frac{T_2}{T_4}. \qquad (2.561a\text{–}d)$$

The efficiency of the Carnot cycle (2.561d) is subtracted from that of the Atkinson cycle (2.560b), leading to (2.562a, 2.501a, b)

$$\eta_a - \eta_c = \frac{T_2}{T_4} \left[\frac{1 - \dfrac{1}{\gamma}\dfrac{T_1}{T_2} - \dfrac{\gamma-1}{\gamma}\dfrac{T_1'}{T_2}}{1 - \dfrac{1}{\gamma}\dfrac{T_3}{T_4} - \dfrac{\gamma-1}{\gamma}\dfrac{T_3'}{T_4}} + 1 \right] = \frac{T_2}{T_4} \frac{2 - \dfrac{1}{\gamma}\left(\dfrac{T_1}{T_2} + \dfrac{T_3}{T_4}\right) - \dfrac{\gamma-1}{\gamma}\left(\dfrac{T_1'}{T_2} + \dfrac{T_3'}{T_4}\right)}{1 - \dfrac{1}{\gamma}\dfrac{T_3}{T_4} - \dfrac{\gamma-1}{\gamma}\dfrac{T_3'}{T_4}}. \qquad (2.562a, b)$$

From (2.556a, b) ≡ (2.256e, g) follows that the denominator in (2.562b) is positive (2.563a, b):

$$1 - \frac{1}{\gamma}\frac{T_3}{T_4} - \frac{\gamma-1}{\gamma}\frac{T_3'}{T_4} > 1 - \frac{1}{\gamma} - \frac{\gamma-1}{\gamma} = 0, \quad \frac{T_1}{T_2} + \frac{T_3}{T_4} < 2 > \frac{T_1'}{T_2} + \frac{T_3'}{T_4}, \qquad (2.563a\text{–}c)$$

the terms in curved brackets in the numerator of (2.562b) are pairs (2.256e–h) with one greater and one smaller than unity. If their sum is less than two (2.563b, c) then the numerator is positive (2.563d) and the Atkinson cycle has higher efficiency than the Carnot cycle (2.563e):

$$2 - \frac{1}{\gamma}\left(\frac{T_1}{T_2} + \frac{T_3}{T_4}\right) - \frac{\gamma}{\gamma-1}\left(\frac{T_1'}{T_2} + \frac{T_3'}{T_4}\right) > 2 - \frac{2}{\gamma} - \frac{2(\gamma-1)}{\gamma} = 0 : \eta_a > \eta_c. \, (2.563d, e)$$

It has been shown that *for the same maximum (2.561a) and minimum (2.561b) temperature the difference between the efficiencies of the Atkinson (2.560b, c) and Carnot (2.561d) cycles is given by (2.562a). In particular if both (2.563b, c): (a) hold as equalities the efficiencies are equal; (b/c) hold with the indicated (reversed) signs the Atkinson (Carnot) cycle is more efficient. In the case (b) when both (2.563b, c) are satisfied the Atkinson cycle is more efficient than the Carnot cycle and thus: (i) produces more work for the same heat input; (ii) for the same work uses less heat.* Both the Carnot (Atkinson) cycle [subsections 2.5.15–2.5.22 (2.5.23–2.5.24)] use ambient air and is suitable for operation in open spaces. For operation in closed spaces, like a submerged submarine, air independent propulsion or power is desirable, leading to a closed cycle engine, with no inlet or exhaust. An example is the Stirling engine and cycle (subsections 2.5.25–2.5.27).

2.5.25   STIRLING (1916) AIR-INDEPENDENT ENGINE

As an example of application consider the propulsion of a submarine: (i) on the sea surface an open cycle engine may be used, with air inlet and exhaust to the atmosphere; (ii) submerged at small depth a "schnorkel" may be used, that is (Figure 2.18) a pair of inlet and exhaust ducts opening above the sea surface; (iii) fully submerged beyond "schnorkel" depth electric motors may be used, with battery power limiting speed, range and endurance; (iv) an alternative is a closed cycle thermal engine, with trapped gas, that does not need inlet or exhaust. A submarine driven by an aerobic engine with schnorkel is limited to periscopic depth (Figure 2.18) whereas with **air independent propulsion** (AIP) based on a Stirling engine it can travel submerged at higher speed and with more endurance than with battery power. The Stirling engine can also be used in other closed spaces without ventilation. The Stirling engine is described with a single piston (Figure 2.19) ordering the four phases in analogy with the four strokes of the Otto engine (Figure 2.14) as seen in the thermodynamic diagram in Figure 2.20: (I) in the isothermal compression phase $1 \rightarrow 2$ at low temperature T_- the piston descends decreasing the volume from V_+ to V_-; (II) in the isochoric heating phase $2 \rightarrow 3$ with the piston close to the lowest position at volume V_- heat is added raising the temperature from low T_- to high T_+; (III) in the isothermal expansion phase $3 \rightarrow 4$ at high temperature T_+ the piston moves up increasing the volume from V_- to V_+; (IV) in the isochoric cooling phase $4 \rightarrow 1$ with the piston near maximum volume V_+ the temperature decreases from T_+ to T_-, restarting the Stirling cycle. Thus the Carnot (Stirling) cycle is a rectangle in the thermodynamic diagram [Figure 2.13 (2.20)] with temperature in the vertical axis and entropy (volume) in the horizonal axis.

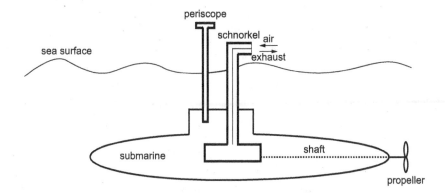

FIGURE 2.18   The use of a piston engine (Figure 2.14) with the Carnot (Figure 2.13) or Atkinson (Figure 2.17) cycle in a closed space requires an inlet to let air in and an exhaust to let gases out, regardless of whether the combustion is: (a) initiated by a spark in an Otto engine burning petrol; (b) caused by compression in a Diesel engine burning oil. In the case of a submarine, the use of an aerobic engine that needs air, requires a "schnorkel" for admission of air and exhaust gases and limits operation to "periscopic" depth. For deep submerged operations the electric motors provide limited speed, endurance and range. An alternative with better performance is Air Independent Propulsion (AIP) using the Stirling cycle (Figures 2.19–2.20).

Stirling engine

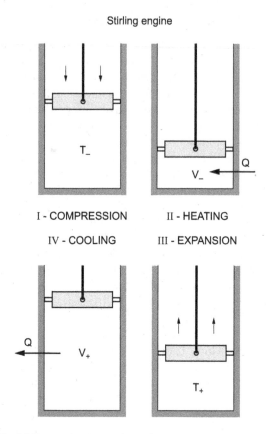

I - COMPRESSION   II - HEATING

IV - COOLING   III - EXPANSION

FIGURE 2.19 The Stirling engine in the simplest single piston, single rod form, has no inlet or exhaust, and operates in four phases: (I) isothermal compression at low temperature T_- as the piston moves down; (II) isochoric heat input with the piston near the lowest position with volume V_-; (III) the heat input pushes the piston up in an isothermal expansion at high temperature T_+; (IV) the isochoric cooling takes place with the piston near the highest position of maximum volume V_+. The cycle of the Stirling engine can restart after one up-and-down motion of the pistons (Figure 2.19) instead of two for the Carnot cycle of piston engine (Figure 2.13).

Thus the simplest thermodynamic diagram for the Stirling cycle (Figure 2.19) is with temperature and volume as coordinates (Figure 2.20b) and consists of straight lines: (I, III) horizontal for compression $1 \to 2$ (expansion $3 \to 4$) at constant low T_- (high T_+) temperature that decreases (increases) the volume; (II, IV) vertical for heating $2 \to 3$ (cooling $4 \to 1$) at constant low V_- (high V_+) volume that increases (decreases) the temperature. The thermodynamic diagram for the Stirling cycle with temperature and entropy as variables (Figure 2.20a): (I, III) retains the horizontal lines for compression $1 \to 2$ (expansion $3 \to 4$) at constant low T_- (high T_+) temperature that decreases (increases) entropy from S_1 (S_3) to S_2 (S_4); (II, IV) the isochoric heating $2 \to 3$ (cooling $4 \to 1$) at constant low V_- (high V_+) volume increases (decreases) the entropy from S_2 (S_4) to S_3 (S_1). The thermodynamic diagram for the Stirling cycle using pressure and volume as variables (Figure 2.20c): (II, IV) retains two vertical

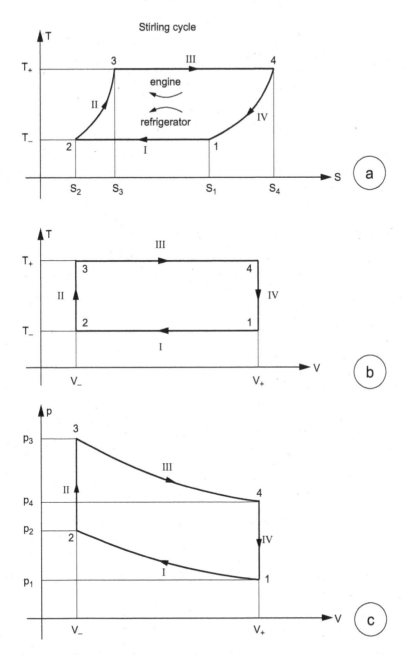

FIGURE 2.20 The operation of the Stirling engine (Figure 2.19) consists of two isothermals linked to two isochorics and thus the simplest representation is a rectangle (Figure 2.20b) in a temperature versus volume diagram. Using a temperature versus entropy diagram (Figure 2.20a) the heating (II) and cooling (IV) phases become curves. Using the pressure versus volume diagram (Figure 2.20c) it is the compression (I) and expansion (III) phases that become curves. The Carnot (Stirling) cycles [Figure 2.15 (2.20) can both be used in opposite directions in (i) either an engine (ii) or a heat pump or refrigerator.

lines for heating $2 \to 3$ (cooling $4 \to 1$) at constant low V_- (high V_+) volume, increasing (decreasing) the pressure from p_2 (p_4) to p_3 (p_1); (I, III) the compression $1 \to 2$ (expansion $3 \to 4$) at constant low T_- (high T_+) temperature increases (decreases) the pressure from p_1 (p_3) to p_2 (p_4). The heat and work exchanges during the Stirling cycle (subsection 2.5.25) are considered next (subsection 2.5.26) for a perfect gas.

2.5.26   Heat and Work in the Stirling Cycle

The four phases (Figure 2.19) of the Stirling cycle (Figure 2.20) are: (I) compression $1 \to 2$ at constant low temperature (2.564a) with heat loss (2.522b) $\equiv$ (2.564b) and work gain (2.523e) $\equiv$ (2.564c):

$$I - \text{compression}: T = T_-: \quad Q_{12} = T_-\left(S_2 - S_1\right) < 0, \quad W_{12} = RT_- \log\left(\frac{V_+}{V_-}\right) > 0; \quad (2.564a\text{-}c)$$

(II) heating $2 \to 3$ at constant low volume (2.565a), without work (2.515c) $\equiv$ (2.565b) and with heat gain (2.517d) $\equiv$ (2.565c):

$$II - \text{heating}: V = V_- \quad W_{23} = 0, \quad Q_{23} = \frac{R}{\gamma-1}\left(T_+ - T_-\right) > 0; \quad (2.565a\text{-}c)$$

(III) expansion $3 \to 4$ at constant high temperature (2.566a) for which the heat gain is (2.522b) $\equiv$ (2.566b) and the work loss is (2.523e) $\equiv$ (2.566c):

$$III - \text{expansion}: T = T_+: \quad Q_{34} = T_+\left(S_4 - S_3\right) > 0, \quad W_{34} = RT_+ \log\left(\frac{V_-}{V_+}\right) < 0; \quad (2.566a\text{-}c)$$

(IV) cooling $4 \to 1$ at constant high volume (2.567a), without work (2.515c) $\equiv$ (2.567b) and with heat loss (2.517d) $\equiv$ (2.567c):

$$IV - \text{cooling}: V = V_+: \quad W_{41} = 0, \quad Q_{41} = \frac{R}{\gamma-1}\left(T_- - T_+\right) < 0. \quad (2.567a\text{-}c)$$

The description of the Stirling cycle (subsection 2.5.26) leads to its power and efficiency (subsection 2.5.27).

2.5.27   Power and Efficiency of the Stirling Cycle

The Stirling cycle performs work in the expansion (2.566c) and compression (2.564c) for a total work (2.568a-c):

$$W^S = W_{34} + W_{12} = -R\left(T_+ - T_-\right)\log\left(\frac{V_+}{V_-}\right) < 0. \quad (2.568a\text{-}c)$$

There is heat exchange in all four phases with cancelation (2.569a) for cooling (2.567c) and heating (2.565c), so that only the expansion (2.566b) and compression (2.565b) contribute to the total heat (2.569b, c):

$$Q_{41} + Q_{23} = 0; \quad Q^S = Q_{12} + Q_{34} = T_- \left(S_2 - S_1 \right) + T_+ \left(S_4 - S_3 \right). \quad (2.569a\text{--}c)$$

For a perfect gas in an isothermal process (2.522c) implies (2.570a, 2.501a, b):

$$S_2 - S_1 = R \, \log\left(\frac{V_2}{V_1}\right) = R \, \log\left(\frac{V_-}{V_+}\right), \quad S_4 - S_3 = R \, \log\left(\frac{V_4}{V_3}\right) = R \, \log\left(\frac{V_+}{V_-}\right), \quad (2.570a, b)$$

that substituted in (2.569c) specify the total heat (2.571a):

$$Q^S = R\left(T_+ - T_-\right)\log\left(\frac{V_+}{V_-}\right) = -W^S, \quad (2.571a, b)$$

that equals (2.571b) minus the work (2.568c) because the internal energy is conserved. The heat supplied in the Stirling cycle during the expansion phase is (2.572a) $\equiv$ (2.566b) leading to (2.572b, c):

$$Q_{34}^S = T_+ \left(S_4 - S_3 \right) = T_+ \left[S(T_+, V_+) - S(T_+, V_-) \right] = RT_+ \log\left(\frac{V_+}{V_-}\right), \quad (2.572a\text{--}c)$$

using (2.477a).

The efficiency of the Stirling engine is the ratio of the total work (2.568c) to the heat supplied during the heating expansion phase (2.572c) and is given by (2.573a, b)

$$\eta_S = \frac{-W^S}{Q_{34}^S} = 1 - \frac{T_-}{T_+} = \eta_e, \quad (2.573a\text{--}c)$$

and coincides with that of the Carnot engine (2.573c) $\equiv$ (2.546e). The power delivered by the Stirling engine equals (2.574) the work (2.571b) multiplied by the angular velocity:

$$\dot{W}^S = -W^S \Omega = \Omega R\left(T_+ - T_-\right)\log\left(\frac{V_+}{V_-}\right). \quad (2.574)$$

The Carnot (Stirling) cycles [subsections 2.5.15–2.5.22 (2.5.25–2.5.27)] are next compared [subsection 2.5.28 (2.5.29)] as concerns work output, heat input, efficiency and power (pressure at the four corners of the thermodynamic diagram).

2.5.28   MORE WORK IN THE CARNOT THAN IN THE STIRLING CYCLE

The work of the engine using the Carnot (2.543e) [Stirling (2.568c)] cycle involves the high and low temperatures and entropies (volumes). In order to make a comparison the three pairs of thermodynamic variables are related by (2.477a) $\equiv$ (2.575a) for an ideal gas:

$$S_+ - S_- = C_V \log\left(\frac{T_+}{T_-}\right) + R \, \log\left(\frac{V_+}{V_-}\right) = R\left[\log\left(\frac{V_+}{V_-}\right) + \frac{1}{\gamma - 1}\log\left(\frac{T_+}{T_-}\right)\right], \quad \text{(2.575a, b)}$$

using (2.264d) in (2.575b). Substituting (2.575b) in (2.548c) leads to the work of the Carnot cycle (2.576) in terms of temperatures and volumes:

$$W^e = -R\left(T_+ - T_-\right)\left[\log\left(\frac{V_+}{V_-}\right) + \frac{1}{\gamma - 1}\log\left(\frac{T_+}{T_-}\right)\right], \quad \text{(2.576)}$$

that exceeds the work of the Stirling cycle (2.568c) by (2.577a, 2.501a, b)

$$\left|W^e\right| - \left|W^S\right| = R\frac{T_+ - T_-}{\gamma - 1}\log\left(\frac{T_+}{T_-}\right) > 0. \quad \text{(2.577a, b)}$$

The heat supply in the Carnot cycle during combustion (2.544a; 2.575b) $\equiv$ (2.578):

$$Q_+^e = RT_+\left[\log\left(\frac{V_+}{V_-}\right) + \frac{1}{\gamma - 1}\log\left(\frac{T_+}{T_-}\right)\right], \quad \text{(2.578)}$$

also exceeds the heat supply to the Stirling engine during the thermal expansion phase (2.572c) by (2.579a, 2.501a, b):

$$Q_{12}^r - Q_+^e = \frac{RT_+}{\gamma - 1}\log\left(\frac{T_+}{T_-}\right) > 0. \quad \text{(2.579a, b)}$$

The same factor, namely the same efficiency (2.573a–c) appears in the work (2.577a, b) [heat supply (2.579a, b)] differences between the Carnot (2.543a, f) $\equiv$ (2.576) [(2.544a) $\equiv$ (2.578)] and Stirling (2.568a–c) [(2.572a–c)] cycles. The pressures are different at each of the four corners of the thermodynamic diagram, both for the Carnot (Stirling) cycle [Figure 2.13 (2.20)] and are compared next (subsection 2.5.29).

2.5.29   CONDITIONS AT THE FOUR CORNERS OF THE THERMODYNAMIC DIAGRAM

The pressures are compared in the Carnot (Stirling) cycles [Figures 2.13–2.15 (2.19–2.20)] starting with the lowest p_{1C} (p_{1S}) that is assumed to be the same (2.580a). The second pressure p_{2C} (p_{2S}) in the Carnot (Stirling) cycle in $1 \rightarrow 2$ the admission (compression) phase I results from an isothermal compression from entropy S_+ to S_- (volume V_+ to V_-) in (2.568b) $\equiv$ (2.580b) [(2.472a) $\equiv$ (2.580c)]

$$p_{1C} = p_{1S}: \quad S_- - S_+ = -R \log\left(\frac{p_{2C}}{p_{1C}}\right), \quad p_{1S} V_+ = p_{2S} V_-, \qquad \text{(2.580a–c)}$$

implying (2.580b) $\equiv$ (2.581a) [(2.580c) $\equiv$ (2.581b)] and hence (2.581c):

$$\frac{p_{2C}}{p_{1C}} = \exp\left(\frac{S_+ - S_-}{R}\right) > 1, \quad \frac{p_{2S}}{p_{1S}} = \frac{V_+}{V_-} > 1, \qquad \text{(2.581a, b)}$$

$$\frac{p_{2C}}{p_{2S}} = \frac{V_-}{V_+} \exp\left(\frac{S_+ - S_-}{R}\right) = \left(\frac{T_+}{T_-}\right)^{1/(\gamma-1)} > 1, \qquad \text{(2.581c–e)}$$

from (2.581c; 2.575b) follows (2.581d) proving that the pressure is higher in the Carnot cycle than in the Stirling cycle.

The third pressure p_{3C} (p_{3S}) in the Carnot (Stirling) cycle in $2 \rightarrow 3$ the compression (heating) phase II is the result of adiabatic (isochoric) heating from the temperature T_- to T_+ in (2.478f) $\equiv$ (2.582a) [(2.472a) $\equiv$ (2.582b)] implying (2.582c):

$$\frac{p_{3C}}{p_{2C}} = \left(\frac{T_+}{T_-}\right)^{\gamma/(\gamma-1)} > 1, \quad \frac{p_{3S}}{p_{2S}} = \frac{T_+}{T_-} > 1, \qquad \text{(2.582a, b)}$$

$$\frac{p_{3C}}{p_{3S}} = \left(\frac{T_+}{T_-}\right)^{1/(\gamma-1)} \quad \frac{p_{2C}}{p_{2S}} = \left(\frac{T_+}{T_-}\right)^{2/(\gamma-1)} > 1, \qquad \text{(2.582c–e)}$$

that simplifies (2.581d) to (2.582d) showing that the maximum pressure is higher in the Carnot cycle compared with the Stirling cycle (2.582e).

The fourth pressure p_{4C} (p_{4S}) in the Carnot (Stirling) cycle in $3 \rightarrow 4$ the combustion (expansion) phase III is an isothermal expansion from the entropy S_- to S_+ (volume V_- to V_+) in (2.477b) $\equiv$ (2.583a) [(2.472a) $\equiv$ (2.583b)]

$$S_+ - S_- = -R \log\left(\frac{p_{4C}}{p_{3C}}\right), \quad p_{3S} V_- = p_{4S} V_+, \qquad \text{(2.583a, b)}$$

implying (2.584a, b) and (2.584c):

$$\frac{p_{4C}}{p_{3C}} = \exp\left(\frac{S_- - S_+}{R}\right) < 1, \quad \frac{p_{4S}}{p_{3S}} = \frac{V_-}{V_+} < 1: \quad \frac{p_{4C}}{p_{4S}} = \frac{V_+}{V_-} \exp\left(\frac{S_- - S_+}{R}\right) \frac{p_{3C}}{p_{3S}}; \qquad \text{(2.584a–c)}$$

using (2.582c) in (2.584c) leads to (2.585a) and substitution of (2.575b) yields (2.585b):

$$\frac{p_{4C}}{p_{4S}} = \frac{V_+}{V_-} \exp\left(\frac{S_- - S_+}{R}\right)\left(\frac{T_+}{T_-}\right)^{2/(\gamma-1)} = \left(\frac{T_+}{T_-}\right)^{1/(\gamma-1)} > 1, \qquad (2.585\text{a--c})$$

again showing a larger pressure in the Carnot than in the Stirling cycle.

The $4 \to 1$ phase IV in the Carnot (Stirling) cycle is an adiabatic (isochoric) cooling from the temperature T_+ to T_- in (2.478f) $\equiv$ (2.586a) [(2.472a) $\equiv$ (2.586b)] implying (2.586c):

$$\frac{p_{1C}}{p_{4C}} = \left(\frac{T_-}{T_+}\right)^{\gamma/(\gamma-1)}, \qquad \frac{p_{1S}}{p_{4S}} = \frac{T_-}{T_+}, \qquad (2.586\text{a, b})$$

$$\frac{p_{1C}}{p_{1S}} = \frac{p_{1C}}{p_{4C}}\frac{p_{4C}}{p_{4S}}\frac{p_{4S}}{p_{1S}} = \left(\frac{T_+}{T_-}\right)^{-\gamma/(\gamma-1)+1/(\gamma-1)+1} = 1. \qquad (2.586\text{c, d})$$

Substituting (2.585b) in (2.586c) leads to (2.586d) $\equiv$ (2.580a) confirming that the cycle has been completed at the original value.

It has been shown that *the ratio of pressures at the four corners of the thermodynamic diagram for the Carnot (Stirling) cycle [Figure 2.13 (2.20)] is given by (2.581a) $\equiv$ (2.587a), (2.582a; 2.575b) $\equiv$ (2.587b) and (2.586a) $\equiv$ (2.587c) [(2.581b) $\equiv$ (2.588a), (2.582b) $\equiv$ (2.588b), (2.586b) $\equiv$ (2.588c)]:*

$$\frac{p_{2C}}{p_{1C}} = \exp\left(\frac{S_+ - S_-}{R}\right) > 1, \quad \frac{p_{3C}}{p_{1C}} = \frac{V_+}{V_-}\left(\frac{T_+}{T_-}\right)^{(\gamma+1)/(\gamma-1)} > 1, \quad \frac{p_{4C}}{p_{1C}} = \left(\frac{T_+}{T_-}\right)^{\gamma/(\gamma-1)} > 1, \; (2.587\text{a--c})$$

$$\frac{p_{2S}}{p_{1S}} = \frac{V_+}{V_-} > 1, \quad \frac{p_{3S}}{p_{1S}} = \frac{T_+ V_+}{T_- V_-} > 1, \quad \frac{p_{4S}}{p_{1S}} = \frac{T_+}{T_-} > 1. \qquad (2.588\text{a--c})$$

The ratio of the pressures at the corresponding corners of the thermodynamic diagrams for the Carnot and Stirling cycles follows from (2.587a--c; 2.588a--c) and is given by (2.581d; 2.582d; 2.585b) $\equiv$ (2.589b--d):

$$\theta \equiv \left(\frac{T_+}{T_-}\right)^{1/(\gamma-1)} > 1; \quad \left\{\frac{p_{2C}}{p_{2S}}, \frac{p_{3C}}{p_{3S}}, \frac{p_{4C}}{p_{4S}}\right\} = \left\{\theta, \theta^2, \theta\right\}\frac{p_{1C}}{p_{1S}}, \qquad (2.589\text{a--d})$$

involving the same parameter (2.589a). For the same value of the pressure at one corner (2.580a) the Carnot cycle has higher pressures than the Stirling cycle at all other corners (2.581e; 2.582e; 2.585c) explaining why it has larger work (2.577a,b). The proof of (2.587b) follows using (2.582a, 2.581a) in (2.590a), and then (2.575b) in (2.590b) leading to (2.590c) $\equiv$ (2.587b):

$$\frac{p_{3C}}{p_{1C}} = \frac{p_{3C}}{p_{2C}}\frac{p_{2C}}{p_{1C}} = \left(\frac{T_+}{T_-}\right)^{\gamma/(\gamma-1)}\exp\left(\frac{S_+ - S_-}{R}\right) = \frac{V_+}{V_-}\left(\frac{T_+}{T_-}\right)^{(\gamma+1)/(\gamma-1)}. \qquad (2.590\text{a--c})$$

The larger pressures imply larger forces, accelerations and stresses in the Otto engine relative to the Stirling engine, requiring a more robust design. Other types of engine exist using thrust, such as the turbojet, ramjet, rocket and ionic (subsection 2.4.15) engines. In these cases the high flow velocities require consideration of fluid dynamics of compressible flows (Sections 2.6–2.21), that is both an application and an extension of thermodynamics (Sections 2.1–2.5). In the context of compressible fluid mechanics, the reversible (irreversible) processes [Sections 2.1–2.2 (2.3–2.4)] are illustrated [Section(s) 2.6 (2.7–2.8)] by adiabatic flow (shock waves), that can coexist in a nozzle (Section 2.9), and apply to real/ideal and imperfect/perfect gases (Section 2.5).

2.6   ADIABATIC COMPRESSIBLE FLUID FLOW

Thermodynamics applies not only to thermal machines (subsections 2.5.15–2.5.29) but also to the constitutive (Sections 2.1–2.2) and diffusive (Sections 2.3–2.4) properties of matter, implying it affects many physical processes, for example, the flow of a compressible fluid without (with) dissipation [Sections 2.6–2.9 (Notes 2.5–2.21)]; the simplest case excludes external forces and mass and energy sources. Fluid flow must satisfy besides the equation of state (subsections 2.5.2–2.5.9), the conservation of mass (subsection 2.6.1), momentum (subsection 2.6.2) and energy (subsection 2.6.4). The equation of energy is an alternative to the equation of entropy, which, for a non-dissipative fluid corresponding to an isentropic flow, is conserved along streamlines. The equation of state may be combined (subsection 2.6.3) with the equations of entropy and continuity. The set of three equations of mass, momentum and energy conservation can be put in a similar conservation form (subsection 2.6.5). A conservation equation for a flow variable states that it is constant in a reference frame moving locally with the flow, and hence its convected derivative is zero (subsection 2.6.7). For example, if the vorticity or curl of the velocity is zero in a potential flow, it remains so for all time in homentropic conditions, that is for constant entropy (subsection 2.6.6), implying also that the circulation is conserved along streamlines (subsection 2.6.8).

The simplest solutions of the equations of non-dissipative fluid mechanics are for irrotational flows (subsection 2.6.9) leading to the Bernoulli equation in its simplest form for an incompressible fluid (subsection 2.6.10) with extension to compressible homentropic flow (subsection 2.6.11) in the absence of heat input, using the stagnation enthalpy (subsection 2.6.12). The properties of irrotational homentropic steady flow are considered in some detail including: (i) the relation of the temperature, sound speed, mass density and pressure with the velocity (subsection 2.6.13); (ii) the critical or sonic condition for which the flow velocity equals the sound speed (subsection 2.6.14); (iii) three Mach numbers defined as the ratios of the flow velocity to the local, stagnation and critical sound speeds (subsection 2.6.15) and their upper limits for non-negative flow variables (subsection 2.6.16).

The Mach number can be used to classify speed into seven ranges (subsections 2.6.17–2.6.18), namely incompressible, subsonic, transonic, supersonic, hypersonic, orbital and escape and identify four barriers (subsection 2.6.19), namely sound, heat, ionization and dissociation. These speed ranges affect the choice of propulsion systems including piston engines, jets with or without turbines, rockets and electric propulsion (subsections 2.6.20–2.6.22) and hybrid combinations (Section 2.6.23).

Besides the Carnot/Atkinson (Stirling) thermodynamic cycles [subsections 2.5.15–2.5.22/2.5.23–2.5.24 (2.5.26–2.5.29)] that is open (closed), and provides aerobic (air independent) propulsion for low speeds, is also considered the Barber-Brayton cycle (subsections 2.6.24–2.6.28) that applies to turbojets suitable for high-speed flight (subsections 2.6.30–2.6.35) in the atmosphere (subsection 2.6.29). Numerical examples are given for the Barber-Brayton (Carnot) cycle of a turbojet (petrol or diesel piston) engine [subsections 2.6.27–2.6.34 (example 10.5)].

2.6.1   EQUATION OF CONTINUITY OR MASS CONSERVATION

The equations of fluid mechanics are considered for a flow in the absence of external forces in two cases: (i) non-dissipative for an isentropic flow for which the entropy is conserved along streamlines (subsections 2.6.1–2.6.5); (ii) dissipative allowing for thermal conduction, viscous stresses, mass diffusion and chemical reactions (Notes 2.5–2.21). Both for non-dissipative and dissipative flows, the equation of continuity states that the output Q of mass sources in a domain equals (2.591a) the increase of total mass in the volume dV plus the mass flux $\vec{I}$ crossing the boundary ∂D area $d\vec{A}$:

$$\int_V Q\,dV = \frac{d}{dt}\int_D \rho\,dV + \int_{\partial D} \vec{I}.d\vec{A}. \qquad (2.591\text{a})$$

Using the divergence theorem (III.5.163a–c) in (2.591a) gives (2.591b):

$$\int_D \left[\frac{\partial \rho}{\partial t} + \nabla.\vec{I} - Q \right] dV = 0, \qquad (2.591\text{b})$$

and since the domain is arbitrary the integrand must be zero, leading to the *equation of continuity stating that the output of mass sources $Q > 0$ (sinks $Q < 0$) per unit volume and time equals (2.591c) the local time rate of change of the mass density plus the divergence of the mass flux:*

$$\frac{\partial \rho}{\partial t} + \nabla.\vec{I} = Q. \qquad (2.591\text{c})$$

In the absence of dissipation there is no diffusion of mass, and the mass flux is convective, that is equals (2.592a) the product of the mass density by the velocity; in the absence of mass sources or sinks (2.592b) the continuity Equation (2.591c) becomes (2.592c):

$$\vec{I} = \rho\vec{v}, \quad Q = 0: \quad \frac{\partial \rho}{\partial t} + \nabla.\left(\rho\vec{v}\right) = 0. \qquad (2.592\text{a–c})$$

In a **co-moving reference frame**, dragged along by the fluid, the fluid velocity is zero (2.593a) and the mass density is conserved (2.593b):

$$\bar{v}=0: \quad \frac{\partial \rho \, \bar{v}}{\partial t}=0; \quad \bar{v}\neq 0: \quad \frac{\partial \rho}{\partial t}+\frac{L}{\bar{v}}\{\rho\}=0 \quad \Rightarrow \quad \frac{L}{\bar{v}}\{\rho\}=\nabla.(\rho\bar{v}), \quad (2.593\text{a--e})$$

in a reference frame at rest the flow velocity is not zero (2.593c) and the **convection derivative (Lie 1888)** of the mass density must be added to (2.592b) leading to (2.593d). Comparing (2.592c) $\equiv$ (2.593d) follows that the *convective derivative of the mass density with regard to the velocity equals* (2.593e) *the divergence of their product. Since the mass density is a function of position (time) for a **non-uniform (unsteady) flow** (2.594a) the **material derivative** or total time derivative (2.594b) is given in index notation by (2.594c, d):*

$$\rho=\rho\left(x_i,t\right): \quad \frac{d\rho}{dt}=\frac{\partial \rho}{\partial t}+\frac{\partial \rho}{\partial x_j}\frac{dx_j}{dt}=\frac{\partial \rho}{\partial t}+v_j\frac{\partial \rho}{\partial x_j}, \quad \frac{d\rho}{dt}\equiv\frac{\partial \rho}{\partial t}+\bar{v}.\nabla\rho, \quad (2.594\text{a--e})$$

or alternatively in vector notation by (2.594e) showing that it is the sum of: (i) the local variation with time at a fixed position; (ii) the convective effect of transport by the fluid at a fixed time.

The material derivative (2.594e) can be used in the equation of continuity (2.592c) $\equiv$ (2.595b, c) without mass sources or sinks (2.592b) $\equiv$ (2.595a):

$$Q=0: \quad 0=\frac{\partial \rho}{\partial t}+\bar{v}.\nabla\rho+\rho\nabla.\bar{v}=\frac{d\rho}{dt}+\rho\nabla.\bar{v}, \quad (2.595\text{a--c})$$

also in the alternative form (2.595d, e):

$$-\nabla.\bar{v}=\frac{1}{\rho}\frac{d\rho}{dt}=\frac{d}{dt}\left(\log\rho\right). \quad (2.595\text{d, e})$$

*The mass density is constant along streamlines for an **incompressible flow**, that is the material derivative of the mass density is zero (2.596a):*

$$incompressible\ flow: \frac{d\rho}{dt}=0 \Leftrightarrow \nabla.\bar{v}=0, \quad (2.596\text{a, b})$$

*implying that the **dilatation** (2.596b) that is the divergence of the velocity is zero.*

2.6.2 EQUATION OF MOMENTUM FOR AN INVISCID FLUID (EULER 1755)

In the absence of external volume forces, the equation of momentum balances the inertia force in a domain against the surface forces on the boundary:

$$\int_D \rho \frac{d\bar{v}}{dt}dV=-\int_{\partial D} p\,d\bar{A}=-\int_D \nabla p\,dV; \quad (2.597\text{a, b})$$

the latter consists of an inward pressure (2.597a) in the normal direction in the absence of viscous stresses; the viscous stresses are a dissipative effect included in the Navier-Stokes equations (subsections 2.4.9–2.4.11; Notes 2.4–2.7). Applying the gradient theorem (III.5.170a–c) to the r.h.s. of (2.597a) leads to (2.597b) for an arbitrary domain, thus implying the *inviscid momentum equation (Euler 1755) which shows (2.598a–c) that the fluid is accelerated in the direction of decreasing pressure:*

$$-\frac{1}{\rho}\nabla p = \frac{d\bar{v}}{dt} = \frac{\partial\bar{v}}{\partial t} + (\bar{v}.\nabla)\bar{v} \equiv \bar{a}, \qquad (2.598a\text{–}c).$$

The acceleration (2.598c) ≡ (2.599a–c) is the material derivative (2.594a–e) of the velocity with regard to time:

$$a_i \equiv \frac{dv_i}{dt} = \frac{\partial v_i}{\partial t} + \frac{\partial v_i}{\partial x_j}\frac{dx_j}{dt} = \frac{\partial v_i}{\partial t} + v_j\frac{\partial v_i}{\partial x_j}, \qquad (2.599a\text{–}c)$$

and includes local and convective terms, respectively, the first and second on the r.h.s. of (2.598b) ≡ (2.599b, c). Thus a constant pressure (2.600a) corresponds to uniform flow (2.600b):

$$p = const \quad \Leftrightarrow \quad \bar{v} = const, \qquad (2.600a, b)$$

that is constant velocity, in the inviscid case and in the absence of external volume forces.

2.6.3 EQUATION OF STATE FOR AN ISENTROPIC FLOW

The thermodynamic state of the fluid is specified by the equation of state that is a relation among any three of four thermodynamic variables: (i) pressure p; (ii) temperature T; (iii) mass density (2.601a) or specific volume (2.601b); (iv) entropy per unit mass (2.601c):

$$\rho \equiv \frac{dm}{dV} = \frac{1}{\vartheta}, \quad s \equiv \frac{dS}{dm}; \quad u \equiv \frac{dU}{dm} \qquad (2.601a\text{–}d)$$

all four thermodynamic variables appear together with the internal energy per unit mass (2.601d) specified by the first principle of thermodynamics (2.221d) ≡ (2.602a, b):

$$du = T\,ds - p\,d\vartheta = T\,ds + \frac{p}{\rho^2}d\rho. \qquad (2.602a, b)$$

The equation of state, expressing the pressure in terms of the mass density and entropy (2.603a), holds in a convected frame (2.603b):

$$p = p(\rho,s): \quad \frac{dp}{dt} = \left(\frac{\partial p}{\partial \rho}\right)_s\frac{d\rho}{dt} + \left(\frac{\partial p}{\partial s}\right)_\rho\frac{ds}{dt}. \qquad (2.603a, b)$$

The two thermodynamic derivatives in (2.603b) are: (i) the adiabatic sound speed (2.479b) ≡ (2.604a) that simplifies to (2.479c) ≡ (2.604b) [(2.479d) ≡ (2.604c)] for an ideal (IG) [perfect (PG)] gas; (ii) the **non-adiabatic coefficient** (2.494a) ≡ (2.604d) that simplifies to (2.494b) ≡ (2.604e, f) [(2.604g) using (2.428a)] for an ideal (IG) [perfect (PG)] gas:

$$\left(c_s\right)^2 \equiv \left(\frac{\partial p}{\partial \rho}\right)_s \overset{IG}{=} \gamma \frac{p}{\rho} \overset{PG}{=} \gamma \bar{\bar{R}} T; \quad \left(b_p\right)^2 \equiv \left(\frac{\partial p}{\partial s}\right)_\rho \overset{IG}{=} \frac{p}{C_V} = p\frac{\gamma-1}{R} \overset{PG}{=} (\gamma-1)\rho T. \quad (2.604a\text{-}g)$$

Thus *the **equation of state** (2.603a) ≡ (2.603b) ≡ (2.605) involves the adiabatic sound speed (2.604a) [non-adiabatic coefficient (2.604d)] that is given by (2.604b) [(2.604e, f)] for an ideal gas, and by (2.604c) [(2.604g)] for a perfect gas:*

$$\frac{dp}{dt} = \left(c_s\right)^2 \frac{d\rho}{dt} + \left(b_p\right)^2 \frac{ds}{dt}. \qquad (2.605)$$

*The **adiabatic Equation** (2.605a) for an **isentropic flow** whose entropy is conserved along streamlines (2.606a) leads to (2.606b):*
isentropic flow:

$$\frac{ds}{dt} = 0: \quad \frac{dp}{dt} = \left(c_s\right)^2 \frac{d\rho}{dt} = -\left(c_s\right)^2 \rho\left(\nabla.\bar{v}\right) \overset{IG}{=} -\gamma\, p\left(\nabla.\bar{v}\right), \qquad (2.606a\text{-}d)$$

that may be combined with the continuity equation (2.595c) leading to (2.606c), and by (2.604b) also leads to (2.606d) for an ideal gas. The properties of adiabatic flows have been considered before (Sections I.12.1–I.12.6). By the first principle of thermodynamics the equation of entropy is equivalent to the equation of energy, that is derived for a non–dissipative (dissipative) fluid next (subsection 2.6.4) [in the Note 2.7], based on entropy conservation (2.606a) [entropy production (Note 2.6)].

2.6.4 EQUATION OF ENERGY FOR A NON-DISSIPATIVE FLUID

The conservation equation is similar for the mass (2.591c) or energy (2.607a), involving the energy density E and flux $\bar{G}$, and in the absence of dissipation or other energy sources or sinks the l.h.s. of (2.607a) is zero. The energy equation follows from the preceding equations of continuity, momentum and entropy (respectively, subsections 2.6.1/2.6.2/2.6.3). The **energy density** of a fluid, dissipative or not, is (2.607b) the sum of: (i) the internal energy per unit volume, where u is the internal energy per unit mass (2.601d); (ii) the kinetic energy. Thus the energy density of a fluid is the **stagnation internal energy** per unit volume (2.607c):

$$\frac{\partial E}{\partial t} + \nabla.\bar{G} = 0: \quad E = \rho u + \frac{\rho}{2}v^2 = \rho u_0, \quad u_0 \equiv u + \frac{v^2}{2}; \qquad (2.607a\text{-}d)$$

the stagnation internal energy per unit mass (2.607d) equals the internal energy $u = u_0$ for a fluid at rest $v = 0$. To determine the energy flux in (2.607a) it is sufficient to write, in the form of the divergence of a vector, the time derivative of the energy density (2.607a), given by (2.608):

$$\frac{\partial E}{\partial t} = \left(u + \frac{v^2}{2} \right) \frac{\partial \rho}{\partial t} + \rho \frac{\partial u}{\partial t} + \rho \vec{v} \cdot \frac{\partial \vec{v}}{\partial t}. \tag{2.608}$$

For a non-dissipative fluid the inviscid momentum equation (2.598b) may be used in the last term on the r.h.s. of (2.608) in the equivalent form with inner vector product by the velocity (2.609a, b):

$$\rho \vec{v} \cdot \frac{\partial \vec{v}}{\partial t} = -\vec{v} \cdot \nabla p - \rho \vec{v} (\vec{v} \cdot \nabla) \vec{v} = -\nabla \cdot (p \vec{v}) + p (\nabla \cdot \vec{v}) - \rho \left[(\vec{v} \cdot \nabla) \left(\frac{v^2}{2} \right) \right]. \tag{2.609a, b}$$

Substituting besides (2.609b) also the equation of continuity (2.592c) in the first term on the r.h.s. of (2.608) leads to (2.610):

$$\frac{\partial E}{\partial t} + \nabla \cdot (p \vec{v}) = -u \nabla \cdot (\rho \vec{v}) - \frac{v^2}{2} \nabla \cdot (\rho \vec{v}) + \rho \frac{\partial u}{\partial t} + p (\nabla \cdot \vec{v}) - \rho (\vec{v} \cdot \nabla) \frac{v^2}{2}; \tag{2.610}$$

a divergence term has been passed to the l.h.s. of (2.610) as in the energy equation (2.607a).

The r.h.s. of (2.610) involves two more divergence terms (2.611):

$$\frac{\partial E}{\partial t} + \nabla \cdot (p \vec{v}) = -\nabla \cdot (\rho u \vec{v}) + \rho (\vec{v} \cdot \nabla) u + \rho \frac{\partial u}{\partial t} + p (\nabla \cdot \vec{v}) - \nabla \cdot \left(\frac{\rho}{2} v^2 \vec{v} \right); \tag{2.611}$$

all divergence terms in (2.611) are collected on the l.h.s. of (2.612a):

$$\frac{\partial E}{\partial t} + \nabla \cdot \left[\left(p + \rho u + \frac{\rho}{2} v^2 \right) \vec{v} \right] = \rho \frac{du}{dt} - \frac{p}{\rho} \frac{d\rho}{dt} = T \frac{ds}{dt} = 0, \tag{2.612a–c}$$

and on the r.h.s.: (i) is used the equation of continuity (2.595d) in (2.612b); (ii) the internal energy (2.602b) shows that the r.h.s. of (2.612b):

$$\rho \frac{du}{dt} - \frac{p}{\rho^2} \frac{d\rho}{dt} = T \frac{ds}{dt} = 0, \tag{2.613a, b}$$

vanishes (2.613a) in isentropic conditions (2.613b) ≡ (2.606a).

Thus (2.612a–c) is the **equation of energy for a non-dissipative fluid** (2.607a), where: (i) the energy density (2.607b) per unit volume is the sum of internal and kinetic energies and equals (2.607c) the product of the mass density by (2.607d) the stagnation internal energy per unit mass; (ii) the **energy flux**:

$$h = u + \frac{p}{\rho} \quad h_0 = h + \frac{v^2}{2} = u + \frac{p}{\rho} + \frac{v^2}{2} = u_0 + \frac{p}{\rho},$$

$$\bar{G} = p\bar{v} + \rho u \bar{v} + \frac{\rho}{2} v^2 \bar{v} = (E + p)\bar{v} = \rho h \bar{v} + \frac{\rho}{2} v^2 \bar{v} = \rho h_0 \bar{v}, \tag{2.614a–i}$$

*is the flux (2.614e–i) of the **stagnation enthalpy** (2.614b–d), that is the sum of the kinetic energy per unit mass with the enthalpy (2.614a). The enthalpy (2.614a) ≡ (2.615a–d) [internal energy (2.602a, b)] per unit mass:*

$$h \equiv \frac{dH}{dm}: \quad dh = d\left(u + \frac{p}{\rho}\right) = du + \frac{dp}{\rho} - \frac{p}{\rho^2} d\rho = T\,ds + \frac{dp}{\rho}, \tag{2.615a–d}$$

appears in the energy density (2.607b) [flux (2.614b)], because (2.607a; 2.614f) the total rate of change with time of the total energy in a region (2.616a–c):

$$\frac{d}{dt}\int_D E\,dV = \int_D \frac{\partial E}{\partial t}\,dV = -\int_D (\nabla.\bar{G})dV = -\int_{\partial D} \bar{G}.d\bar{A}$$

$$= -\int_{\partial D} E\bar{v}.d\bar{A} - \int_{\partial D} p\bar{v}.d\bar{A}, \tag{2.616a–d}$$

consists (2.616d) of: (i) the flux of the energy density through the boundary; (ii) plus the work of the pressure forces.

The equation of energy can be extended to a dissipative fluid including heat conduction and viscosity (Note 2.7), mass diffusion (Note 2.10) and chemical reactions (Note 2.13). *The fundamental equations of a non-dissipative flow in the absence of external forces or mass sources/sinks are: (i) one vector equation, namely the equation of momentum (2.598a–c); (ii/iii/iv) three scalar equations, namely continuity (2.595a–e), isentropic state (2.606a–d) and energy (2.607a–d; 2.614a–i). They involve one vector (the velocity) and three scalar (mass density, pressure and entropy) variables, on which depend other quantities like internal energy, enthalpy, temperature and sound speed.* The equations of (ii) continuity, (i) momentum and (iv) energy are written next (subsection 2.6.5) in a similar conservative form.

2.6.5 CONSERVATIVE EQUATIONS FOR NON–DISSIPATIVE OR ISENTROPIC FLOWS

The equations of non-dissipative fluid mechanics appear in conservative form for the conservation of mass (2.592c) ≡ (2.617a) and energy (2.607a, b; 2.614e) ≡ (2.617b):

$$\frac{\partial \rho}{\partial t} + \frac{\partial}{\partial x_i}(\rho v_i) = 0, \quad \frac{\partial}{\partial t}\left[\rho\left(u + \frac{v^2}{2}\right)\right] + \frac{\partial}{\partial x_i}\left\{\left[p + \rho\left(u + \frac{v^2}{2}\right)\right]v_i\right\}. \tag{2.617a, b}$$

The inviscid momentum equation (2.598b; 2.599a) $\equiv$ (2.618a), or (2.618b) using the identity matrix:

$$\rho \frac{\partial v_i}{\partial t} + \rho v_j \frac{\partial v_i}{\partial x_j} = -\frac{\partial p}{\partial x_i} = -\frac{\partial}{\partial x_j}(p\delta_{ij}),$$ (2.618a, b)

can be put in a conservative form similar to the equation of the continuity (2.617a), replacing the mass flux (2.592a) $\equiv$ (2.619a) by the **momentum flux** (2.619b, c):

$$I_i = \rho v_i: \quad m_{ij} = I_i v_j = \rho v_i v_j,$$ (2.619a–c)

that appears in (2.620a):

$$\frac{\partial}{\partial t}(\rho v_i) = \rho \frac{\partial v_i}{\partial t} + v_i \frac{\partial \rho}{\partial t} = -\rho v_j \frac{\partial v_i}{\partial x_j} - \frac{\partial p}{\partial x_i} - v_i \frac{\partial}{\partial x_j}(\rho v_i)$$

$$= -\frac{\partial}{\partial x_j}(\rho v_i v_j + p\delta_{ij}) = -\frac{\partial}{\partial x_j}(p\delta_{ij} + m_{ij}),$$ (2.620a–d)

where substitution of the momentum (2.618a) and continuity (2.617a) equations leads to (2.620b) $\equiv$ (2.620c) involving (2.620d) the isotropic pressure and momentum flux (2.619c).

Thus has been obtained the *conservative form of the equations of non-dissipative fluid mechanics specifying the conservation of mass (2.617a) $\equiv$ (2.621a), momentum (2.620d) $\equiv$ (2.621b) and energy (2.617b) $\equiv$ (2.621c):*

$$0 = \frac{\partial}{\partial t}\left\{\rho, \rho v_i, \rho\left(u+\frac{v^2}{2}\right)\right\} + \frac{\partial}{\partial x_j}\left\{\rho v_i \delta_{ij}, \rho v_i v_j + p\delta_{ij}, \left[\rho\left(u+\frac{v^2}{2}\right)+p\right]v_j\right\},$$ (2.621a–c)

involving the mass density ρ, velocity v_i and its modulus v, pressure p, internal energy u per unit mass, and the identity matrix δ_{ij}. The system of 5 Equations (2.621a–c) can be written in matrix form (2.622a–c) in cartesian coordinates:

$$0 = \frac{\partial}{\partial t}\begin{bmatrix} \rho \\ \rho v_x \\ \rho v_y \\ \rho v_z \\ \rho u_0 \end{bmatrix} + \begin{bmatrix} \frac{\partial}{\partial x} \\ \frac{\partial}{\partial y} \\ \frac{\partial}{\partial z} \end{bmatrix}\begin{bmatrix} \rho v_x & \rho v_y & \rho v_z \\ p+\rho v_x^2 & \rho v_x v_y & \rho v_x v_z \\ \rho v_x v_y & p+\rho v_y^2 & \rho v_y v_z \\ \rho v_x v_z & \rho v_y v_z & p+\rho v_z^2 \\ \rho h_0 v_x & \rho h_0 v_y & \rho h_0 v_z \end{bmatrix},$$ (2.622)

involving the stagnation internal energy (2.607d) [enthalpy (2.614b–d)] per unit mass in the last line, that corresponds to the energy equation.

For a perfect gas the internal energy (2.471a) [enthalpy (2.474b)] per unit mass (2.601d) [(2.615a)] equals (2.623a) [(2.623b)] the temperature multiplied by the specific heat per unit mass at constant volume (2.623a–c) [pressure (2.623d–f)] that are both constant:

$$\{u,h\} = T\{c_V,c_p\}; \quad \{c_V,c_p\} = \frac{d}{dm}(C_V,C_p) = \frac{\overline{\overline{R}}}{\gamma-1}(1,\gamma) = const:$$

$$\frac{\partial}{\partial t}\left[\rho\left(v^2 + 2c_V T\right)\right] + \frac{\partial}{\partial x_i}\left[\rho\left(v^2 + 2c_p T\right)v_i\right] = 0,$$

(2.623a–g)

and the equation of energy, that is the last line of (2.622) becomes (2.623g). The conservative form of the equations of fluid mechanics is well suited to specify the matching conditions across surfaces of discontinuity (subsections 2.7.1–2.7.3). The conservative form of the equations of fluid mechanics can be extended to include heat conduction and viscous dissipation (Note 2.8), mass diffusion (Note 2.11) and chemical reactions (Note 2.15).

The fundamental equations of non-dissipative fluids (subsections 2.6.1–2.6.5) imply isentropic flow (2.606a) $\equiv$ (2.624a), that is entropy constant along streamlines:

$$adiabatic\ flow \quad \begin{cases} isentropic: ds/dt = 0, \\ homentropic: s = const. \end{cases}$$

(2.624a–b)

In the more restrictive case of **homentropic flow** (2.624b), that is entropy constant everywhere, can be obtained as a consequence an equation of conservation of vorticity (subsection 2.6.6). Thus *the **adiabatic flow**, excluding heat exchanges, can be interpreted in a less (more) restrictive sense, as isentropic (2.624a) [homentropic (2.524b)] flow with entropy constant along streamlines (everywhere).* The isentropic flow (subsections 2.6.1–2.6.5) is reconsidered next (subsections 2.6.6–2.6.8) with the restriction to homentropic flow.

2.6.6 CONSERVATION OF VORTICITY IN A HOMENTROPIC FLOW

Using the identity in vector (2.625a) [index (2.625b,c)] notation:

$$\nabla\left(\frac{v^2}{2}\right) = (\vec{v}.\nabla)\vec{v} - (\nabla \wedge \vec{v}) \wedge \vec{v} \Leftrightarrow \frac{\partial}{\partial x_i}\left(\frac{1}{2}v_j v_j\right) = v_j\frac{\partial v_j}{\partial x_i} - v_j\left(\frac{\partial v_i}{\partial x_j} - \frac{\partial v_j}{\partial x_i}\right),$$

(2.625a–c)

the inviscid momentum equation in the absence of external forces (2.598b, c) is rewritten (2.626a, b) involving the **vorticity** (2.625c) defined as the curl of the velocity, and related to local rotation of the flow:

$$\frac{1}{\rho}\nabla p + \frac{\partial \vec{v}}{\partial t} + \nabla\left(\frac{v^2}{2}\right) = -(\nabla \wedge \vec{v}) \wedge \vec{v} = -\vec{\varpi} \wedge \vec{v}, \quad \vec{\varpi} \equiv \nabla \wedge \vec{v}.$$

(2.626a–c)

The proof of (2.625a) can be made (2.627b–e) using separate indices (2.627a):

$$\{i,j,k\} = cyclic \; \{1,2,3\} :$$

$$\left[\left(\nabla \wedge \vec{v}\right) \wedge \vec{v}\right]_i = \left(\nabla \wedge \vec{v}\right)_j v_k - \left(\nabla \wedge \vec{v}\right)_k v_j$$

$$= v_k \left(\partial_k v_i - \partial_i v_k\right) - v_j \left(\partial_i v_j - \partial_j v_i\right)$$

$$= v_j \partial_j v_i + v_k \partial_k v_i - \frac{1}{2}\left(\partial_i v_j^2 + \partial_i v_k^2\right)$$

$$= \left(v_i \partial_i + v_j \partial_j + v_k \partial_k\right) v_i - \frac{1}{2}\partial_i \left(v_i^2 + v_j^2 + v_k^2\right)^2 .$$

(2.627a–e)

The internal energy per unit mass (2.602b) implies that the enthalpy per unit mass (2.615a) satisfies (2.615d), that also holds for gradient (2.628a) and derivatives with regard to a parameter (2.628b):

$$\nabla h = T\nabla s + \frac{1}{\rho}\nabla p, \qquad \frac{dh}{dt} = T\frac{ds}{dt} + \frac{1}{\rho}\frac{dp}{dt},$$

(2.628a, b)

for example, the material derivative (2.594a–e) ≡ (2.629a, b) that is the total derivative with regard to time:

$$\frac{d}{dt} = \frac{\partial}{\partial t} + v_i \frac{\partial}{\partial x_i} = \frac{\partial}{\partial t} + \vec{v}.\nabla.$$

(2.629a, b)

Substituting (2.628a) in (2.625a) ≡ (2.630a) leads to the inviscid momentum equation in **Crocco's form** (2.630b), where was introduced (2.630c) the stagnation enthalpy (2.614b–d):

$$\frac{\partial \vec{v}}{\partial t} + \vec{\omega} \wedge \vec{v} = -\nabla\left(\frac{v^2}{2}\right) - \frac{\nabla p}{\rho} = -\nabla\left(\frac{v^2}{2}\right) - \nabla h + T\nabla s = -\nabla h_0 + T\nabla s.$$

(2.630a–c)

Thus *the inviscid momentum equation can be written in alternative forms involving the velocity and: (i) mass density and pressure (2.598a–c); (ii) the vorticity (2.626c) in (2.626a, b); (iii/iv) the enthalpy (2.614a) [stagnation enthalpy (2.614b–d) in (2.630b) [(2.630c)].* In the case of a homentropic flow, that is constant entropy (2.624b) ≡ (2.631a), the last term on the r.h.s. of (2.630c) is zero, and the curl (2.630b) leads to the vorticity equation (2.631c):

$$\nabla s = 0: \; 0 = -\nabla \wedge \left(\nabla h_0\right) = \frac{\partial}{\partial t}\left(\nabla \wedge \vec{v}\right) + \nabla \wedge \left(\vec{\omega} \wedge \vec{v}\right) = \frac{\partial \vec{\omega}}{\partial t} + \nabla \wedge \left(\vec{\omega} \wedge \vec{v}\right).$$

(2.631a–c)

Thus in a homentropic flow (2.624b) ≡ (2.631a) ≡ (2.632a) the vorticity (2.626c) ≡ (2.631b) satisfies the conservation equation (2.632c):

$$s = const, \quad \vec{\omega} = \nabla \wedge \vec{v}: \quad \frac{\partial \vec{\omega}}{\partial t} + \nabla \wedge \left(\vec{\omega} \wedge \vec{v}\right) = 0.$$

(2.632a–c)

The conservation of vorticity (2.632c) like the conservation of mass (2.592a–c), and the acceleration (2.599a–c) can be interpreted (subsection 2.6.7) in terms of convected derivatives like (2.593a–e).

2.6.7 CONVECTION DERIVATIVE (LIE 1888) OF THE DENSITY, VELOCITY AND VORCITICY

Considering a flow variable (2.633a) such as the mass density, velocity or vorticity, the conservation along streamlines is specified by: (i) the vanishing of the local time derivative (2.633c) in a **co-moving reference frame** for which the velocity is zero (2.633b); (ii) by the addition of the convection or Lie derivative (2.633d) with regard to the velocity in a fixed reference frame in which the velocity is $\vec{v}$:

$$X \equiv \left\{ \rho, \vec{v}, \vec{\varpi} \right\}: \quad \vec{v} = 0 \quad \Rightarrow \quad \frac{\partial X}{\partial t}\bigg|_{\vec{v}=0}; \quad \frac{\partial X}{\partial t} + \underset{\vec{v}}{L}\{X\} = 0. \qquad (2.633a\text{–}d)$$

The **convection derivative (Lie 1888)** of a fluid variable X with regard to the velocity is (2.634a) the difference between the total time derivative in a reference frame at rest for which the flow velocity is $\vec{v}$ and the local time derivative in a reference frame co-moving with the flow in which the velocity is zero:

$$\underset{\vec{v}}{L}\{X\} \equiv \frac{dX}{dt} - \frac{\partial X}{\partial t}: \quad \underset{\vec{v}}{L}\{\vec{v}\} = \vec{a} - \frac{\partial \vec{v}}{\partial t} = \frac{d\vec{v}}{dt} - \frac{\partial \vec{v}}{\partial t} = (\vec{v}.\nabla)\vec{v}, \qquad (2.634a\text{–}d)$$

for example, *the convection derivative of the velocity with regard to itself is (2.633b, c) the difference between the acceleration or material derivative (2.599a–c) and the local time derivative, and is thus given by (2.634d).*

The convection derivative of the mass density relative to the velocity (2.593e) was obtained from the equation of continuity (subsection 2.6.1). The convection derivative of the vorticity is obtained next from its conservation (2.635b) along streamlines for a homentropic flow (2.635a):

$$s = const; \quad \frac{\partial \vec{\varpi}}{\partial t} + \underset{\vec{v}}{L}\{\vec{\varpi}\} = 0; \qquad (2.635a, b)$$

comparing (2.633a, b) with the **equation of conservation of vorticity** in a homentropic flow (2.632a–c) ≡ (2.636a–c):

$$s = const, \quad \vec{\varpi} \equiv \nabla \wedge \vec{v}: \quad 0 = \frac{\partial \vec{\varpi}}{\partial t} + \nabla \wedge (\vec{\varpi} \wedge \vec{v}) \Rightarrow \underset{\vec{v}}{L}\{\vec{\varpi}\} = \nabla \wedge (\vec{\varpi} \wedge \vec{v}), \quad (2.636a\text{–}d)$$

shows that the convection derivative of the vorticity equals (2.636d) the curl of its outer product by the velocity.

Thus has been obtained *the convection derivatives with regard to: velocity (2.599a–c; 2.634b, c), that is a contravariant vector (Note III.9.7); (ii) the mass* density (2.592a–c; 2.593a–c) that is a scalar density (Note III.9.14), like the energy (2.607a–d) and enthalpy (2.614a–d) densities; (iii) the vorticity (2.635a, b; 2.636a–d) that is a bivector (Note III.9.20) or axial vector in three dimensions (Note III.9.22). A viscous flow would not change the convection derivative of the vorticity (2.636d) but would imply that the vorticity is no longer conserved (Note 2.9). This is similar to: (i) the acceleration being non-zero in the presence of pressure gradients (2.598a–c); (ii) the mass conservation (2.592c) not holding in the presence of flow sources/sinks (2.591c). The conservation (2.632c) of vorticity (2.632b) in a homentropic flow implies that:(i) if the vorticity is zero at one time (2.637a, b) it will be zero for all time; (ii) if the flow is potential at one time it will remain potential for all time (2.637b):

$$\text{irrotational flow:} \quad \vec{\omega} \equiv \nabla \wedge \vec{v} = 0 \Leftrightarrow \vec{v} = \nabla \Phi; \qquad (2.637a\text{–}c)$$

(iii) the circulation of velocity is conserved (Kelvin 1869) along streamlines (subsection 2.6.8).

2.6.8   CONSERVATION OF CIRCULATION ALONG STREAMLINES (KELVIN 1869)

For an irrotational flow (2.638a) the Stokes theorem (1.89a, b) ≡ (2.638c) implies that the **circulation** (2.638b) of the velocity along a closed loop (2.638c) is zero (2.638d):

$$\nabla \wedge \vec{v} = 0 \quad : \Gamma \equiv \oint_L \vec{v}.d\vec{x} = \int_B \left(\nabla \wedge \vec{v} \right).d\vec{A} = 0, \qquad (2.638a\text{–}d)$$

where B is any regular surface supported on the regular curve L. An alternative is to note that for an irrotational flow (2.637a, b) ≡ (2.638a) the velocity is the gradient of a potential (2.637c) ≡ (2.639a) and thus the circulation along a loop (2.639b, c) is the difference of the potential at coincident starting and ending points (2.639d) that is zero (2.639e) because the velocity potential is a single-valued function:

$$\vec{v} = \nabla \Phi : \quad \Gamma = \oint_L \vec{v}.d\vec{x} = \oint_L \nabla \Phi.d\vec{x} = \oint_L d\Phi = 0, \qquad (2.639a\text{–}e)$$

A **loop** is a closed curve, that is a curve whose initial and final points coincide. A **regular surface (curve)** has unique continuous normal (tangent) vector at all points, and thus cannot have cusps (edges). For example, an ellipsoid (ellipse) is a regular surface (regular closed curve), but a cone (square) is not a regular surface (curve) because of the vertex (corners); a wedge is also not a regular surface.

The material derivative (2.629a, b) of the circulation (2.638b) ≡ (2.639b) along the streamlines of the flow (2.640a, b) of an inviscid fluid (2.626a–c) implies (2.640c):

$$\frac{d\Gamma}{dt} = \frac{d}{dt}\oint \vec{v}.d\vec{x} = \oint \frac{\partial \vec{v}}{\partial t}.d\vec{x} = -\oint \left[\nabla \left(\frac{v^2}{2} \right) + \vec{\omega} \wedge \vec{v} + \frac{\nabla p}{\rho} \right].d\vec{x}$$

$$= -\oint \left(\vec{\omega} \wedge \vec{v} + \frac{\nabla p}{\rho} \right).d\vec{x}, \qquad (2.640a\text{–}d)$$

where the gradient term involving the kinetic energy per unit mass leads to an exact differential that takes the same value at the start and end points and can be omitted in (2.640d). For an irrotational flow (2.641a) only the second term in (2.641d) remains in (2.641b) where can be used (2.628a) the enthalpy leading to (2.641c) involving the heat input (2.641d–e):

$$\vec{\omega} = 0: \quad \frac{d\Gamma}{dt} = \oint_L \frac{\nabla p}{\rho} . d\vec{x} = \oint_L \left(\nabla h - T \nabla s \right) . d\vec{x} = -\oint_L T \, ds = -\oint_L dq. \qquad (2.641a\text{--}e)$$

Thus *in order for the flow of an inviscid fluid to remain potential (2.642a) the circulation along streamlines must be conserved (2.642b) and this is ensured by a homentropic flow (2.642c):*

$$\vec{\omega} = 0 \Leftarrow \frac{d\Gamma}{dt} = 0 \Leftarrow s = const. \qquad (2.642a\text{--}c)$$

*The **Kelvin (1869) circulation theorem** (2.641a–c) applies also to an irrotational (2.642b) barotropic flow for which the pressure depends only on the mass density (2.643a)*

$$\text{Barotropic flow:} \quad p = p(\rho): \quad \oint \frac{dp}{\rho} = 0, \quad dh = \frac{dp}{\rho}, \qquad (2.643a\text{--}c)$$

since in this case the loop integral (2.641c) ≡ (2.643b) is zero as would be the case for a homentropic enthalpy (2.643c). *A homentropic flow (2.644b) is always barotropic, because in the equation of state with pressure as function of mass density and entropy (2.644a) only the dependence on mass density appears (2.644c):*

$$p = p(\rho, s): \quad s = const \Rightarrow p = p(\rho). \qquad (2.644a\text{--}c)$$

Thus a potential flow is compatible with homentropic conditions, but may not be with flows involving heat input. This suggests a more detailed consideration of barotropic potential flow leading (subsection 2.6.9) to the Bernoulli (1738) equation.

2.6.9 POTENTIAL FLOW OF A BAROTROPIC FLUID (BERNOULLI 1738)

It has been shown that: (i) for an inviscid fluid, neglecting viscous stresses (subsections 2.4.9–2.4.11) the momentum equation in Navier-Stokes form (2.386) reduces to Euler Equation (2.598a–c) in the absence of external forces; (ii) the neglect of viscous stresses and other dissipation mechanisms, such as mass (heat) diffusion [subsection(s) 2.4.18 (2.4.1–2.4.2)] leads to isentropic flow (2.625a), that is conservation of entropy along streamlines, so that any heat flux must be perpendicular to them; (iii) in the more restrictive case of homentropic flow (2.625b), that is constant entropy and no heat exchanges, a flow that is potential at one time will remain potential for all time (2.641a–c), and the result extends to a barotropic fluid (2.643a–c). Thus the potential flow of a barotropic fluid is considered next.

The inviscid momentum equation in the absence of external forces in the form (2.626a) balances the pressure gradient divided by the mass density against the total acceleration (2.599a–c) consisting of a local and a convective acceleration; the convective acceleration is due to the change in the modulus of the velocity and to the change of direction of the velocity, with the latter term on the r.h.s. of (2.626b) involving the vorticity, that is the curl of the velocity (2.626c). In an **irrotational flow** *(2.637a–c), that is with zero vorticity (2.645a, b), the velocity is the gradient (2.645c) of a* **velocity potential**, *and thus the inviscid momentum equation (2.626a) simplifies to (2.645d, e):*

$$0 = \vec{\varpi} = \nabla \wedge \vec{v}; \quad \vec{v} = \nabla \Phi, \quad -\frac{1}{\rho} \nabla p = \nabla \left(\frac{\partial \Phi}{\partial t} + \frac{v^2}{2} \right) = \nabla \left[\frac{\partial \Phi}{\partial t} + \frac{\nabla \Phi . \nabla \Phi}{2} \right]. \qquad (2.645a\text{–}e)$$

From (2.645e) follows various generalizations of the Bernoulli equation (1738) for an incompressible (homentropic) flow [subsection 2.6.10 (2.6.11)].

2.6.10 IRROTATIONAL INCOMPRESSIBLE UNSTEADY FLOW

In the case of an incompressible flow (2.596a, 2.501a, b) with constant mass density (2.646a) the inviscid momentum equation (2.645e) becomes (2.646b):

$$\rho = const : \quad \nabla \left(\frac{p}{\rho} + \frac{v^2}{2} + \frac{\partial \Phi}{\partial t} \right) = 0, \qquad (2.646a, b)$$

implying the **incompressible Bernoulli theorem**: *in an irrotational (2.647a) incompressible (2.647b) flow is conserved (2.647c, d) the sum of: (i) the stagnation enthalpy per unit mass, that equals the enthalpy plus kinetic energy, for a steady or unsteady flow; (ii) the time derivative of the potential for an unsteady flow:*

$$\vec{v} = \nabla \Phi, \quad \rho = const : \quad const = \frac{\partial \Phi}{\partial t} + \frac{v^2}{2} + \frac{p}{\rho} = \frac{\partial \Phi}{\partial t} + \frac{1}{2} \left(\nabla \Phi . \nabla \Phi \right) + \frac{p}{\rho}. \qquad (2.647a\text{–}d)$$

If in addition (2.647a, b) ≡ *(2.648a, c) the flow is steady (2.648b) the original* **Bernoulli (1738) equation** *is regained (2.648d):*

$$\nabla \wedge \vec{v} = 0 = \frac{\partial \vec{v}}{\partial t}, \quad \rho = const : \quad p + \frac{1}{2} \rho v^2 = p_0, \qquad (2.648a\text{–}d)$$

stating that the sum of the pressure and dynamic pressure or kinetic energy per unit volume is a constant, namely the **stagnation pressure**, *that is the pressure for zero velocity, since $p = p_0$ for $v = 0$.* This implies that a higher pressure is associated with a lower velocity and vice-versa. In the case of a compressible fluid, like a gas, the pressure depends on the mass density, and some additional condition is needed (subsection 2.6.11) for the l.h.s. of (2.645d, e) to be the gradient of a scalar.

2.6.11 IRROTATIONAL HOMENTROPIC UNSTEADY FLOW

A homentropic flow is defined by a constant entropy (2.625b) ≡ (2.649a), and is a particular case of an isentropic flow (2.625a) ≡ (2.649b) for which the entropy is constant along streamlines but may vary from one streamline to another:

$$\text{homentropic:} \quad s = const \quad \Rightarrow \quad \frac{ds}{dt} = 0 \quad \text{isentropic.} \qquad (2.649a, b)$$

Thus *a homentropic flow is always isentropic, and thus non-dissipative; an isentropic flow is non-dissipative but may be non-homentropic if the entropy is not constant and only varies across streamlines.* In a potential (2.650a) homentropic flow (2.649a) ≡ (2.650b) the inviscid momentum equation (2.630b) becomes the gradient of a scalar (2.650c):

$$\vec{v} = \nabla \Phi, \quad s = const : \quad \nabla \left(\frac{\partial \Phi}{\partial t} + \frac{v^2}{2} + h \right) = 0, \qquad (2.650a\text{–}c)$$

implying the **homentropic Bernoulli theorem**:

$$\vec{v} = \nabla \Phi, \quad s = const \quad : const = \frac{\partial \Phi}{\partial t} + h + \frac{v^2}{2} = \frac{\partial \Phi}{\partial t} + h_0. \qquad (2.651a\text{–}d)$$

stating that in an irrotational (2.651a) homentropic (2.651b) flow there is conservation (2.651c, d) of the sum of the stagnation enthalpy (2.615b–d) with the time derivative of the unsteady potential. If in addition (2.651a, b) ≡ (2.652a, c) the flow is steady (2.652b) the stagnation enthalpy is conserved (2.652d, e):

$$\nabla \wedge \vec{v} = 0 = \frac{\partial \vec{v}}{\partial t}, \quad s = const : \quad h + \frac{v^2}{2} = h_0 = const. \qquad (2.652a\text{–}e)$$

The homentropic flow may be extended to a barotropic flow (2.643a, b) taking (2.643c) for the enthalpy. The enthalpy is considered next for: (i) an incompressible flow leading back to previous results (subsection 2.6.10); (ii) for a barotropic flow (subsection 2.6.12) that includes the polytropic and homentropic as particular cases.

2.6.12 ENTHALPY FOR INCOMPRESSIBLE, BAROTROPIC, POLYTROPIC AND ADIABATIC FLOWS

The enthalpy for a homentropic (2.649a) ≡ (2.653a) flow of a compressible fluid depends (2.653b) on the relation between pressure and mass density:

$$s = const \quad h = \int \frac{dp}{\rho} ; \quad \rho = const \quad h = \frac{p}{\rho}, \qquad (2.653a\text{–}d)$$

for an incompressible fluid (2.647b) ≡ (2.653c) the enthalpy is the ratio of pressure to mass density (2.653d) and the unsteady (2.651c, d) [steady (2.652d, e)] homentropic Bernoulli equation reduces to the unsteady (2.647c, d) [steady (2.648c, d)] incompressible Bernoulli equation. In order to make more explicit the pressure and mass density in the homentropic (2.649a) unsteady (2.651d) [steady (2.652e)] Bernoulli equation the relation between them must be used to specify the enthalpy (2.653b). An example is a **polytropic law** specifying the pressure as proportional to a power of the mass density (2.654a), with **polytropic exponent** n, and **polytropic constant** k specified (2.654c) by the stagnation values of pressure and density (2.654b):

$$\text{polytropic:} \quad p = k\rho^n \quad \Leftrightarrow \quad \frac{p}{p_0} = \left(\frac{\rho}{\rho_0}\right)^n, \quad k \equiv p_0\,\rho_0^{-n}. \qquad (2.654a\text{--}c)$$

The homentropic flow (2.655a) of a perfect gas is an adiabatic flow (2.478d) ≡ (2.655b) for which the polytropic exponent coincides with the adiabatic exponent (2.655c):
ideal gas:

$$s = const \quad \Rightarrow \quad p = k\rho^\gamma, \quad n = \gamma = \frac{c_p}{c_V} = 1 + \frac{2}{f}. \qquad (2.655a\text{--}e)$$

that: (i) equals (2.264b) ≡ (2.655d) the ratio of specific heats at constant pressure and volume; (ii) is given by (2.475a–c) ≡ (2.655e) for molecules with f degrees-of-freedom. The equation of state of a perfect gas (2.473d) ≡ (2.656a) for an isothermal flow (2.656b) leads to a polytropic law (2.656c) with exponent unity (2.656d) and coefficient (2.656e):

$$p = \rho\bar{\bar{R}}T, \quad T = const: \quad p = k\rho, \quad n = 1, \quad k \equiv \bar{\bar{R}}T. \qquad (2.656a\text{--}e)$$

An isothermal flow is not homentropic and thus would not preserve the potential flow of a perfect gas (subsection 2.6.7).

Next is considered the enthalpy (2.653b) for a homentropic (2.653a) polytropic (2.654a) gas; in the particular case of a perfect gas the polytropic exponent coincides with the adiabatic exponent (2.655c–e). The enthalpy (2.653b) of a homentropic (2.653a) polytropic (2.654a) gas is given by (2.657c–e):

$$n \neq 1 \quad h = \int \frac{1}{\rho} d(k\rho^n) = kn\int \rho^{n-2}\, d\rho = \frac{kn}{n-1}\rho^{n-1} = \frac{n}{n-1}\frac{p}{\rho} = \frac{c^2}{n-1}. \qquad (2.657a\text{--}e)$$

where is introduced in (2.657e) the **sound speed** (2.658a–d):

$$c^2 = \frac{dp}{d\rho} = \frac{d}{d\rho}(k\rho^n) = kn\rho^{n-1} = n\frac{p}{\rho}. \qquad (2.658a\text{--}d)$$

From (2.657b–e; 2.658a–d) is excluded (2.657a) the case of an isothermal perfect gas (2.656a–e). Thus *the enthalpy (2.653b) of a homentropic (2.653a) polytropic (2.654a–c) flow is given by (2.657a–e) ≡ (2.659b) involving the pressure and mass density, or (2.659c) involving the sound speed (2.659d) ≡ (2.658d):*

$$n \neq 1: \quad h = \frac{n}{n-1} \frac{p}{\rho} = \frac{c^2}{n-1}, \quad c^2 = n \frac{p}{\rho}, \tag{2.659a–d}$$

and excluding (2.659a) the isothermal case (2.656a–e). For a perfect gas the homentropic flow is an adiabatic flow, and the polytropic exponent is replaced by the adiabatic exponent in the enthalpy and thus in the Bernoulli equation (2.652e), with implications discussed in the sequel (subsections 2.6.13–2.6.18).

2.6.13   IRROTATIONAL HOMENTROPIC FLOW OF A PERFECT GAS

Substituting the enthalpy (2.657d) with polytropic exponent replaced by the adiabatic exponent in (2.651c) leads to:

$$\vec{v} = \nabla\Phi, \quad s = const: \quad \frac{\partial\Phi}{\partial t} + \frac{v^2}{2} + \frac{\gamma}{\gamma-1} \frac{p}{\rho} = const, \tag{2.660a–c}$$

the Bernoulli equation (2.660c) for the irrotational (2.660a) homentropic (2.660b) flow of a perfect gas. If in addition (2.660a, b) ≡ (2.661a, c) the flow is steady (2.661b) the Bernoulli equation simplifies to (2.661d):

$$\nabla \wedge \vec{v} = 0 = \frac{\partial\vec{v}}{\partial t}, \quad s = const: \quad \frac{v^2}{2} + \frac{c^2}{\gamma-1} = const = \frac{c_0^2}{\gamma-1}, \tag{2.661a–d}$$

*relating the flow velocity v to the local c and stagnation c_0 sound speeds, that has two alternative forms (2.662b, c) [(2.663b, c)] involving the **stagnation (local) Mach number** (2.662a) [(2.663a)]:*

$$M_0 \equiv \frac{v}{c_0}: \quad \left(\frac{c}{c_0}\right)^2 = 1 - \frac{\gamma-1}{2} \frac{v^2}{c_0^2} = 1 - \frac{\gamma-1}{2} M_0^2, \tag{2.662a–c}$$

$$M \equiv \frac{v}{c}: \quad \left(\frac{c_0}{c}\right)^2 = 1 + \frac{\gamma-1}{2} \frac{v^2}{c^2} = 1 + \frac{\gamma-1}{2} M^2. \tag{2.663a–c}$$

Defining the ***critical or sonic flow condition*** as the points where the flow velocity equals the sound speed (2.664a) relates (2.661d) the stagnation c_0 and critical c_* sound speeds (2.664b, c):

$$v_* = c_*: \quad c_0^2 = \frac{\gamma-1}{2} c_*^2 + c_*^2 = \frac{\gamma+1}{2} c_*^2. \tag{2.664a–c}$$

Substituting (2.664c) in (2.662b) *relates the velocity to the stagnation and local sound speeds (2.665b–e) and to the critical Mach number (2.665a):*

$$M_* \equiv \frac{v}{c_*} : \quad \left(\frac{c}{c_*}\right)^2 = \left(\frac{c_0}{c_*}\right)^2 \left(\frac{c}{c_0}\right)^2 = \frac{\gamma+1}{2}\left[1 - \frac{\gamma-1}{2}\frac{v^2}{c_0^2}\right]$$

$$= \frac{\gamma+1}{2} - \frac{\gamma-1}{2}\frac{v^2}{c_*^2} = \frac{\gamma+1}{2} - \frac{\gamma-1}{2}M_*^2.$$

$$(2.665a-e)$$

The homentropic relations (2.481a–d) together with (2.662a–c)/(2.663a–c)/(2.665a,d,e) imply, respectively:

$$1 - \frac{\gamma-1}{2}M_0^2 = 1 - \frac{\gamma-1}{2}\frac{v^2}{c_0^2} = \left(\frac{c}{c_0}\right)^2 = \frac{T}{T_0} = \left(\frac{\rho}{\rho_0}\right)^{\gamma-1} = \left(\frac{p}{p_0}\right)^{1-\frac{1}{\gamma}}, \quad (2.666a-e)$$

$$1 + \frac{\gamma-1}{2}M^2 = 1 + \frac{\gamma-1}{2}\frac{v^2}{c^2} = \left(\frac{c_0}{c}\right)^2 = \frac{T_0}{T} = \left(\frac{\rho_0}{\rho}\right)^{\gamma-1} = \left(\frac{p_0}{p}\right)^{1-1/\gamma}, \quad (2.667a-e)$$

$$\frac{\gamma+1}{2} - \frac{\gamma-1}{2}M_*^2 = \frac{\gamma+1}{2} - \frac{\gamma-1}{2}\frac{v^2}{c_*^2}$$

$$= \left(\frac{c}{c_*}\right)^2 = \frac{T}{T_*} = \left(\frac{\rho}{\rho_*}\right)^{\gamma-1} = \left(\frac{p}{p_*}\right)^{1-\frac{1}{\gamma}}, \quad (2.668a-e)$$

that *for the irrotational (2.661a), steady (2.661b), homentropic (2.661c) flow of a perfect gas the Mach number and velocity are related to the sound speed/tempera-ture/mass density/ pressure by (2.666a–e)/(2.667a–e)/(2.668a–e involving, respec-tively, the stagnation/local/critical sound speeds and Mach numbers (2.662a)/(2.663a)/(2.665a). Setting (2.669a, b) in (2.668a–e) specifies the relations between the values of the stagnation and critical sound speeds (2.669c), temperatures (2.669d), mass densities (2.669e) and pressures (2.669f):*

$$v = 0 = M_* : \quad \frac{\gamma+1}{2} = \left(\frac{c_0}{c_*}\right)^2 = \frac{T_0}{T_*} = \left(\frac{\rho_0}{\rho_*}\right)^{\gamma-1} = \left(\frac{p_0}{p_*}\right)^{1-1/\gamma}, \quad (2.669a-f)$$

that are detailed next (Section 2.6.14).

2.6.14   RELATION BETWEEN CRITICAL AND STAGNATION VARIABLES

The critical and stagnation values may be related in the conditions of validity of (2.661a–c): (i) by (2.669c) for the sound speed (2.670a–d):

$$\frac{c_*}{c_0} = \sqrt{\frac{2}{\gamma+1}} = \sqrt{\frac{f}{f+1}} = \left\{ \sqrt{\frac{6}{7}}, \sqrt{\frac{5}{6}}, \sqrt{\frac{3}{4}} \right\} = \{ 0.92582, 0.91287, 0.86603 \}; \quad (2.670a\text{--}d)$$

(ii) by (2.669d) for the temperature (2.671a–e):

$$\frac{T_*}{T_0} = \left(\frac{c_*}{c_0} \right)^2 = \frac{2}{\gamma+1} = \frac{f}{f+1} = \left\{ \frac{6}{7}, \frac{5}{6}, \frac{3}{4} \right\} = \{ 0.85714, 0.83333, 0.75000 \}; \quad (2.671a\text{--}e)$$

(iii) by (2.669e) for the mass density (2.672a–g):

$$\frac{\rho_*}{\rho_0} = \left(\frac{c_*}{c_0} \right)^{2/(\gamma-1)} = \left(\frac{T_*}{T_0} \right)^{1/(\gamma-1)} = \left(\frac{T_*}{T_0} \right)^{f/2} = \left(\frac{2}{\gamma+1} \right)^{1/(\gamma-1)} = \left(\frac{f}{1+f} \right)^{f/2}$$

$$\quad (2.672a\text{--}g)$$

$$= \left\{ \left(\frac{6}{7} \right)^3, \left(\frac{5}{6} \right)^{2.5}, \left(\frac{3}{4} \right)^{1.5} \right\} = \{ 0.62974, 0.63394, 0.64952 \};$$

(iv) by (2.669f) for the gas pressure (2.673a–h):

$$\frac{p_*}{p_0} = \left(\frac{\rho_*}{\rho_0} \right)^{\gamma} = \left(\frac{c_*}{c_0} \right)^{2\gamma/(\gamma-1)} = \left(\frac{T_*}{T_0} \right)^{\gamma/(\gamma-1)} = \left(\frac{T_*}{T_0} \right)^{1+f/2} = \left(\frac{2}{\gamma+1} \right)^{\gamma/(\gamma-1)}$$

$$\quad (2.673a\text{--}h)$$

$$= \left(\frac{f}{f+1} \right)^{1+f/2} = \left\{ \left(\frac{6}{7} \right)^4, \left(\frac{5}{6} \right)^{3.5}, \left(\frac{3}{4} \right)^{2.5} \right\} = \{0.53978, 0.52828, 0.48714\}.$$

In all cases considered the values of the adiabatic exponent (2.674a–c) correspond (2.655d) ≡ (2.674d) to (2.674e–g):

$$\gamma = \left\{ \frac{4}{3}, \frac{7}{5}, \frac{5}{3} \right\} \Leftrightarrow f = \frac{2}{\gamma-1} = \{ 6, 5, 3 \}, \quad (2.674a\text{--}g)$$

to the irrotational (2.660a) ≡ (2.661a) steady (2.661b) homentropic (2.660b) ≡ (2.661c) flow of a perfect gas with, respectively, polyatomic/diatomic/monoatomic molecules (2.475a–c). The ratio of the flow velocity to three sound speeds, namely local, stagnation and critical, leads to the three corresponding Mach numbers (Section 2.6.15).

2.6.15 LOCAL, STAGNATION AND CRITICAL MACH NUMBERS

The local/stagnation/critical Mach numbers are defined by the ratio of the flow velocity, respectively, to the local (2.662a) ≡ (2.675a), stagnation (2.663a) ≡ (2.675b) and critical (2.665a) ≡ (2.675c) sound speeds:

$$M \equiv \frac{v}{c}, \quad M_0 = \frac{v}{c_0}, \quad M_* = \frac{v}{c_*}. \quad (2.675a\text{--}c)$$

The stagnation (2.675b) and critical (2.675c) Mach numbers are related (2.664c) by (2.676a):

$$M_* = M_0 \sqrt{\frac{\gamma+1}{2}}, \quad \frac{\gamma-1}{2} + \frac{1}{M^2} = \frac{1}{M_0^2} = \frac{\gamma+1}{2}\frac{1}{M_*^2}, \qquad (2.676\text{a–c})$$

and are related to the local Mach number (2.675a) by (2.661d) in (2.676b, c). Solving (2.676b) [(2.676c)] for M_0 (M_*) leads to (2.677a, b) [(2.677c, d)], and solving for M leads to (2.678a, b) [(2.678c, d)]:

$$M_0^2 = \frac{2M^2}{2+(\gamma-1)M^2} = \frac{f M^2}{f+M^2}, \quad M_*^2 = \frac{(\gamma+1)M^2}{2+(\gamma-1)M^2} = \frac{(1+f)M^2}{f+M^2}, \qquad (2.677\text{a–d})$$

$$M^2 = \frac{2M_0^2}{2-(\gamma-1)M_0^2} = \frac{f M_0^2}{f-M_0^2} = \frac{2M_*^2}{\gamma+1-(\gamma-1)M_*^2} = \frac{f M_*^2}{1+f-M_*^2}, \qquad (2.678\text{a–d})$$

Thus *the local (2.675a), stagnation (2.675b) and critical (2.675c) Mach numbers are related by (2.676a–c; 2.677a–d; 2.678a–d), among which is singled out (2.676a) = (2.679a–d)*

$$\frac{M_*}{M_0} = \sqrt{\frac{\gamma+1}{2}} = \sqrt{1+\frac{1}{f}} = \left\{ \sqrt{\frac{7}{6}}, \sqrt{\frac{6}{5}}, \sqrt{\frac{4}{3}} \right\}$$

$$= \{1.08012, 1.09545, 1.15470\} = \frac{c_0}{c_*}, \qquad (2.679\text{a–e})$$

respectively, for polyatomic/diatomic/monoatomic perfect gases, that are inverses of (2.670a–d) $\equiv$ *1/(2.679a–e)*. The ratio of sound speed, temperature, mass density and pressure to their stagnation values (2.481a–d) $\equiv$ (2.669a–f) can be expressed (2.666a–e / 2.667a–e / 2.668a–e) in terms of the local (2.675a), stagnation (2.675b) and critical (2.675c) Mach numbers, specifying their maximum possible values (Section 2.6.16) for the irrotational, homentropic, steady flow of a perfect gas.

2.6.16   Sound Speed, Temperature, Mass Density and Pressure

The ratios of sound speed (2.666b), temperature (2.666c), mass density (2.666d) and pressure (2.666e) to their stagnation values can be expressed in terms of the stagnation (2.675b), critical (2.675c) or local (2.675a) Mach numbers by (2.690a-f) using (2.676a; 2.677a):

$$\frac{T}{T_0} = \left(\frac{c}{c_0}\right)^2 = \left(\frac{\rho}{\rho_0}\right)^{\gamma-1} = \left(\frac{p}{p_0}\right)^{1-1/\gamma}$$

$$= 1 - \frac{\gamma-1}{2}M_0^2 = 1 - \frac{\gamma-1}{\gamma+1}M_*^2 = \frac{2}{2+(\gamma-1)M^2}. \qquad (2.680\text{a–f})$$

For the irrotational (2.681a), steady (2.681b), homentropic (2.681c) flow of a perfect gas the velocity has (2.652d) an upper limit (2.681e) specified by the lower limit for the stagnation enthalpy (2.681d):

$$\nabla \wedge \bar{v} = 0 = \frac{\partial \bar{v}}{\partial t}, \quad s = const: \quad h = 0 \quad \Rightarrow \quad v_{max} = \sqrt{2\,h_0}. \qquad (2.681a\text{–}e)$$

The lower limits (2.682a–d) lead in (2.680d, e, f) to the maximum values of the stagnation (2.682e), critical (2.682f) and local (2.682g) Mach numbers:

$$T = c = \rho = p = 0: \quad M_0 \leq \sqrt{\frac{2}{\gamma - 1}}, \quad M_* \leq \sqrt{\frac{\gamma + 1}{\gamma - 1}}, \quad M \equiv \frac{v}{c} \leq \infty. \qquad (2.682a\text{–}g)$$

Thus *the local Mach number (2.675a) is unlimited (2.682g) because the local sound speed can vanish (2.682b), whereas the stagnation (critical) Mach numbers (2.675b) [(2.675c)] have upper limits (2.682f) $\equiv$ (2.447a) [(2.682g) $\equiv$ (2.447b–h)]:*

$$M_0 \equiv \frac{v}{c_0} \leq \sqrt{\frac{2}{\gamma - 1}} = \sqrt{f} = \left\{ \sqrt{6}, \sqrt{5}, \sqrt{3} \right\} = \left\{ 2.44949, 2.23607, 1.73205 \right\}, \qquad (2.683a\text{–}e)$$

$$M_* \equiv \frac{v}{c_*} \leq \sqrt{\frac{\gamma + 1}{\gamma - 1}} = \sqrt{1 + f} = \left\{ \sqrt{7}, \sqrt{6}, 2 \right\} = \left\{ 2.64575, 2.44949, 2.00000 \right\}, \qquad (2.683f\text{–}j)$$

respectively, for polyatomic, diatomic, monoatomic perfect gases.

As can be seen in particular in (2.683a–e) and (2.683f–j) *the critical Mach number is always larger than the stagnation Mach number (2.684b) because (2.676a) $\equiv$ (2.664c) the stagnation sound speed is larger than the critical sound speed (2.684a):*

$$c_0 > c_* \Rightarrow M_* \equiv \frac{v}{c_*} > \frac{v}{c_0} \equiv M_0, \qquad (2.684a, b)$$

$$v < c_* \Rightarrow c > c_* \Rightarrow M \equiv \frac{v}{c} < \frac{v}{c_*} \equiv M_*, \qquad (2.684c\text{–}e)$$

$$v > c_* \Rightarrow c < c_* \Rightarrow M \equiv \frac{v}{c} > \frac{v}{c_*} \equiv M_*. \qquad (2.684f\text{–}h)$$

Also if the velocity is smaller (2.684c) [larger (2.684f)] than the critical sound speed, the local sound speed is larger (2.684d) [smaller (2.684g)] than the critical sound speed, and the local Mach number smaller (2.684e) [larger (2.684h)] than the critical Mach number. The results (2.684c–h) follow from (2.661d) $\equiv$ (2.684i) [(2.665d) $\equiv$ (2.684j)]:

$$\left| c_0^2 - \frac{\gamma - 1}{2} v^2 \right|^{\frac{1}{2}} = c = \left| \frac{\gamma + 1}{2} c_*^2 - \frac{\gamma - 1}{2} v^2 \right|^2, \qquad (2.684i, j)$$

that show that *the local sound speed decreases with increasing velocity and vice-versa*. The limits on the stagnation (2.683a–e) [critical (2.683f–j)] Mach numbers can be exceeded only if the flow is unsteady or rotational or non-homentropic. A non-homentropic flow involves heat transfer, and is incompatible with preserving a potential flow (subsection 2.6.8).

The preceding results (subsections 2.6.11–2.6.16) on the irrotational homentropic steady flow of a perfect gas are summarized in Table 2.13 with four blocks with 15 lines: (i–iv) gas properties, namely number of degrees-of-freedom of a molecule, specific heats at constant volume and pressure and adiabatic exponent; (v–viii) ratio of sonic to stagnation values of sound speed, temperature, mass density and gas pressure; (ix–xiii) relations between stagnation, critical and local Mach numbers; (xiv–xv) maximum values of the stagnation and critical Mach numbers. There are two blocks, 6 columns: (i–iii) set of properties, specific variables and equations; (iv–vi) values for monoatomic/diatomic/polyatomic gases. The principles of thermodynamics without (with) flow [subsections 2.5.1–2.5.14 (2.6.1–2.6.16)] apply to the Carnot, Atkinson and Stirling (Barber-Brayton) cycles of piston (jet) engines [subsections 2.5.15–2.5.29 (2.6.24–2.6.35)] that are used in the propulsion of aerospace vehicles (subsections 2.6.20–2.6.23) over a wide range of speed (subsections 2.6.17–2.6.19).

2.6.17 INCOMPRESSIBLE, SUBSONIC, TRANSSONIC, SUPERSONIC AND HYPERSONIC FLOWS

Henceforth the stagnation Mach number (2.675b) will be taken as the reference based on the stagnation sound speed. It follows from (2.680b) that there is approximately incompressible flow $\rho \sim \rho_0$ if the Mach number is small $M_0^2 \ll 1$ but not necessarily negligible $M_0 \ll 1$ (which would imply absence of flow). A small Mach number is (2.685a) that for a sea level sound speed (IV.5.133c) ≡ (2.685b) corresponds an airspeed (2.685c) above the landing and take-off speed of typical airliners (2.685d), but below cruise Mach number at higher altitudes (2.685e):

$$M_0 \leq 0.3; \quad c_0 = 340\,m\,s^{-1}, \quad v_0 = M_0\,c_0 = 102\,m\,s^{-1} > 70 - 90\,m\,s^{-1},$$
$$M_0 \sim 0.8 - 0.9. \tag{2.685a–e}$$

The change in mass density for the Mach number (2.685a) ≡ (2.686a) for a perfect gas (2.686b) is, respectively (2.686c, d) in the monoatomic, diatomic and polyatomic cases:

$$M_0^2 = 0.09: \quad \frac{\rho}{\rho_0} = \left(1 - \frac{0.09}{f}\right)^{f/2} = \left(1 - \frac{0.09}{3}\right)^{1.5}, \left(1 - \frac{0.09}{5}\right)^{2.5}, \left(1 - \frac{0.09}{6}\right)^{3} \tag{2.686a–d}$$
$$= 0.95534, 0.95561, 0.95567,$$

and hence less than 5% in all cases.

TABLE 2.13
Properties of the Irrotational, Homentropic and Steady Flow of a Perfect Gas

Molecule			Monoatomic	Diatomic or Linear Polyatomic	Three-dimensional Polyatomic
Number of Degrees of Freedom		f	3	5	6
Gas properties	Specific heat at constant volume	$C_V = \dfrac{fR}{2}$	$\dfrac{3R}{2}$	$\dfrac{5R}{2}$	$3R$
	Specific heat at constant pressure	$C_p = \dfrac{f+2}{2}R$	$\dfrac{5R}{2}$	$\dfrac{7R}{2}$	$4R$
	Adiabatic exponent	$\gamma = \dfrac{C_p}{C_V} = \dfrac{f+2}{f}$	$\dfrac{5}{3} = 1.667$	$\dfrac{7}{5} = 1.400$	$\dfrac{4}{3} = 1.333$
Ratio critical to stagnation value	Sound speed	$\dfrac{c_*}{c_0} = \sqrt{\dfrac{f}{f+1}}$	0.86603	0.91287	0.95282
	Temperature	$\dfrac{T_*}{T_0} = \dfrac{f}{f+1}$	0.75000	0.83333	0.85714
	Mass density	$\dfrac{\rho_*}{\rho_0} = \left(\dfrac{f}{f+1}\right)^{f/2}$	0.64952	0.63394	0.62974
	Pressure	$\dfrac{p_*}{p_0} = \left(\dfrac{f}{f+1}\right)^{1+f/2}$	0.48714	0.52828	0.53987
Relation between Mach numbers	Ratio critical to stagnation	$\dfrac{M_*}{M_0} = \sqrt{1+\dfrac{1}{f}}$	1.15470	1.09545	1.08012
	Local to stagnation	$M_0^2 = \dfrac{fM^2}{f+M^2}$	$M_0^2 = \dfrac{3M^2}{3+M^2}$	$M_0^2 = \dfrac{5M^2}{5+M^2}$	$M_0^2 = \dfrac{6M^2}{6+M^2}$
	Local to critical	$M_*^2 = \dfrac{(1+f)M^2}{f+M^2}$	$M_*^2 = \dfrac{4M^2}{3+M^2}$	$M_*^2 = \dfrac{6M^2}{5+M^2}$	$M_*^2 = \dfrac{7M^2}{6+M^2}$
	Critical to local	$M^2 = \dfrac{fM_*^2}{1+f-M_*^2}$	$M^2 = \dfrac{3M_*^2}{4-M_*^2}$	$M^2 = \dfrac{5M_*^2}{6-M_*^2}$	$M^2 = \dfrac{6M_*^2}{7-M_*^2}$
	Stagnation to local	$M^2 = \dfrac{fM_0^2}{f-M_0^2}$	$M^2 = \dfrac{3M_0^2}{3-M_0^2}$	$M^2 = \dfrac{5M_0^2}{5-M_0^2}$	$M^2 = \dfrac{6M_0^2}{6-M_0^2}$
Maximum Mach number	Stagnation	$M_0^{max} = \sqrt{f}$	1.73205	2.23607	2.44949
	Critical	$M_*^{max} = \sqrt{1+f}$	2.00000	2.44949	2.64575

Note: Properties of the irrotational, homentropic, steady flow of a monoatomic/diatomic/polyatomic perfect gas: (i–iii) gas properties: specific heats at constant pressure and volume and adiabatic exponent as their ratio; (iv–vii) ratio of critical to stagnation values of four flow variables: sound speed, temperature, mass density and pressure; (viii–xiii) six relations between pairs of three Mach numbers: stagnation, critical and local; (xiv–xv) maximum values of stagnation and critical Mach numbers.

Above the Mach number (2.685a) there is **subsonic flow** up to $M_0 = 0.8$; the flow is accelerated around a body like an aircraft, and above a free stream Mach number 0.8 there may be regions of supersonic flow around the body, leading to the appearance of **shock waves** (Section 2.7). The **transonic flow** ranges up to $M_0 = 1.2$ and is the regime in which regions of subsonic and supersonic flow coexist. Beyond Mach number 1.2 the shock waves remain for **supersonic flow**. The **sound barrier** corresponds to the Mach number unity and at about Mach 3 there is the **heat barrier** since flow temperature increases rapidly. It has been shown (2.683c–e) that an irrotational homentropic steady flow is limited to a Mach number below 2.5, showing that, beyond, heat exchanges are significant. From (2.680a) for a free stream temperature (2.687a) in the stratosphere in air taken as a diatomic perfect gas (2.687b) follows a stagnation temperature (2.687d, e)

$$T = -56.5\,C = 216.5\,K, \quad \gamma = 1.4, \quad M_0 = \{1.0, 2.0\},$$

$$T_0 = \frac{T}{1-0.2\,M_0^2} = \{270, 1082\}\,K = \{-3, \ 810\}c \tag{2.687a–e}$$

that increases significantly between Mach 1 and 2 in (2.687c). For Mach numbers above 3 corresponding to **hypersonic flow** heat transfer by conduction and radiation affects significantly the flow and reduces the stagnation temperature. Higher Mach numbers lead to ionization and dissociation that are significant at orbital and escape speeds (subsection 2.6.18).

2.6.18 ORBITAL/ESCAPE SPEEDS AND IONIZATION/DISSOCIATION

An order of magnitude of the velocity of a satellite at a low orbit around the earth is given by the balance of the acceleration of gravity at the surface of the earth (2.688a) against the centrifugal acceleration (2.688c) for a tangential velocity v_s in a circular orbit with radius (2.688b) equal to that of the Earth:

$$g = 9.8\,m\,s^{-2}, \quad R = 6400\,km = 6.4\times10^6\,m: \quad g = \frac{(v_s)^2}{R}. \tag{2.688a–c}$$

The resulting **orbital speed of satellization** (2.688d) would correspond to a stagnation Mach number (2.688e):

$$v_s = \sqrt{gR} = \sqrt{9.8\times6.4\times10^6\,m^2\,s^{-2}} = 7.9\times10^3\,m\,s^{-1} = 7.9\,km\,s^{-1},$$

$$\frac{v_s}{c_0} = \frac{7.9\times10^3}{3.4\times10^2} = 23. \tag{2.688d–f}$$

The **escape velocity** v_e for a spacecraft to exit the gravity field of the Earth is such that the kinetic energy equals the potential energy per unit mass (2.689a) leading to (2.689b, c):

$$\frac{1}{2}(v_e)^2 = g\,R; \quad v_e = \sqrt{2\,g\,R} = v_s\sqrt{2} = 1.41\,v_s = 1.12\times10^4\,m\,s^{-1}.$$

$$= 11.2\,km\,s^{-1}, \quad \frac{v_e}{c_0} = 33, \tag{2.689a–c}$$

that shows that the escape velocity exceeds the satellization velocity by 41% and cor-responds to a stagnation Mach number (2.689c). In space there is no air or it is very rarefied, and the values of the Mach number (2.688e) [(2.689c)] are indicative of the velocity (2.688d) [(2.689b)] that a rocket must attain to put a satellite in low Earth orbit (to send a spacecraft beyond the Earth gravity field to outer space).

The very high speeds and temperatures lead to **ionization**, that is the electrons separate from the atoms, forming an electrically charged flow with positive and nega-tive charges, and possibly neutrals. Thus fluid mechanics must be extended to **mag-netohydrodynamics (plasmas)** if the density is high enough to justify a continuum treatment (too low requiring the consideration of particles). At even higher tempera-tures and velocities there is **dissociation** of atoms into their constituent elements. A satellite falling from orbit (a spacecraft returning from a mission to another planet or asteroid) enters the atmosphere of the earth at orbital (2.688d) [escape (2.689b)] speed at the high atmosphere, with a heat release per kilogram of the order of the kinetic energy (2.458a, 2.445a, b) [(2.459a–f)]:

$$W_s = \frac{(v_s)^2}{2} = \frac{1}{2}(7.9\times10^3\,m\,s^{-1})^2 = 3.12\times10^7\,J\,kg^{-1} = 31.2\,MJ\,kg^{-1}, \tag{2.690a}$$

$$W_e = \frac{(v_e)^2}{2} = (v_s)^2 = (7.9\times10^3\,m\,s^{-1})^2 = 6.24\times10^7\,J\,kg^{-1} = 62.4\,MJ\,kg^{-1}. \tag{2.690b}$$

This large heat input causes almost total ionization (plus partial dissociation) of atmospheric air in the case of fall of a satellite (a spacecraft on a sample return mis-sion from another planet). The intense heating associated requires a **thermal protec-tion**, that may be an **ablative coating** that evaporates absorbing heat. The **angle of entry** in the atmosphere is important to avoid an excessive heating rate and keep the **heat flux** within acceptable limits. The atmospheric entry conditions in other planets, like Venus or Mars, involve similar phenomena with other values. The fall of a meteor into a planetary atmosphere, for example, the Earth, involves similar heating rates, leading to vaporization, so that only the larger meteors can reach the surface of the Earth. The speed ranges (subsections 2.6.17–2.6.18) may be classified by Mach num-ber (subsection 2.6.19) and are associated with sound, heat, ionization and dissocia-tion barriers.

2.6.19 SOUND, HEAT, IONIZATION AND DISSOCIATION BARRIERS

The speed regimes considered before (subsections 2.6.17–2.6.18) may be classified by stagnation Mach number as: (i) incompressible (2.691a) when free stream values are close to stagnation values (2.453a, 2.445a, b); (ii) this is no longer the case for subsonic flow (2.691b) that is compressible but generally devoid of shock waves around a vehicle like an aircraft; (iii) shock waves (Section 2.7) cause a significant drag increase for transonic flow in the Mach number range (2.691c) around the sonic condition; (iv) for higher Mach numbers shock waves persist in supersonic flow (2.691d); (v) heat exchanges significantly affect the hypersonic flow (2.691e) beyond the heat barrier; (vi) the rate of ionization increases with Mach number becoming almost total at orbital speeds (2.691f); (vii) there is dissociation before, at or beyond the escape speed (2.691g) of exit from the Earth's gravity field:

$$
\text{flow stagnation Mach numbers}
\begin{cases}
incompressible: M_0 \le 0.3, \\
subsonic: 0.3 < M_0 \le 0.8, \\
transonic: 0.8 < M_0 \le 1.2, \\
supersonic: 1.2 < M_0 \le 3, \\
hypersonic: 3 < M_0 \le 23, \\
orbital: 23 < M_0 \le 33, \\
escape: 33 < M_0.
\end{cases}
\quad (2.691a\text{--}g)
$$

The separation of seven flow regimes by Mach number involves 4 barriers: (i) the **sound barrier** (2.692a) at Mach one associated with the formation of shock waves; (ii) the **heat barrier** (2.692b) at about Mach 3 associated with intense heat exchanges by conduction and radiation affecting the flow; (iii) the **ionization barrier** starting with partial ionization in hypersonic flow to almost total ionization at orbital speeds (2.692c); (iv) the **dissociation barrier** at speeds above the orbital speed with significant effect at escape speed (2.692d):

$$
barriers
\begin{cases}
sound: M_0 = 1, \\
heat: M_0 \ge 3, \\
ionization: M_0 \ge 5, \\
dissociation: M_0 \ge 25.
\end{cases}
\quad (2.692a\text{--}d)
$$

Of all the values in (2.691a–g; 2.692a–d): (i) only the sound barrier at Mach one is precise; (ii) all the remaining values are approximate indications; (iii) the ionization and dissociation rates are the most variable with increasing temperature and different gases. To achieve very high speeds the piston engine driving a propeller is not feasible, and propulsion is provided by jet engines with (without) turbine [subsections 2.6.20–2.6.21 (2.6.22–2.6.23)].

2.6.20 PISTON-ENGINED PROPELLER-DRIVEN VERSUS JET-POWERED AIRCRAFT

A piston engine can drive a propeller (Figure 2.21) whose maximum velocity is tangential at the tip of the blades and equals to $V_t = \Omega a$ where Ω is the angular velocity and a the radius. In a piston-engined propeller-driven aircraft the forward speed V_f is orthogonal to the tangential tip speed of the propeller V_t; thus the total velocity of the

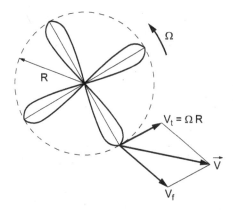

FIGURE 2.21   A very efficient means of propulsion is a propeller for which the total velocity $\bar{v}$ is the sum of two perpendicular components: (i) the forward velocity v_r of the vehicle; (ii) the tangential velocity $v_t = \Omega R$ proportional to the angular velocity of rotation Ω, highest at the tip radius R. Thus the tip of the propeller approaches the sound speed $v = c$ or Mach number unity $M = v/c = 1$ when the forward velocity is about 70% of the sound speed $v_1 = 0.7\,c$ or Mach number $M_1 = 0.7$. Beyond this Mach number the propeller becomes inefficient due to the formation of shock waves near the blade tips, that increase drag and dissipate energy.

tips of propeller blades relative to the air is given by Pythagoras theorem as the square root of the sum of squares (2.693a) and should not approach the sound speed (2.693b) to avoid shock waves and high drag:

$$\bar{V} = \left| \left(V_f\right)^2 + \left(V_t\right)^2 \right|^{1/2} \leq c_0; \quad \Omega\, a = V_t = V_f \leq \frac{c_0}{\sqrt{2}} = 240\, m\,s^{-1} = 865\, km\,s^{-1},$$

$$\frac{V_f}{c_0} = \frac{V_t}{c_0} = \frac{1}{\sqrt{2}} = 0.707.$$

(2.693a–g)

assuming equal aircraft forward speed and propeller tip speed (2.693c, d) they cannot exceed (2.693e) corresponding (2.693f) to a Mach number (2.693g). Above about Mach number 0.7 jet propulsion is needed, using turbine (non-turbine) engines [subsection 2.6.20 (2.6.21)].

The simplest turbojet engine (Figure 2.22) operates in four phases: (I) atmospheric air is admitted through the **inlet** and further compressed in an axial or radial **compressor**; (II) the air passes to an annular or cannular **combustion chamber** where fuel is injected and air is burned; (III) the hot combustion gases expand through a **turbine** that drives the compressor through a **shaft**; (IV) the jet **exhausts** through a nozzle providing thrust. From the nozzle edges on tips issue **shear layers** separating the hot combustion gas jet from the ambient atmosphere until **mixing** occurs further downstream restoring background atmospheric state back to inlet conditions.

The compressor may be: (2.519a–e) **radial** single stage using the centrifugal force to compress air at the periphery; (2.520a–e) **axial** using one or more stages with bladed disks to compress air along the axis of the engine. The axial compressor is

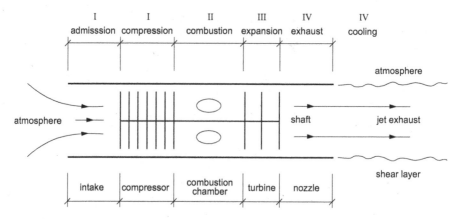

FIGURE 2.22   A propulsion system not limited to subsonic speeds is the turbojet engine consisting of: (i) an air intake admitting atmospheric air; (ii) a one-stage centrifugal compressor or multistage axial compressor with compression ratio up to 40:1; (iii) an annular combustion chamber or a set of cans burning fuel to increase gas temperature up to 2000K; (iv) the gases expand through a turbine that drives the compressor through a shaft; v) the colder but still hot combustion gases exhaust to the atmosphere through a nozzle providing thrust. The turbojet (piston) engine [Figure 2.14 (2.22)] cannot self-start, and need a starter motor to turn the compressor-turbine shaft (crankshaft) to a rotation speed high enough to enable continuous operation.

more suitable than the radial compressor for high-speed propulsion because: (i) it has a smaller frontal area, and can be installed in an aircraft with less drag; (ii) it can achieve a higher overall **compression ratio** that improves efficiency by using more stages. The combustion chamber can be: (a) **annular** if it occupies all the space between the shaft and the casing; (b) **cannular** if the same space is occupied by separate "cans" with the combustion occurring only inside the cans.

The most critical parameter of a turbojet engine is the **turbine entry temperature** (TET) because the turbine is subject to the harshest combination of conditions: (i) high temperature of the combustion gases; (ii) dynamic pressure of the high-speed flow; (iii) centrifugal force due to the high-rotational speed. Due this combination of operating conditions (i) to (iii) turbine blades may be: (i) single-chrystal to avoid cracks; (ii) internally-cooled by bleed air from the compressor to reduce temperature; (iii) made of metals or alloys with high-strength at high-temperature like titanium, tungsten, nickel and molybdenum or alloys combining several materials. High compression ratio and high TET are the main contributors to high efficiency, high thrust and low fuel consumption of turbojet engines in their various forms (subsection 2.6.21).

2.6.21   Turbojet, Turboprop, Turboshaft, Turbofan and Propfan

A turbojet may be: (i) **single-spool**, that is, has a single shaft connecting the turbine to the compressor (Figures 2.22 and 2.23); (ii) **twin-spool** (Figures 2.23) with two concentric shafts, for example, the inner (outer) shaft connecting the low (high) pressure turbine behind (ahead) to the low (high) pressure compressor first (second); (iii)

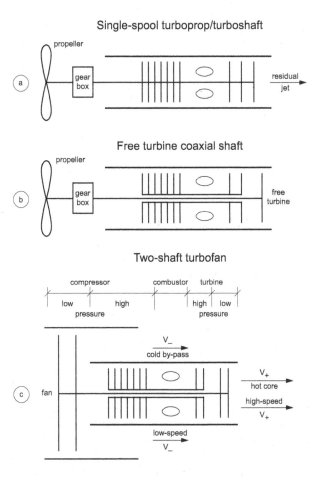

FIGURE 2.23 There are several variations of the jet engine, and three are mentioned. The propeller of an aircraft [rotor(s) of an helicopter] can be connected (Figure 2.23a) to the shaft of a turbojet (Figure 2.22) through a gearbox to reduce the rotation speed from thousands to hundreds of revolutions per minute (R.P.M.) in a turboprop (turboshaft). Compared with a piston engine driving a propeller (rotor) the turboprop (turboshaft) has much fewer moving parts and provides higher power for the same weight and volume. The turboprop/turboshaft may be: (a) single spool if the propeller (rotor) is connected through the gearbox to the turbine – compressor shaft; (b) free turbine if there is a coaxial shaft inside the compressor – turbine shaft linking the propeller (rotor) through the gearbox to a second free turbine operating behind the first at lower pressure. Another two-shaft variation of the turbojet engine (Figure 2.22) is the turbofan (Figure 2.23c): (i) all air admitted through the inlet passes through a large diameter low pressure compressor or fan; (ii) a fraction of the inlet flow called the "by-pass flow" goes around the rest of the engine, is not burned, and downstream of the engine mixes in the exhaust with the "core flow"; (iii) the core flow goes through the high-pressure compressor, combustion chamber, and high and low pressure turbines; (iv) the high (low) pressure turbine drives the high (low) pressure compressor through the outer (inner) of two concentric shafts rotating at different angular velocities, supported on bearings. The by-pass ratio (BPR) turbofan is the ratio of by-pass to core mass flow rate. As the BPR increases, the specific fuel consumption (SFC) decreases (Table 2.14) and also the noise.

TABLE 2.14
Fuel Consumption of Jet Engine

Type	Specific Fuel Consumption	Specific Impulse
Unit	$kgf. kg^{-1} h^{-1}$	s
Propeller* BPR: 20–40	0.4	9 000
Propfan* BPR: 10–20	0.5	7 200
High BPR: Turbofan BPR: 5–10	0.6	6 000
Low BPR: Turbofan BPR: 1	0.8	4 500
Turbojet BPR: 0	1.0	3 600
Ramjet	2–4	900–1800
Ionic	3–6	600–1200
Rocket**	8–12	300–450

B.P.R. – By-pass ratio
* equivalent B.P.R.
** total propellent consumption: fuel + oxidizer

Note: Typical fuel consumption of several classes of engine specified alternatively as: (a) specific fuel consumption (SFC) of kilograms of fuel, kgf, spent to provide one kilogram of thrust kg for one hour, h; (b) the inverse, that is specific impulse (SI) that is the time in seconds that one kilogram of fuel or propellant can produce one kilogram of thrust. The fuel consumption reduces, increasing the by-pass ratio (BPR) from a turbojet BPR = 0 to a turbofan to an open rotor like a propfan or propeller. The fuel consumption is higher for a ramjet, and the propellant consumption is highest for a rocket engine that carries oxidizer in addition to fuel.

the **triple-spool** jet engine is less widely used and a higher number of spools is unheard of. The multiple spools can run at different rotational speeds and increase efficiency at the expense of complexity.

A variation of the two (or three) spool turbojet is the **turbofan** (Figure 2.23): (i) the low-pressure compressor is replaced by a large **fan**; (ii) part of the flow goes through a **by-pass duct** around the engine and is not burned; (iii) the remaining **core flow** goes through the high-pressure compressor, combustor, high and low pressure turbines; (iv) the low-pressure turbine drives the fan; (v) the core and by-pass flows mix in the exhaust. The by-pass ratio of a turbofan is the ratio of the mass flux of the by-pass and core flows: (i) zero 0:1 for a turbojet; (ii) unity 1:1 for a low by-pass ratio turbofan with equal mass flux in the by-pass and core flows; (iii) up to ten 10:1 for a **high by-pass ratio turbofan** (HBPRT) with a large fan at the front sending 10 times more air to the by-pass duct than to the core flow, that drives the whole.

The turbofan has several advantages over the turbojet that increase with the BPR: (i) larger mass flow rate leads to higher thrust; (ii) lower average jet velocity increases efficiency; (iii) lower fuel consumption since only air in the core flow is burned; (iv) less noise since the cold slow by-pass flow acts as an "acoustic shield" scattering noise from the hot high-speed core flow. For higher by-pass ratios beyond ten B.P.R.$\geq$10 the fan and case become very large and heavy, and it is preferable to dispense with the casing, going for an **open rotor** or propeller, that entrains air freely without the constraint of an air intake, leading to B.P.R~20 − 40. Thus the propeller is the most efficient propulsor until it reaches sonic conditions at the blade tips.

The **fuel consumption** of a jet engine can be measured in two inverse ways: (a) the **specific fuel consumption** (S.F.C.) is the mass of fuel needed to produce a kilogram force of thrust during one hour, in units h^{-1}; (b) the inverse is the **specific impulse** (S.I.) that is the time during which 1 kilogram of fuel produces a kilogram of thrust, in seconds. For example, an S.F.C. 0.5 means that 0.5 kilogram of fuel produces one kilogram of thrust during one hour, and thus the specific impulse is $1 : 0.5 = 2$ hours or 7200 seconds.

Table 2.14 indicates the order of magnitude of the S.F.C. and S.I. of jet engines: the S.F.C. of about 1 for a turbojet decreases to 0.8 for a low B.P.R. turbofan with B.P.R. of 1, to 0.6 for a high B.P.R. turbofan with a B.P.R. up to 10; the corresponding S.I. increases from 1 to 1.25 to 1.67 hours, that is 3600 to 4500 to 6000 seconds. The **propeller** entrains the largest mass flow, and corresponds to the highest B.P.R. 30–40, and hence lowest S.F.C. 0.4 and highest S.I. of 2.5 hours = 9000 seconds.

Since the propeller is the most efficient propulsor for low speed flight, the turbojet engine has been adapted as a **turboprop**, with the shaft driving the propeller through a **gearbox** to reduce the r.p.m. from thousands (to hundreds) at the turbine (propeller). As an alternative to the piston engine to drive the propeller the turboprop has two advantages: (i) much smaller number of moving parts; (ii) ability to generate higher power. The **turboshaft** (turboprop) are similar in that they use a jet engine to drive the rotor(s) of a helicopter (propeller(s) of an aircraft) through a **reduction gearbox** consuming most power, and additional **residual thrust**. The propeller (rotor) is driven by the turboprop (turboshaft) in one of two ways: (a) either from the compressor-turbine **power shaft** (Figure 2.23); (b) or from a **free turbine** (Figure 2.23) on an inner concentric shaft.

The **propfan** is an intermediate between the high B.P.R. turbofan and propeller: (i) the high B.P.R. 10–20 excludes the weight and bulk of the casing of a turbofan, leading to an **open rotor** like a propeller; (ii) the propfan uses two **contrarotating propellers** to balance torque with distinct blade numbers or configurations to reduce noise; (iii) the blades have twist and swept tips to reduce compressibility effects and delay the appearance of shock waves until cruise Mach number about 0.8, higher than for a conventional propeller (2.693g). The turbofan, turboprop and propfan are the most efficient engines for subsonic flight, but for supersonic flight other types of turbine or non-turbine engines are needed (subsection 2.6.22).

2.6.22 AFTERBURNING, PULSE JETS, RAMJETS, ROCKETS AND IONIC PROPULSION

For supersonic flight high jet exhaust speeds are required, that may exceed the TET the turbine, leading to the use of an **afterburner**, that is an extension of the exhaust duct where additional fuel is injected, to burn the remaining oxygen in the exhaust gas, increasing speed and thrust. The afterburner (Table 2.15) can increase the thrust of a turbojet engine by up to 50% but doubles the S.F.C. leading to a triple total fuel consumption. A turbofan can also be used for supersonic aircraft, providing lower fuel consumption than a turbojet in subsonic flight, with the supersonic flight capability retained for low B.P.R. up to 1. Afterburning on a low B.P.R. turbofan: (i) increases thrust by up to 70%, more than for a turbojet, by burning the cold by-pass air; (ii) the specific fuel consumption also increases up to 2.5, leading to a quadruple

TABLE 2.15

Comparison of a Turbojet and Turbofan Engine

Engine	Type	Turbojet	Turbofan
By-pass ratio	BPR	1:0	1:1
Reference Thrust	Without AB	100	100
	With AB	150	170
Specific fuel consumption $kgf.\ kg^{-1}\ h^{-1}$	Without AB	1.0	0.8
	With AB	2.0	2.5
Total fuel consumption $kg.\ h^{-1}$	Without AB	100	80
	With AB	300	425

AB - Afterburning

Note: Comparison of a turbojet (turbofan) with by-pass ratio BPR = 0 (BPR = 1) with and without after-burning (AB) in terms of: (i) reference thrust relative to 100 for turbojet without afterburning; (ii) specific fuel consumption (SFC); (iii) their product as the total fuel consumption.

total fuel consumption. Afterburning allows supersonic flight for limited periods due to the high fuel consumption and high temperature. Sustained supersonic flight in Concorde and some modern fighters is achieved without afterburning. Concorde used afterburning only on take-off and landing and modern fighters use afterburning only for acceleration and manoeuvring.

Beyond Mach 2 the compressor of a jet engine becomes unnecessary since shock waves in the air inlet provide sufficient compression of atmospheric air; the compressor may become a hindrance due to **compressor stall**, that is flow separation from the compressor blades. For supersonic flight a suitable engine is the **ramjet**, that has no moving parts (Figure 2.24): (i) air is compressed in the air intake by shock waves; (ii) it is burned with fuel injection in the combustion chamber; (iii) the hot jet expands through a nozzle to provide thrust. The turbojet engine cannot self-start: it needs an electric motor or compressed air charge to start the compressor and then operates at any speed from static to supersonic. The ramjet can operate only at a sufficiently high speed to have significant ram air compression in the inlet, say a minimum Mach number 0.6 to 1.2. The ramjet can operate up to Mach 5 when flow in the combustion chamber becomes supersonic. The **scramjet** or supersonic combustion ramjet can operate up to Mach 10 if stable supersonic combustion can be assured.

A precursor of the turbojet and ramjet was the **pulse jet** (Figure 2.24) used in the German V-1 (Vergeltungswaffen) flying bomb in the second world war, that had only one moving part: a **shutter** in the air intake: (i) ram air pressure opened the shutter letting air into the engine; (ii) fuel was injected and combustion increased the internal pressure, closing the shutter and forming a jet exhaust providing thrust; (iii) when the pressure in the combustion chamber decayed enough the shutter would open again re-starting the pulse jet engine cycle. The pulse jet, like the **detonation engine**, with cyclic controlled explosions, has discontinuous combustion, in contrast with continuous combustion in turbojet and ramjet engines. The piston, turbojet, ramjet, pulse jet and detonation engines are all **aerobic**: they use the oxygen from ambient air to burn

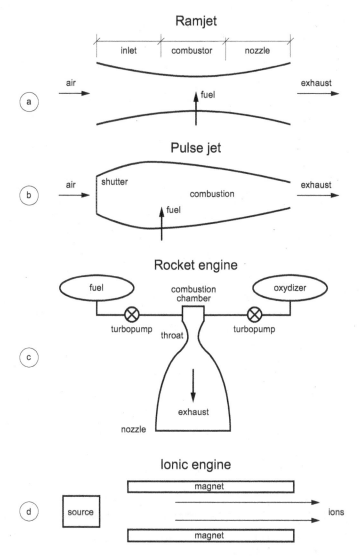

FIGURE 2.24   Besides the piston (Figure 2.14), turbojet (Figure 2.22), and turboprop/turbo-shaft/turbofan (Figure 2.23) engines, there are other means of propulsion of aerospace vehicles of which four examples are given. The ramjet (a) has no moving parts, since the air is compressed by dynamic pressure or shock waves in the inlet, burned in the combustion chamber and exhausts through a nozzle providing thrust; it operates only at high speed for sufficient ram-air compression in the inlet, and needs a booster engine to reach that speed. The pulse jet (b) has a set of shutters in the air inlet as the only moving parts; the shutters open by dynamic pressure to let air in, the combustion increases the pressure inside and closes the shutters, until the jet exhaust weakens, and the shutters open again. The rocket engine (c) carries not only fuel but also oxidizer for combustion, and does not need air, so it can operate in the atmosphere or in vacuo; it can have hight thrust, but also high propellant consumption (Table 2.14) and short burn time. The ionic engine (d) accelerates charged particles using electromagnetic fields generated by magnets, operates only in vacuum of space, and provides low thrust but much better SFC than rocket engines.

fuel, so that only fuel need be carried on board. The aerobic engines cannot function in the vacuum of space, or the very rarefied air in the high atmosphere and are limited in the thrust they can provide by the amount of air available. To launch a satellite into space a **rocket engine** carrying both fuel and oxidizer is needed. The **propellants** of a rocket engine may be liquid or solid.

An example of **liquid rocket propulsion** is to burn hydrogen with oxygen to form water (2.694a):

$$O_2 + 2H_2 \rightarrow 2OH_2 : \quad m_{O_2} = 32, \quad m_{H_2} = 2, \quad m_{O_2}/(2m_{H_2}) = 8:1. \quad (2.694a\text{--}d)$$

The molecular mass of oxygen (hydrogen) is (2.694b) [(2.694c)], so the proportion is (2.694d) implying that most of the propellant mass is oxidizer. Thus a rocket engine must carry a much larger propellant mass than the fuel of an aerobic engine; this leads to a much lower specific impulse of about 300–400 s in Table 2.14. The rocket engine has almost unlimited thrust provided that the **turbopumps** can supply (Figure 2.24) a high enough flow rate from the tanks to the combustion chamber for jet expansion through the nozzle. The oxygen (2.694b) [hydrogen (2.694c)] are gases with low density at ambient temperature and would require very large tanks causing high drag. For rocket propulsion may be used as fuel (oxidizer) **liquid hydrogen (oxygen)**, at temperatures below 20 K (80 K); the LH (LOX) are **cryogenic propellants** that must be kept at very low temperatures and cannot stay much time in the tanks before firing the rocket motor, due to evaporation. **Solid propellants** have generally lower specific impulse than the best combination of liquid propellants (specific impulse of 430 s for LH – LOX) but can be stored for long periods. Rocket propellants can be hazardous: flammable, explosive, corrosive, toxic and polluting.

Another form of anaerobic propulsion is the **ionic engine** (Figure 2.24) that accelerates charged particles using a magnetic field (subsection 2.4.15). Whereas the rocket engine can provide high thrust and short endurance anywhere, in the atmosphere or space, the ionic engine has low thrust and long endurance and: (i) cannot operate in the atmosphere because the thrust is insufficient to overcome drag and the ions are quickly absorbed; (ii) in the vacuum of space, with no drag, the ionic propulsion can lead to very high speeds over a long time. As the thrust of ionic engines increases to a few Newtons they can be used to stabilize satellites in orbit for longer periods than chemical thrusters using small rocket engines; the useful life of a satellite may be shorter than the life expectancy of on-board systems if the stabilization system runs out of propellant and cannot keep the satellite pointed in the right direction for Earth or astronomical observation, telecommunications or other missions. Ionic motors with higher thrust may allow orbital changes possibly taking weeks or months where a more powerful rocket motor could take hours or days. The different propulsion methods, like turbojets, ramjets, rockets and electric can be combined in hybrid propulsion systems.

2.6.23 TURBORAMJET, RAMROCKET, TURBOELECTRIC AND VARIABLE-CYCLE ENGINES

Since the ramjet cannot operate at low speeds it may be associated with a turbojet (rocket engine) in a turboramjet (ramrocket). A **turboramjet** (Figure 2.25) consists of a turbojet inside the duct of a ramjet: (i) at low speed the inlet of the turbojet is open and it provides thrust to accelerate up to the speed of operation of the ramjet; (ii) then the turbojet intake is closed and the engine operates as a ramjet. The turbo-ramjet is a bulky engine and has a wide speed range from zero to hypersonic, up to Mach 5 (10) for the ramjet (scramjet). Another method to start a ramjet is the **rocket boosted ramjet** (Figure 2.25) that uses a rocket engine as a first stage to accelerate

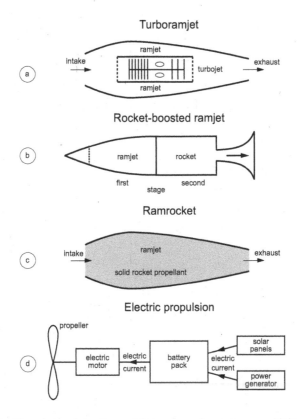

FIGURE 2.25 There is a variety of propulsion systems for aerospace vehicles including thermal (Figures 2.14 and 2.22–2.24) and electric (Figure 2.25) propulsion and hybrids and combinations of which seven examples are given. The turboramjet (a) uses a turbojet (Figure 2.22) or turbofan (Figure 2.23) inside the ramjet duct (Figure 2.24a) for acceleration to high speed, when the turbo engine is shut-off and the ramjet takes over; when returning to lower speeds, the ramjet is de-activated by shutting off the fuel supply and the turbo engine is reopened and relighted. The rocket boosted ramjet (Figure 2.24b) is used for missiles with a rocked first stage accelerating the vehicle to the high-speed of ramjet operation, and then dropping-off leaving the second-stage ramjet to continue the flight. The ramrocket (Figure 2.24c) is a more compact solution, in which the ramjet duct is filled with solid propellants for the rocket booster phase, and when the rocket propellants are totally spent the ramjet duct is free and the speed high enough for the ramjet to take over propulsion. Completely different from the preceding thermal propulsion systems, is electric propulsion (Figure 2.24d) using solar panels or other power generators to charge batteries to drive electric motors connected to propellers

(Continued)

Solar-electric propulsion

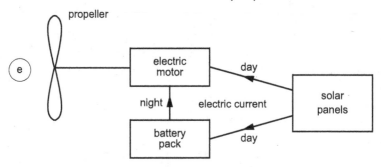

Turbo-electric propulsion

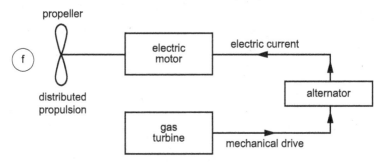

Variable cycle turbofan

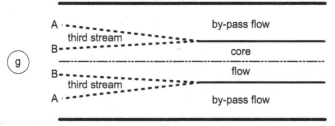

flight: A - supersonic, low bpr; B - subsonic, high bpr

FIGURE 2.25 **(Continued)** In the case of solar-electric propulsion (e) the solar panels by day charge a battery pack and drive electric motors connected to propellers; by night the energy stored in the batteries is used to drive the electric motors connected to the propellers. The high weight and volume of batteries and limited power of solar panels can be overcome by hybrid turbo-electric propulsion (f) with a turbine driving mechanically an alternator, supplying electric current to electric motor(s) connected to propeller(s), possibly in a distributed propulsion system. The development potential of thermal propulsion is far from exhausted, as shown by the variable-cycle engine (g) that is a turbofan (Figure 2.23c) with three air streams to vary the by-pass ratio between the core and by-pass flows, to achieve higher efficiency at different speeds and flight regimes.

up to the operating speed of the ramjet in the second stage, after which the first stage separates. A more compact solution is the **ramrocket** (Figure 2.25) that is a ramjet filled with solid rocket propellant: the rocket propellants burn first until they are exhausted, leaving the ramjet duct open and at the same time accelerating to ramjet operating speed.

The ionic engine (Figure 2.24) is a form of electric propulsion suited for space; other forms of electric propulsion suitable for atmospheric flight include solar (battery) powered aircraft. In a **battery-powered aircraft** (Figure 2.25) batteries provide the electric current to an electric motor driving a propeller or a distributed propulsion system with several propellers, say along the span of a wing. Although the electric motors are efficient (over 90%) and distributed propulsion also, the batteries are heavy and deplete rapidly leading to low speed and short endurance. A kilogram of fossil fuel may provide as much energy as tens of kilograms of batteries while occupying a much smaller volume; the progress in power-to-weight and power-to-volume ratios of batteries is gradual, and some technologies more favourable in this respect, like lithium-ion batteries; they have fire and explosion risks, requiring dedicated control and safety systems.

The **solar powered aircraft** (Figure 2.25) uses wings with large area covered with solar cells to: (i) provide power for flight in daylight; (ii) store energy in batteries for night flying. Although endurance is in principle unlimited the power available is limited, the speed and payload small with vulnerability to strong winds and storms. Ideally solar powered aircraft should fly above all cloud cover for permanent sunlight by day, but the speed may be exceeded by strong winds at higher altitudes. Also the solar powered aircraft needs to reach high altitudes after take-off which may require hours or days with limited interruptions of sunlight, also avoiding severe storms.

For large high-speed aircraft with significant passenger numbers or payload weight the choice could be **turbo-electric propulsion** (Figure 2.25): (i) a gas turbine drives alternator(s) to generate electric energy; (ii) the electric energy is sent to electric motors; (iii) the electric motors drive propellers in a distributed propulsion system. The main challenges relate to the large power (tens of megawatts) needed to power an airliner, leading to large currents (thousands of amperes) and high voltage (hundreds or thousands of volts). Thus there are two matters of concern: (i) high voltages that can cause electric discharges especially at high-altitude and electromagnetic interference if equipment is not widely separated; (ii) large electric currents implying significant energy losses and of greater concern large amounts of heat that are difficult to dissipate in an aircraft. A solution to the challenge (ii) could be superconducting electric cables with nearly zero electrical resistance. Superconductivity has been demonstrated for small currents close to zero absolute temperature and is difficult to achieve at ambient temperatures and for large electric currents that deviate considerably from the ground state of matter. The problems of transport of large amounts of electrical energy could be avoided by integrated power-propulsion units that have their own design problems.

The turbine engine will remain for some time the choice for fast and efficient transport of large numbers of passengers and large payloads over long distances, and its development potential is far from exhausted. The **variable cycle turbofan** uses (Figure 2.25) three flow streams: (i) the core flow that is burned to provide power; (ii) the by-pass flow for propulsion efficiency; (iii) a third flow stream to adjust the

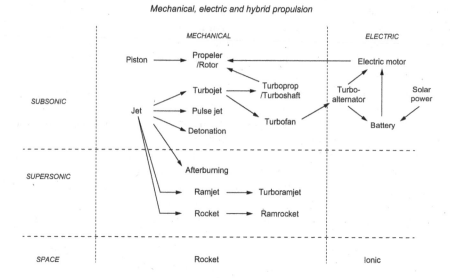

Mechanical, electric and hybrid propulsion

DIAGRAM 2.3   For the propulsion of aerospace vehicles, such as aeroplanes, helicopters, drones, missiles, satellite launchers and spacecraft, there are several choices of engines including air breathing (piston and jets), rockets and electric, with variants, combinations and hybrids.

by-pass ratio. In the variable cycle engine the third flow stream may: (i) augment the core flow for high jet velocity in supersonic flight; (ii) or alternatively augment the by-pass flow for low fuel consumption cruise flight and low noise at take-off and landing. The Diagram 2.3 summarizes the propulsion methods for aerospace vehicles, that also apply in other fields, distinguishing: (i) mechanical and electric propulsion; (ii) in space and in the atmosphere; (iii) for the atmosphere at low and high-speed. The Carnot/Atkinson/Stirling cycles (subsections 2.5.15–2.5.22/2.5.23–2.5.24/2.5.25–2.5.29) are described by thermodynamics (subsections 2.5.1–2.5.14) with low air velocities, whereas the Barber-Brayton cycle (subsections 2.6.24–2.6.35) for a jet engine (subsections 2.6.19–2.6.23) involves higher speeds of a compressible flow (subsections 2.6.1–2.6.18).

2.6.24 THERMODYNAMIC CYCLE OF THE JET ENGINE (BARBER 1791, BRAYTON 1930)

Although a practical piston (jet) engine for aircraft was realized only in the early 1900a (1940s) the relevant thermodynamic cycle was described long before by Carnot (1824) [Barber (1791)]. The four phases of the operation of a jet engine correspond to the Barber-Brayton cycle in the temperature – entropy diagram (Figure 2.26a): (I) the admission of atmospheric air through the inlet and the following compression $1 \to 2$ are adiabatic, increasing temperature from T_1 to T_2, at constant low entropy S_-; (II) the combustion $2 \to 3$ at constant high pressure p_+ increases both temperature and entropy from (T_2, S_-) to (T_3, S_+); (III) the expansion $3 \to 4$ is

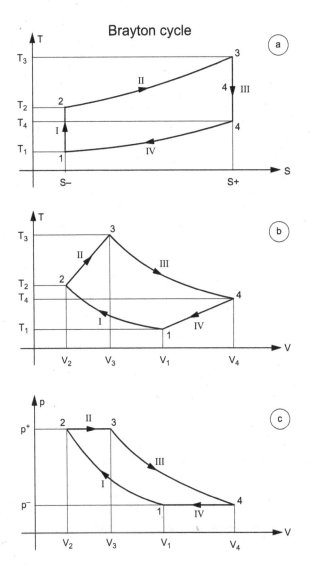

FIGURE 2.26 The Brayton cycle of the turbojet engine (Figure 2.26) consists of adiabatic admission (I) at low entropy S_- and expansion (III) at high entropy S_+, between isobaric combustion (II) at high pressure p_+ and exhaust and atmospheric cooling at low pressure p_-. Thus the Brayton cycle has: (i,ii) two straight lines in the (a) temperature versus entropy and (c) pressure versus volume diagrams; (iii) no straight lines in the temperature versus volume diagram (b).

adiabatic, at constant high entropy S_+ reducing the temperature from T_3 to T_4; (IV) the jet engine cycle is closed by further cooling $4 \rightarrow 1$ reducing the temperature from T_4 to T_1 at constant low atmospheric pressure p_-.

The Barber-Brayton cycle can also be considered (Figure 2.26b) in the temperature – volume diagram: (I) the adiabatic admission and compression $1 \rightarrow 2$ reduces the volume from V_1 to V_2 and increases the temperature from T_1 to T_2; (II) the combustion $2 \rightarrow 3$ at constant pressure increases volume and temperature proportionally to each other from (V_2, T_2) to (V_3, T_3); (III) the adiabatic expansion $3 \rightarrow 4$ increases the volume from V_3 to V_4 and decreases the temperature from T_3 to T_4; (IV) the jet engine cycle is closed by cooling $4 \rightarrow 1$ at the low atmospheric pressure decreasing in proportion the temperature and volume from (T_4, V_4) to the initial values (T_1, V_1).

The Barber-Brayton cycle can also be considered (Figure 2.26c) in the pressure – volume diagram: (I) the adiabatic admission and compression $1 \rightarrow 2$ increases pressure from p_- to p_+ and decreases the volume from V_1 to V_2; (II) the combustion $2 \rightarrow 3$ at constant high pressure p_+ increases the volume from V_2 to V_3; (III) the adiabatic expansion $3 \rightarrow 4$ further increases the volume from V_3 to V_4 and reduces the high pressure p_+ to the low atmospheric value p_-; (IV) the cooling $4 \rightarrow 1$ at the low atmospheric pressure p_- reduces the volume V_4 to the value V_1 at the start of the jet engine cycle. The Carnot, Stirling and Barber-Brayton cycles are compared next (subsection 2.6.25).

2.6.25　COMPARISON OF THE CARNOT, STIRLING AND BARBER-BRAYTON CYCLES

The Carnot (1824) [Stirling (1816)] cycle corresponds to a rectangle in the T – S (T – V) diagram [Figure 2.13 (2.20)], whereas the Barber (1791) – Brayton (1830) cycle (Figure 2.26) does not correspond to a rectangle in any of the T – S, T – V, p – V diagrams used for the Carnot (Stirling) cycles [Figure 2.13 (2.20)]. The Barber-Brayton cycle (Figure 2.26) corresponds to a rectangle in the p – S diagram (Figure 2.27a) that is also used for the Carnot (Stirling) cycle [Figure 2.27b (2.27c)] to compare the three cycles.

The simplest representation of the Barber-Brayton cycle is a rectangle in the pressure – entropy diagram (Figure 2.27a): (I) the adiabatic admission – compression $1 \rightarrow 2$ increases pressure from atmospheric p_- to maximum p_+ at the compressor exit at constant low entropy S_-; (II) the combustion $2 \rightarrow 3$ increases entropy from low S_- to high S_+ at constant high pressure p_+; (III) the expansion $3 \rightarrow 4$ at constant high entropy S_+ decreases the high pressure p_+ to low atmospheric pressure p_-; (IV) the cooling $4 \rightarrow 1$ at constant atmospheric pressure p_- reduces the high entropy S_+ to the low initial value S_-.

The pressure – entropy diagram (Figure 2.27b) can be used as an alternative to others (Figures 2.13) for the Carnot cycle: (I) the admission $1 \rightarrow 2$ reduces entropy from high S_+ to low S_- and increases pressure from p_1 to p_2; (II) the compression $2 \rightarrow 3$ further increases pressure from p_2 to p_3 at constant low entropy S_-; (III) the combustion $3 \rightarrow 4$ at constant temperature increases the entropy from low S_- to high S_+ reducing the pressure from p_3 to p_4; (IV) the exhaust $4 \rightarrow 1$ is adiabatic at high entropy S_+ and reduces the pressure further from p_4 to the initial value p_1.

The pressure – entropy diagram (Figure 2.27c) can be used as an alternative to others (Figure 2.20) for the Stirling cycle: (I) the compression $1 \rightarrow 2$ at constant low temperature increases the pressure from p_1 to p_2 and decreases the entropy from S_1 to S_2; (II) the heating $2 \rightarrow 3$ at constant low volume V_- increases both pressure and entropy from (p_2, S_2) to (p_3, S_3); (III) the expansion $3 \rightarrow 4$ at constant high temperature decreases the pressure from p_3 to p_4 and increases entropy from S_3 to S_4; (IV) the

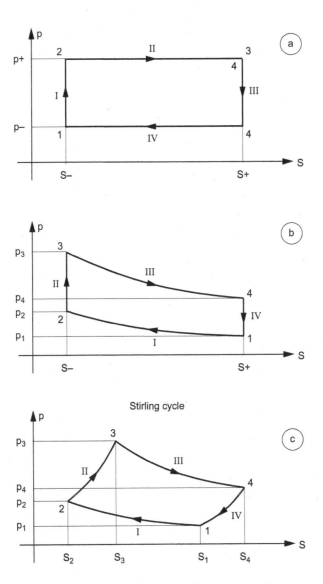

FIGURE 2.27 The simplest representation of the Brayton cycle of the jet engine (Figure 2.26) is a rectangle (a) in the pressure versus entropy diagram. In the same pressure versus entropy diagram the Carnot (Stirling) cycle [Figure 2.13 (2.17)] for aerobic (air independent) propulsion has two (b) [no (c)] straight lines.

cooling $4 \rightarrow 1$ at constant large volume V_+ decreases the entropy and pressure from (S_4, p_4) to (S_1, p_1). After the detailed consideration of the Carnot (Stirling) cycles [subsections 2.5.15–2.5.22 (2.5.25–2.5.29)] and the comparison with the Barber-Brayton cycle (subsections 2.6.24–2.6.25) the latter is considered in more detail (subsections 2.6.26–2.6.28).

2.6.26   FOUR STAGES OF THE BARBER-BRAYTON CYCLE

The first stage of the Barber-Brayton cycle (Figures 2.26, 2.27) is adiabatic (subsection 2.5.10) admission and compression at low entropy (2.695a) hence no heat exchange (2.695b) and work specified by the internal energy (2.514a, c) $\equiv$ (2.695c, d):

$$\text{I - admission - compression:} \quad S = S_-, \quad Q_{12}^b = 0,$$

$$W_{12}^b = U_{12} = \frac{R}{\gamma - 1}(T_2 - T_1) > 0. \qquad (2.695\text{a–d})$$

The second stage is isobaric (subsection 2.5.12) combustion at constant high pressure (2.696a) specifying the work through the volume change (2.696b) and the heat through the change in enthalpy (2.519b, d) $\equiv$ (2.696c, d):

$$\text{II - compression:} \quad p = p_+ : \quad W_{23}^b = p_+(V_2 - V_3) < 0,$$

$$Q_{23}^b = H_3 - H_2 = \frac{\gamma R}{\gamma - 1}(T_3 - T_2) > 0. \qquad (2.696\text{a–d})$$

The third stage is adiabatic expansion at constant high entropy (2.697a) hence no heat exchange (2.697b) and work specified by the internal energy (2.697c, d):

$$\text{III - expansion:} \quad S = S_+ \quad Q_{34}^b = 0,$$

$$W_{34}^b = U_4 - U_3 = \frac{R}{\gamma - 1}(T_4 - T_3) < 0. \qquad (2.697\text{a–d})$$

The final fourth stage is cooling at constant low atmospheric pressure (2.698a) so that the work is specified by the volume change (2.698b) and the heat by the enthalpy change (2.698c, d):

$$\text{IV - atmospheric cooling:}$$

$$p = p_- : \quad W_{41}^b = p_-(V_4 - V_1) > 0,$$

$$Q_{41}^b = H_1 - H_4 = \frac{\gamma R}{\gamma - 1}(T_1 - T_4) < 0. \qquad (2.698\text{a–d})$$

The total heat (2.699a, 2.448a–g) is the sum of heating by combustion (2.696d) and atmospheric cooling (2.698d):

$$Q^b = Q_{23}^b + Q_{41}^b = \frac{\gamma R}{\gamma - 1}(T_3 - T_2 + T_1 - T_4), \qquad (2.699\text{a, b})$$

and the total work (2.700a, 2.467a–e) adds all four phases (2.695d, 2.696b, 2.697d, 2.698b):

$$W^b = W_{12}^b + W_{34}^b + W_{23}^b + W_{41}^b$$
$$= \frac{R}{\gamma - 1}\left(T_2 - T_1 + T_4 - T_3\right) + p_+\left(V_2 - V_3\right) + p_-\left(V_4 - V_1\right). \quad \text{(2.700a, b)}$$

Using the equation of state for a perfect gas (2.460c) ≡ (2.467f–i):

$$p_+\left\{V_2, V_3\right\} = R\left\{T_2, T_3\right\} \quad p_-\left\{V_1, V_4\right\} = R\left\{T_1, T_4\right\}, \quad \text{(2.701a–d)}$$

the total work (2.700b) simplifies to (2.467j–l):

$$W^b = R\left[\left(T_2 - T_3\right) + \left(T_4 - T_1\right)\right]\left(\frac{1}{\gamma - 1} + 1\right) = \frac{\gamma R}{\gamma - 1}\left(T_2 - T_3 + T_4 - T_1\right) = -Q^b, \quad \text{(2.702a–c)}$$

that is minus the total heat (2.702c) ≡ (2.699b) since the variation of the internal energy must be zero over the closed cycle. Next is considered the efficiency of the Barber-Brayton cycle (subsection 2.6.27) and how to maximize the work it supplies (2.6.28).

2.6.27 HEAT, WORK AND EFFICIENCY OF THE BARBER-BRAYTON CYCLE

The efficiency of the Barber-Brayton cycle is the total work (2.702b) divided by the heat input (2.697d) by combustion (2.468a, 2.445a, b):

$$\eta_b = \frac{-W^b}{Q_{23}^b} = \frac{Q^b}{Q_{23}^b} = 1 + \frac{Q_{41}^b}{Q_{23}^b} = \frac{T_3 - T_2 + T_1 - T_4}{T_3 - T_2} = 1 - \frac{T_4 - T_1}{T_3 - T_2}. \quad \text{(2.703a–e)}$$

Since the stage I (III) of admission – compression (exhaust) is adiabatic (2.695a) [(2.697a)] the ratio of high to low pressures (2.478f) satisfies (2.704a) [(2.704b)] implying (2.704c):

$$\frac{T_2}{T_1} = \left(\frac{p_+}{p_-}\right)^{1-1/\gamma} = \frac{T_3}{T_4} \Leftrightarrow \frac{T_4}{T_1} = \frac{T_3}{T_2}: \quad \eta_b = 1 - \frac{T_1\left(\frac{T_4}{T_1} - 1\right)}{T_2\left(\frac{T_3}{T_2} - 1\right)}, \quad \text{(2.704a–d)}$$

that simplifies the efficiency (2.703e) ≡ (2.704d) to (2.705b):

$$r \equiv \frac{p_+}{p_-}: \quad \eta_b = 1 - \frac{T_1}{T_2} = 1 - \frac{T_4}{T_3} = 1 - \left(\frac{p_-}{p_+}\right)^{1-1/\gamma} = 1 - r^{1/\gamma - 1}. \quad \text{(2.705a–e)}$$

Thus *the efficiency of the Barber-Brayton cycle depends only on the ratio of temperatures (2.705b, c), or pressures (2.705d) or pressure ratio (2.705a, e) between: (i) exit of compressor or entry to combustion chamber; (ii) atmospheric pressure at inlet. The efficiency of a turbojet engine increases with the pressure ratio and is given (2.475a–c) by (2.706b, c), respectively, for monoatomic/diatomic/polyatomic perfect gases:*

$$f = 3,5,6: \quad \eta_b = 1 - r^{-\frac{2}{f+2}} = 1 - r^{-\left(\frac{2}{5},\frac{2}{7},\frac{1}{4}\right)}. \tag{2.706a–c}$$

The efficiency of the Barber-Brayton cycle is indicated in Table 2.16 for pressure ratios 10, 20, 30, 40 with the diatomic case applying to air. The maximum efficiency (work) of the Brayton cycle leads [subsection 2.6.27 (2.6.28)] to the minimum fuel consumption (maximum thrust).

2.6.28 MAXIMUM WORK, POWER AND THRUST OF A TURBOJET ENGINE

The work of the Barber-Brayton cycle is given by (2.702b) $\equiv$ (2.707b, c) in terms of the specific heat at constant pressure (2.707a):

$$c_p = \frac{\gamma R}{\gamma - 1}: \quad -W^b = c_p\left(T_3 - T_2 + T_1 - T_4\right). \tag{2.707a–c}$$

The atmospheric temperature is fixed (2.708a) and the turbine entry temperature cannot be exceeded (2.708b) so the work is maximized relative to the compressor exit temperature (2.708c, d)

$$T_1, T_3 = const: \quad \frac{dW^b}{dT_2} = 0 \quad \Rightarrow \quad 1 = -\frac{dT_4}{dT_2} = \frac{T_1 T_3}{\left(T_2\right)^2}, \tag{2.708a–e}$$

TABLE 2.16
Efficiency of the Barber-Brayton Cycle for a Turbojet

Perfect Gas		Monoatomic	Diatomic	Polyatomic
Degrees of freedom of molecule	f	3	5	6
Adiabatic exponent	$\gamma = 1 + \dfrac{2}{f}$	$\dfrac{5}{3}$	$\dfrac{7}{5}$	$\dfrac{4}{3}$
Barber-Brayton cycle efficiency	$\eta_b = 1 - r^{2/(2+f)}$	$1 - r^{2/5}$	$1 - r^{2/7}$	$1 - r^{1/4}$
Compression ratio*	10	0.602	0.482	0.438
	20	0.698	0.575	0.527
$r = \dfrac{p_+}{p_-}$	30	0.743	0.622	0.573
	40	0.771	0.651	0.602
compressor/atmosphere	50	0.791	0.673	0.624

* ratio of pressure at compressor exit p_+ to atmospheric p_-

Note: The efficiency of the Barber-Brayton cycle for a monoatomic/diatomic/polyatomic perfect gas (Tables 2.9–2.11) in a jet engine depends only on the compression ratio between high p_+ and low p_- pressure, and is given for five values = 10, 20, 30, 40, 50.

and (2.704c) leads to (2.707e). Thus *the maximum work of a turbojet engine corresponds (2.708e) = (2.709a) to a compressor outlet or combustion inlet temperature that is the geometric mean of the atmospheric temperature and turbine entry temperature:*

$$T_2 = \sqrt{T_1 T_3}; \quad T_4 = \frac{T_1 T_3}{T_2} = \sqrt{T_1 T_3} = T_2, \qquad (2.709\text{a–d})$$

implying (2.704c) that the exhaust temperature is the same (2.709b–d). Substitution of (2.709a, d) in (2.707c) specifies the **maximum work** *per unit mass of a turbojet engine (2.710a, 2.467a–e):*

$$\left(-W^b\right)_{max} = c_p\left(T_3 + T_1 - 2\sqrt{T_1 T_3}\right) = c_p\left(\sqrt{T_1} - \sqrt{T_3}\right)^2. \qquad (2.710\text{a, b})$$

The maximum power (2.711b, c) is obtained multiplying by the **mass flow rate** *(2.711a) that is the product of mass density ρ, by velocity v, by cross-sectional area S, and is constant through the engine except if there is extraction of bleed air:*

$$\dot{m} = \rho S v: \quad \dot{W} = \dot{m}W = \rho S v W; \quad F = \frac{\dot{W}}{v} = \rho S W, \qquad (2.711\text{a–e})$$

the power equals the velocity multiplied by the **thrust** *(2.711d, e).* The discussion on thermodynamic cycles (subsections 2.5.15–2.5.24 and 2.6.20–2.6.28) is concluded by a numerical example of the Barber-Brayton (Carnot) cycle applied to a turbojet (piston gasoline or diesel) engine [subsections 2.6.29–2.6.35 (example 10.5)] in the atmospheric conditions at sea level (subsections IV.5.5.25 and 2.6.29).

2.6.29   INTERNATIONAL STANDARD ATMOSPHERE (ISA) AT SEA LEVEL

In the numerical examples that follow will be used the **M.K.S. system of units** for which the units of length, mass and time are, respectively, the **meter** m, **kilogram** kg and **second** s. This implies other derived units: (i) **Newton** for force that equals mass multiplied by acceleration (2.712a); (ii) **Joule** for work or energy that equals force multiplied by length (2.712b, c); (iii) **Watt** for power or activity that equals work or energy per unit time (2.477a–e):

$$1N = 1\,kg\,m\,s^{-2}, \quad 1J = 1\,Nm = 1\,kg\,m^2\,s^{-2}, \qquad (2.712\text{a–c})$$

$$1W = 1\,J\,s^{-1} = 1\,N\,m\,s^{-1} = 1\,kg\,m^2\,s^{-3}. \qquad (2.712\text{d–f})$$

The **kilogram force** is (2.713-d) one kilogram mass multiplied by the acceleration of gravity (2.713a) at the surface of the Earth:

$$g = 9.80665\,m\,s^{-2} \quad 1\,kgf = 1\,kg.1g = 9.81\,kg\,m\,s^{-2} = 9.81\,N. \qquad (2.713\text{a–d})$$

The **metric (imperial) horsepower** is 75 (76) kilogram force multiplied by meter per second (2.479a–d) [(2.480a–e)]:

$$1CV = 75\,kgf\,m\,s^{-1} = 75\times 9.81\,N\,m\,s^{-1} = 736\,W = 0.736\,kW, \quad (2.714a\text{–}d)$$

$$1HP = 76\,kgf\,m\,s^{-1} = 76\times 9.81\,N\,m\,s^{-1} = 746\,W = 0.746\,kW. \quad (2.715a\text{–}d)$$

Other units include degrees Celsius C (Kelvin K) for temperature and bar for pressure.

For the **standard atmosphere at sea level** is assumed: (i) a temperature of 15 degrees Celsius (2.716a) adding 273.15 K for the temperature relative to absolute zero leads, neglecting decimals, to 288 Kelvin (2.716b):

$$T_0 = 15C = 288.15K, \quad \rho_{Hg} = 13.6\,g\,cm^{-3} \quad h = 76\,cm:$$

$$
\begin{aligned}
p_0 &= \rho_{Hg}\,h\,g = 13.6\,g\,cm^{-3}\times 76\,cm\times 980.665\,cm\,s^{-2} = 1.013615\times 10^{6}\,g\,cm^{-1}\,s^{-2}\\
&= 1.013615\times 10^{6}\times\left(10^{-3}\,kg\right)\times\left(10^{-2}\,m\right)^{-1}s^{-2} = 1.013615\times 10^{5}\,kg\,m^{-1}\,s^{-2}\\
&= 1.014\times 10^{5}\,N\,m^{-2} \equiv 1\,atm = 1.014\,bar = 101.4\,kPa = 760\,tor;
\end{aligned}
\quad (2.716a\text{–}m)
$$

(ii) the pressure (2.716e–j) of 1 atmosphere corresponding in the **Torricelli (1646)** experiment to the weight per square centimetre of a column of mercury with density (2.716c) and height (2.716d). Besides (i) the atmosphere **atm** another three units of pressure are often used: (ii) **tor** or millimeters of mercury in Torricelli experiment; (iii) **bar** or Newtons per square meter; (iv) Pascal **Pa** $=10^{-5}$ bar.

Atmospheric air consists mostly of 21% of oxygen (2.717a) and 78% of nitrogen (2.717b) plus 1% of argon (2.717c), and I.S.A. neglects other element and components, such as water vapor. The molecular masses of biatomic oxygen (2.712d) and nitrogen (2.717f) then lead to the molecular mass of air (2.717g):

$$\left\{q_{O_2}, q_{N_2}, q_{Al}\right\} = \left\{0.21,\ 0.78,\ 0.01\right\}, \quad (2.717a\text{–}c)$$

$$\left\{\bar{m}_{O_2}, \bar{m}_{N_2}, \bar{m}_{A_1},\right\} = \left\{2\times 16,\ 2\times 14,\ 1\times 40\right\} = \left\{32, 28, 40\right\} : g\ mole^{-1} \quad (2.717d\text{–}f)$$

$$
\begin{aligned}
\bar{m}_a &= q_{O_2}\times \bar{m}_{O_2} + q_{N_2}\times \bar{m}_{N_2} + q_{Al}\times \bar{m}_{Al}\\
&= \left(0.21\times 32 + 0.78\times 28 + 0.01\times 40\right)g\ mole^{-1}\\
&= 28.96\ g\ mole^{-1}.
\end{aligned}
\quad (2.717g)
$$

The gas constant (2.473b) for air (2.717g) is given (2.470f) by (2.718a–e):

$$
\begin{aligned}
R_0 &= \frac{\bar{R}}{\bar{m}_a} = \frac{8.31434\ \ JK^{-1}mole}{28.96\ \ g\ mole^{-1}} = 0.28711\ JK^{-1}g^{-}\\
&= 0.28711\ \ JK^{-1}\times\left(10^{-3}\ kg\right)^{-1} = 287.11\ J\,Kg\ K^{-1}\\
&= 287.11\ m^2 S^{-2} K^{-1}
\end{aligned}
\quad (2.718a\text{–}e)
$$

Considering air as a mostly biatomic perfect gas (2.475b) ≡ (2.719a,b) the specific heats per unit mass at constant volume (2.471d) ≡ (2.719c) [pressure (2.474e) ≡ 2.719d] are given by (2.719e–h):

$$f_a = 5, \; \gamma_a = 1.4 : \{c_{v0}, \; c_{p0}\} = \frac{R_0}{\gamma_a - 1}\{1, \; \gamma_a\} = \frac{R_0}{2}\{f_a, f_a + 2\}$$

$$= \left\{\frac{5}{2}, \frac{7}{2}\right\} R_0 = \{718, \; 1005\} \; m^2 s^2 K^{-1}.$$

(2.719a–h)

The temperature (2.716b), pressure (2.716i) and gas constant (2.718g) lead from the equation of state for a perfect gas (2.718a) ≡ (2.720a) to the mass density at sea level (2.720b, c):

$$\rho_0 = \frac{p_0}{R_0 T_0} = \frac{1.013615 \times 10^5 \; kg \, m^{-1} \, s^{-2}}{287.11 \, m^2 \, s^{-2} \, K^{-1} \times 288.15 \, K} = 1.225 \, kg \, m^{-3}. \quad (2.720a\text{–}c)$$

The adiabatic sound speed (2.479c) ≡ (2.721a, b) at sea level (2.721c, d):

$$c_0 = \sqrt{\gamma_a \frac{p_0}{\rho_0}} = \sqrt{\gamma_a R_0 T_0} = \sqrt{1.4 \times 289 \, m^2 \, s^{-2} \, K^{-1} \times 288 \, K} = 340 \, m \, s^{-1}$$

$$= 340 \, m \, s^{-1} \times 3600 \, s \, h^{-1} = 340 \times 3.6 \times 10^3 \, m \, h^{-1} = 1224 \, km \, h^{-1},$$

(2.721a–g)

was used before (2.685b) ≡ (2.721d) to calculate stagnation Mach numbers (2.675b) and classify flows (subsection 2.6.17–2.6.19). It appears next in the consideration of a numerical example of the Barber-Brayton cycle for a turbojet engine (subsections 2.6.30–2.6.35).

2.6.30   INLET AND EXHAUST OF A TURBOJET ENGINE

It is assumed that the air inlet is circular with radius (2.722a) and area (2.722b, c) and for a typical aircraft velocity on approach to land or during the initial climb after take-off (2.722d), implying the volume (mass) flow rate (2.487a–d) [(2.488a–i)]:

$$a = 0.5 \, m, \quad A = \pi a^2 = 0.785 \, m^2, \quad v_1 = 80 \, m \, s^{-1} : \quad (2.722a\text{–}d)$$

$$\dot{V} = v_1 A = 0.785 \, m^2 \times 80 \, m \, s^{-1} = 62.8 \, m^3 \, s^{-1}, \quad (2.722e\text{–}g)$$

$$\dot{m} = \rho_1 \dot{V} = \rho_1 v_1 A = 1.225 \, kg \, m^{-3} \times 62.8 \, m^3 s^{-1} = 76.7 \, kg \, s^{-1}; \quad (2.722h\text{–}k)$$

the mass flow rate (2.722h–k) is constant through the engine apart from bleed air to cool turbine blades, for air conditioning or other pneumatic systems. The maximum turbine entry temperature possible with current technology is about (2.723a) and for an atmospheric inlet (2.716b) the maximum power (2.709a) ≡ (2.723b) is for a compressor exit temperature (2.723c, d):

$$T_3 = 2000\,K: \quad T_2 = \sqrt{T_1 T_3} = \sqrt{288\,K \times 2000\,K} = 759\,K, \qquad (2.723\text{a–d})$$

$$\eta_b = 1 - \frac{T_1}{T_2} = 1 - \frac{288}{759} = 0.621, \qquad (2.723\text{e–g})$$

and efficiency is (2.705b) ≡ (2.490a–g) for the Barber-Brayton cycle.

The heat supply per unit mass in the combustion phase (2.696d) ≡ (2.724a) is (2.719h; 2.723a, d) given by (2.724b–d):

$$
\begin{aligned}
q_{23}^b &= c_{p_a}\left(T_3 - T_2\right) = 1.005 \times 10^3\, m^2\, s^{-2}\, K^{-1} \times \left(2000 - 759\right) K \\
&= 1.247 \times 10^6\, m^2\, s^{-2},
\end{aligned}
\qquad (2.724\text{a–d})
$$

and multiplying by the mass flow rate (2.722k) specifies the total rate of heat input (2.492a–e):

$$
\begin{aligned}
\dot{Q}_{23}^b &= \dot{m}\, q_{23}^b = 76.7\, kg\, s^{-1}\, K \times 1.247 \times 10^6\, m^2\, s^{-2}\, K^{-1} \\
&= 9.57 \times 10^7\, kg\, m^2\, s^{-3} = 95.7\, MW.
\end{aligned}
\qquad (2.725\text{a–d})
$$

The efficiency (2.723g) multiplied by the heat input per unit mass (2.724d) [total (2.725d)] specifies the maximum work per unit mass (2.493a–e) [total (2.493a–e)] in the expansion phase:

$$w^b = -\eta_b\, q_{23}^b = -0.621 \times 1.247 \times 10^6\, m^2\, s^{-2} = -7.74 \times 10^5\, m^2\, s^{-2}, \quad (2.726\text{a–c})$$

$$\dot{W}^b = -\dot{m}\, w_{34}^b = -0.621 \times 95.7\, MW = -59.4\, MW. \qquad (2.727\text{a–c})$$

The four stages of the Barber-Brayton cycle for the turbojet with maximum power (2.727c) and efficiency (2.723g) are considered next (subsection 2.6.31).

2.6.31 PRESSURE, DENSITY AND TEMPERATURE IN A TURBOJET CYCLE

After the admission the adiabatic compression raises the temperature from (2.716b) to (2.723d) and hence raises the pressure (2.704a) ≡ (2.728b) from (2.716k) ≡ (2.728a) to (2.728b–d):

$$p_- \equiv p_1 = 1\,atm: \quad p_+ = p_-\left(\frac{T_2}{T_1}\right)^{\frac{\gamma}{\gamma-1}} = \left(\frac{759}{288}\right)^{3.5} atm = 29.7. \qquad (2.728\text{a–d})$$

The corresponding mass density for a perfect gas (2.720a) is (2.496a–d):

$$\rho_2 = \frac{p_+}{R_a T_2} = \frac{29.7 \times 1.014 \times 10^5\, kg\, m^{-1}\, s^{-2}}{287\, m^2\, s^{-2}\, K^{-1} \times 759\, K} = 13.8\, kg\, m^{-3}, \qquad (2.729\text{a–c})$$

$$c_2 = c_1 \sqrt{\frac{T_2}{T_1}} = \sqrt{\frac{759}{288}} \times 340\,m\,s^{-1} = 552\,m\,s^{-1}, \qquad (2.730a\text{--}c)$$

and (2.497a–d) is the sound speed.

The pressure (2.728d) and temperature (2.723a) at turbine entry specify the mass density (2.498a–d)

$$\rho_3 = \frac{p_+}{R_a\,T_3} = \frac{29.7 \times 1.014 \times 10^5\,kg\,m^{-1}\,s^{-2}}{287\,m^2\,s^{-2}\,K^{-1} \times 2000\,K} = 5.25\,kg\,m^{-3}, \qquad (2.731a\text{--}c)$$

and the sound speed (2.499a–d) follows (2.717b; 2.718g) from the temperature (2.723a):

$$c_3 = \sqrt{\gamma_a\,R_0\,T_3} = \sqrt{1.4 \times 287\,m^2\,s^{-2}\,K^{-1} \times 2000\,K} = 896\,m\,s^{-1}. \qquad (2.732a\text{--}c)$$

The adiabatic expansion (2.704b) reduces the temperature from (2.723a) to (2.500a–e):

$$T_4 = T_3 \left(\frac{p_+}{p_-}\right)^{1/\gamma-1} = 29.7^{-2/7} \times 2000\,K = 759\,K, \qquad (2.733a\text{--}c)$$

that coincides with (2.723d) in agreement with (2.473a–i) without truncation errors of numerical computation.

The corresponding sound speed is (2.501a, b) $\equiv$ (2.497a–d):

$$c_4 = \sqrt{\gamma_a\,R_0\,T_4} = \sqrt{1.4 \times 287\,m^2\,s^{-2}\,K^{-1} \times 759\,K} = 552\,m\,s^{-1}, \qquad (2.734a\text{--}c)$$

$$\rho_4 = \frac{p_+}{R_0\,T_4} = \frac{1.014 \times 10^5\,kg\,m^{-1}\,s^{-2}}{287\,m^2\,s^{-2}\,K^{-1} \times 759\,K} = 0.465\,kg\,m^{-3}, \qquad (2.735a\text{--}c)$$

and the mass density is (2.502a–e). In the atmospheric cooling phase should be equal (2.503a–e) and (2.504a–e):

$$R_0\,p_- = \rho_4\,T_4 = 0.465\,kg\,m^{-3} \times 759\,K = 353\,kg\,m^{-3}\,K, \qquad (2.736a\text{--}c)$$

$$R_0\,p_- = \rho_1\,T_1 = 1.225\,kg\,m^{-3} \times 288\,K = 353\,kg\,m^{-3}\,K, \qquad (2.737a\text{--}c)$$

thus providing a final check on the accuracy of the chain of calculation over the four stages of the Barber-Brayton cycle, that are reconsidered [subsections 2.6.32 (2.6.33)] in terms of heat (work) per unit mass and heat rate (power).

2.6.32   HEAT RATE AND POWER PER UNIT MASS

There is heat exchange in the Barber-Brayton cycle during the phase II (IV) of heating by combustion (2.696d) [in the atmosphere (2.698d)]) leading (2.719g, h) to the heat per unit mass (2.724a–d) [(2.738a–c)]:

$$q_{41}^b = c_{p_a}\left(T_1 - T_4\right) = 1.005\times 10^3\ m^2\ s^{-2}\ K^{-1}\times\left(288 - 759\right)K$$
$$= -4.73\times 10^5\ m^2\ s^{-2}. \tag{2.738a–c}$$

The heat lost by cooling to the atmosphere (2.738c) is a fraction (b) of the heat input by combustion (2.724d):

$$\frac{q_{41}^b}{q_{23}^b} = -\frac{4.73}{12.47} = -0.379, \tag{2.739}$$

and their sum is the total heat (2.507a–c):

$$q^b = q_{41}^b + q_{23}^b = \left(12.47 - 4.73\right)\times 10^5\ m^2\ s^{-2} = 7.74\times 10^5\ m^2\ s^{-2}. \tag{2.740a–c}$$

The fraction of the total heat (2.740c) corresponding to atmospheric cooling (2.738c) [combustion heating (2.724d)] is (2.741a, 2.467a–e):

$$\frac{\left\{q_{41}^b, q_{23}^b\right\}}{q^b} = \frac{\left\{-4.73, 12.47\right\}}{7.74} = \left\{-0.611, 1.611\right\}. \tag{2.741a, b}$$

Multiplying by the mass flow rate (2.722k) specifies (2.509a–d) the total heat rate due to atmospheric cooling (2.738c), combustion heating (2.724d) and their sum (2.740c):

$$\left\{Q_{41}^b,\ Q_{23}^b,\ Q^b\right\} = \dot{m}\left\{q_{41}^b,\ q_{23}^b,\ q^b\right\}$$
$$= 76.7\ kg\ s^{-1}\times\left\{-4.73, 12.47, 7.74\right\}\times 10^5\ m^2\ s^{-2}$$
$$= \left\{-3.62, 9.56, 5.94\right\}\times 10^7\ kg\ m^2\ s^{-3} \tag{2.742a–d}$$
$$= \left\{-36.2, 95.6, 59.4\right\}MW.$$

Whereas the heat exchanges occur only during the combustion and atmospheric phases, work is also performed in the admission-compression and exhaust phases (subsection 2.6.33).

2.6.33   WORK PER UNIT MASS AND POWER

The work per unit mass is given by (2.695d) $\equiv$ (2.743a–c)/(2.696b) $\equiv$ (2.744a–e)/ (2.697d) $\equiv$ (2.745a–c)/(2.698b) $\equiv$ (2.746a–c) respectively in the admission –

compression/combustion/expansion/atmospheric cooling phases and follows from (2.718g; 2.719g, h) and (2.716b; 2.723a) and (2.723d) $\equiv$ (2.733c):

$$w_{12}^b = c_{V_0}\left(T_2 - T_1\right) = 7.18 \times 10^2 \, m^2 \, s^{-2} \, K^{-1} \times \left(759 - 288\right) K$$
$$= 3.38 \times 10^5 \, m^2 \, s^{-2}, \tag{2.743a–c}$$

$$w_{23}^b = p_+\left(\frac{1}{\rho_2} - \frac{1}{\rho_3}\right) = R_0\left(T_2 - T_3\right) = 287 \, m^2 \, s^{-2} \, K^{-1} \times \left(759 - 2000\right) K$$
$$= -3.56 \times 10^5 \, m^2 \, s^{-2}, \tag{2.744a–c}$$

$$w_{34}^b = c_{V_0}\left(T_4 - T_3\right) = 7.18 \times 10^2 \, m^2 \, s^{-2} \, K^{-1} \times \left(759 - 2000\right) K$$
$$= -8.91 \times 10^5 \, m^2 \, s^{-2}, \tag{2.745a–c}$$

$$w_{41}^b = p_-\left(\frac{1}{\rho_4} - \frac{1}{\rho_1}\right) = R_0\left(T_4 - T_1\right) = 287 \, m^2 \, s^{-2} \, K^{-1} \times \left(759 - 288\right) K$$
$$= 1.35 \times 10^5 \, m^2 \, s^{-2}. \tag{2.746a–c}$$

The total work per unit mass is (2.747a, 2.467a–e):

$$w^b = w_{12}^b + w_{23}^b + w_{34}^b + w_{41}^b = -7.74 \times 10^5 \, m^2 \, s^{-2}, \tag{2.747a, b}$$

in agreement with (2.747b) $\equiv$ (2.740c) without computational truncation errors. The fraction of the total work (2.747b) in the admission – compression (2.743c), combustion (2.747c), expansion (2.745c) and cooling (2.746c) phases is given by (2.514a–d):

$$\frac{\left\{w_{12}^b, w_{23}^b, w_{34}^b, w_{41}^b\right\}}{w^b} = \frac{\left\{3.38, -3.56, -8.91, 1.35\right\}}{-7.74}$$
$$= \left\{-0.437, 0.460, 1.151, -0.174\right\}. \tag{2.748a–c}$$

Multiplying by the mass flow rate (2.722k) specifies the power (2.749a–d) associated with the admission – compression (2.743c), combustion (2.744c), expansion (2.745c) and cooling (2.746c) phases:

$$\left\{\dot{W}_{12}^b, \dot{W}_{23}^b, \dot{W}_{34}^b, \dot{W}_{41}^b, \dot{W}^b\right\} = \dot{m}\left\{w_{12}^b, w_{23}^b, w_{34}^b, w_{41}^b, w^b\right\}$$
$$= 76.7 \, kg \, s^{-1} \times \left\{3.36, -3.56, -8.91, 1.35, -7.74\right\} \times 10^5 \, m^2 \, s^{-2} \tag{2.749a–e}$$
$$= \left\{25.8, -27.3, -68.8, 10.4, -59.4\right\} MW,$$

with the total power (2.749e) agreeing with (2.727c). Table 2.17 summarizes the physical conditions at the four stages of the Barber-Brayton cycle for a turbojet

TABLE 2.17

Four Phases of Flow in a Turbojet Engine

Stage	I: $1 \to 2$	II: $2 \to 3$	III: $3 \to 4$	IV: $4 \to 1$
At the start of	admission + compression	combustion	expansion	atmospheric cooling
Temperature (K)	288	759	2000	759
Pressure (atm)	1.00	29.7	29.7	1.00
Mass density $(kg\ m^{-3})$	1.225	13.8	5.25	0.465
Sound speed $(m\ s^{-1})$	340	552	896	552
Heat per unit mass $(\times 10^2\ s^{-2})$	0.000	12.47	0.000	−4.73
Fraction of total heat	0.000	1.611	0.000	−1.611
Total heat rate (MW)	0.00	95.6*	0.00	−36.2
Work per unit mass $(\times 10^2\ m^2\ s^{-2})$	3.38	−3.56	−8.91	1.35
Fraction of total work	−0.437	.0.460	1.151	−0.174
Power (MW)	25.8	−27.3	−68.3	10.4

* Total heat input - volume flow rate: $62.8\ m^2\ s^{-1}$ - efficiency: 0.621- power: $59.8\ MW$ - mass flow rate: $76.7\ kg\ s^{-1}$ - exhaust velocity: $210\ m\ s^{-2}$

Note: An example of the four phases of flow in a turbojet, namely (I) admission and compression, (II) combustion, (III) expansion and (IV) atmospheric heating, indicating for each phase: (i–iv) the four flow variables: temperature, pressure, mass density and sound speed; (v–vii) [(ix–xi)] heat (work) per unit mass, fraction of total heat (work) and total heat (work) rate.

engine in terms of temperature, pressure, mass density and sound speed, work and heat per unit mass, total heat rate and power. The power specifies the thrust through the exhaust velocity (subsection 2.6.34).

2.6.34   EXHAUST SPEED AND THRUST OF A TURBOJET ENGINE

It is assumed that the exit nozzle has the same radius (2.722a) and area (2.722c) as the inlet, so that the same mass flow rate (2.722k) with the density (2.735c) leads to the exhaust speed (2.516a–d):

$$v_4 = \frac{\dot{m}}{\rho_4 A} = \frac{76.7\ kg\ s^{-1}}{0.465\ kg\ m^{-3} \times 0.785\ m^2} = 210\ m\ s^{-1}, \qquad (2.750a\text{–}c)$$

corresponding to stagnation (2.721d) [local (2.734c)] Mach numbers (2.751a–c) [(2.751d–f)]:

$$M_{40} = \frac{v_4}{c_1} = \frac{210}{340} = 0.618, \quad M_{44} = \frac{v_4}{c_4} = \frac{210}{552} = 0.380, \qquad (2.751a\text{–}f)$$

using (2.721d) [(2.734c)] in (2.751b) [(2.751e)]. Dividing the power (2.727c) by the exhaust velocity (2.750c) specifies the thrust (2.518a–e):

$$F = -\frac{W^b}{v_4} = \frac{5.94 \times 10^7 \, N \, m \, s^{-1}}{210 \, m \, s^{-1}} = 2.83 \times 10^5 \, N = 283 \, kN$$

$$= \frac{283}{9.81} \times 10^3 \, kgf = 28.8 \times 10^4 \, kgf, \qquad (2.752a\text{--}f)$$

that is about 30 tons force of thrust.

If the flow were irrotational, homentropic and steady the conservation of the stag-nation sound speed (2.661d) would lead to v2.519a–e):

$$v_1^2 + \frac{2}{\gamma-1} c_1^2 = \frac{2}{\gamma-1} c_0^2 = v_4^2 + \frac{2}{\gamma-1} c_4^2, \qquad (A)$$

implying that the exhaust velocity could be calculated from (2.520a–e):

$$v_4^2 = v_1^2 - \frac{2}{\gamma-1}\left(c_4^2 - c_1^2\right). \qquad (B)$$

The values (2.722d; 2.717b; 2.721d; 2.734c) lead to (2.521a–e):

$$v_4^2 = 80^2 - 5 \times \left(552^2 - 340^2\right) m^2 \, s^{-2} = -9.39 \times 10^5 \, m^2 \, s^{-2}, \qquad (C)$$

that shows the conditions of irrotational, homentropic and steady flow cannot be met since there has been a heat input between the inlet and the exhaust; the large heat addition during combustion increases the exhaust relative to the inlet velocity, accounts alone for the unphysical negative sign in (2.521a–e). The preceding theory provides a simplified account of the complex rotational, non-isentropic, unsteady flow in turbomachinery (subsection 2.6.35).

2.6.35 ROTATIONAL, NON-ISENTROPIC, UNSTEADY FLOW

The flow in the air inlet of a turbojet has boundary layers near the walls and thus is not uniform over the cross-section and is rotational. The friction against the walls of the inlet and the casing of the compressor causes heating and pressure losses. The increasing temperature of the flow through successive compressor stages leads to increasing heat losses by conduction to the walls. The compressor stages consist of blades that leave trailing vortex wakes that are a rotational unsteady flow.

The flow in the combustion chamber has pressure fluctuations associated with combustion. The vorticity in the combustion chamber or combustion cans leads to a rotational flow aiding in the mixing of fuel and air and a more complete combustion of the fuel. The high temperatures of combustion cause heat losses by conduction to the walls and also by radiation.

The flow through the turbine can cause high thermal stresses, reduced by inter-nal cooling of the blades, which loses heat by conduction. The rows of turbine stages have blades with vortex wakes and their rotation causes a strong swirl that is a vortical flow.

The expansion through the nozzle involves a flow with two kinds of vorticity: (i) shear flow due to the boundary layers of the wall; (ii) swirling flow due to the high-speed rotation of the turbine. The high-speed jet issuing from the nozzle gives rise to shear layers (subsection 2.7.2), that broaden downstream as the hot jet mixes with the cold ambient air.

For all these reasons the detailed design of a jet engine goes far beyond the very preliminary consideration of a thermodynamic cycle. Detailed models of vortical and compressible flows, heat conduction and radiation, and chemical reactions and combustion are used along the successive phases of the engine cycle.

In the case of supersonic flight several new phenomena arise: (i) the supersonic inlet flow forms a normal shock wave (Section 2.7) in the inlet duct that causes irreversible compression and heating; (ii) the hot high-speed jet may form oblique shock waves (Section 2.8) from the nozzle lips, that may be reflected between the shear layers, forming shock cells; (iii) to accelerate the exhaust flow to supersonic speeds a throated convergent–divergent nozzle is needed (Section 2.9) that may contain or not shock waves. The set of three phenomena (i–iii) is considered next (Sections 2.7–2.9).

2.7   VORTEX SHEET AND THE NORMAL SHOCK (RANKINE 1870; HUGINOT 1887)

The equations of fluid mechanics (subections 2.6.1–2.6.5) have: (i) continuous solutions, for example, the flow of a compressible polytropic fluid (subsections 2.6.6–2.6.16), including as a particular case the adiabatic flow; (ii) surfaces of discontinuity (Note 1.9) across which must be conserved the mass, momentum and energy (subsection 2.7.1). The latter three conditions allow (subsection 2.7.1) three possibilities: (i) a tangential discontinuity, with tangential velocity and other flow quantities discontinuous, except for pressure equilibrium, that is a vortex sheet (subsection 2.7.2); (ii) a normal shock (subsection 2.7.3) for which the tangential velocity is zero and other quantities are discontinuous; (iii) the combination of (i) and (ii) leads to an oblique shock (subsection 2.7.4) for which all flow variables are discontinuous except the tangential velocity. A normal shock wave (subsection 2.7.5) causes an irreversible decrease in velocity and increase in mass density and pressure associated with entropy production, both for weak (strong) shocks [subsection 2.7.6 (2.7.8)]. Thus there is a significant difference between the static (dynamic) or Poisson (Rankine – Huginot) loci relating pressure to specific volume associated with a reversible process (subsection 2.7.7) [normal shock wave (subsection 2.7.9)]. The relations between the flow variables across the normal shock front can be put in several forms, using as independent variables, for example, the ratio of pressures and temperatures for an ideal gas (subsection 2.7.10). In the case of a perfect gas the geometric (arithmetic) mean (subsection 2.7.13) of upstream and downstream velocities [subsection 2.7.11 (2.7.12)] is the critical sound speed (speed of advance of the shock front). The comparison of the adiabatic and shock loci of pressure – specific volume relations (subsection 2.7.14) is a preliminary step to relate the conditions across a plane shock of a perfect gas (subsection 2.7.15) in terms of the: (i) upstream and downstream pressure ratio (subsection 2.7.16); (ii/iii) upstream (downstream) local Mach number [subsection 2.7.17 (2.7.18)]. The numerical examples of the normal shock of a perfect gas with a supersonic (subsonic) stagnation Mach number (subsection 2.7.19) are used to: (i) calculate the subsonic downstream local, stagnation and critical Mach numbers (subsection 2.7.20), (ii–iv) the upstream (downstream) flow variables before (after) the shock [subsection 2.7.21 (2.7.22)] and their ratios and differences (subsection 2.7.23). The entropy difference (subsection 2.7.24) is used to distinguish the physical (unphysical) cases of entropy increase (decrease) that are compatible with (contradict) the second principle of thermodynamics. Since a normal shock is an irreversible thermodynamic process some stagnation flow variables must decrease and none can increase (subsection 2.7.25).

2.7.1 MATCHING CONDITIONS ACROSS FLOW DISCONTINUITIES

The equations of fluid mechanics, in spite of being partial differential with continuous coefficients, admit discontinuous solutions (Note 1.9). The flow on each side of the surface of discontinuity is specified by variables (2.753a) like the velocity, pressure, mass density, temperature, sound speed, internal energy and enthalpy:

$$X \equiv \{\bar{v},p,\rho,T,c,u,h\}: \quad [X] \equiv X^+ - X^- \begin{cases} = 0 & \text{continuous,} \\ \neq 0 & \text{jump,} \end{cases} \quad (2.753a, b)$$

and the jump across the surface (2.753b) is the difference between downstream and upstream values (Figure 2.28). The jump is zero (non-zero) for flow variables that are continuous (discontinuous) across the surface; is positive (negative) for an increase (decrease) across the surface of discontinuity. Choosing a reference frame moving

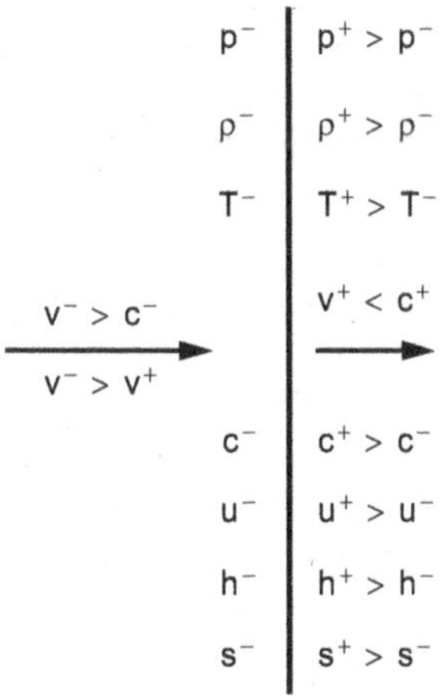

FIGURE 2.28 A normal shock wave has an upstream (downstream) velocity supersonic (subsonic) relative to the sound speed, and across the shock front: (i) the velocity decreases; (ii–vii) the pressure, mass density, temperature, sound speed, internal energy and enthalpy all increase. Since it is an irreversible process the entropy also increases.

with the surface the flow is steady, and in the equations of non-dissipative fluid mechanics in conservative form (2.621a–c): (i) the time derivatives vanish (2.754a); (ii) the spatial derivatives projected on the unit normal to the surface specify the jump between the two sides (2.754b, c):

$$\frac{\partial X}{\partial t} = 0: \quad n_i \frac{\partial X}{\partial x_i} = [X] = X^+ - X^-. \tag{2.754a–c}$$

This leads to the following **conservation conditions** *across a surface of discontinuity of a non-dissipative fluid: (i/ii) from the conservation of mass (2.621a) [energy (2.621b)] the continuity of the normal component of the mass (2.755a) [energy (2.755c)] flux, that involves the normal velocity only (also the total velocity and enthalpy):*

$$[\rho v_n] = 0, \quad \left[\rho\left(h + \frac{v^2}{2}\right)v_n\right] = 0; \tag{2.755a, b}$$

(iii) the conservation of momentum (2.621b) leads to the continuity of the normal (2.756a) [tangential (2.756b)] component that involves (does not involve) the pressure and involves the normal velocity only (also the tangential velocity):

$$[p + \rho(v_n)^2] = 0, \quad [\rho v_n \vec{v}_t] = 0. \tag{2.756a, b}$$

Then (2.755a, b; 2.756a, b) lead to a classification of flow discontinuities (Diagram 2.4) into three types (Figure 2.29) whose properties (Table 2.18) are considered next (subsections 2.7.2–2.7.4).

2.7.2 TANGENTIAL DISCONTINUITY BETWEEN TWO JETS

The conservation of the mass flux (2.755a) ≡ (2.757a, b) leads in the conservation of the normal (2.756a) [tangential (2.756b)] momentum to (2.757c) [(2.757d)]:

$$\rho^- v_n^- = \rho^+ v_n^+ \equiv j, \quad [p + j v_n] = 0 = j[\vec{v}_t], \tag{2.757a–d}$$

$$0 = j[2h + v^2] = 2j[h_0] = j\left[2h + (v_n)^2 + (v_t)^2\right], \tag{2.757e–g}$$

and in (2.755b) to the conservation of the stagnation enthalpy (2.757e–g). Case I is a tangential velocity on one side of the surface (2.758a) that implies: (i) from (2.757a) that the velocity is also tangential on the other side (2.758a) and vice-versa (2.758b):

vortex sheet: $v_n^- = 0 \neq \vec{v}_t^- \quad \Leftrightarrow \quad v_n^+ = 0 \neq \vec{v}_t^+;$

$$j = 0, \quad p^+ = p^-, \quad \vec{v}_t^- \neq \vec{v}_t^+, \quad \rho^+ \neq \rho^-, \quad h^+ \neq h^-; \tag{2.758a–g}$$

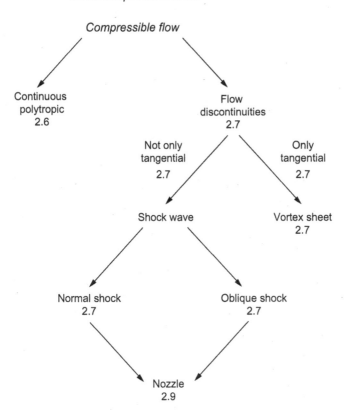

Some compressible flows

Compressible flow

Continuous
polytropic
2.6

Flow
discontinuities
2.7

Not only
tangential
2.7

Only
tangential
2.7

Shock wave

Vortex sheet
2.7

Normal shock
2.7

Oblique shock
2.7

Nozzle
2.9

DIAGRAM 2.4 The compressible flows can be continuous or discontinuous. The tangential discontinuities, like a vortex sheet, can separate incompressible or compressible flows, and can issue from nozzle lips. The discontinuities of the normal velocity lead to normal or oblique shock waves, that are always compressible and irreversible, and can occur inside ducts or at nozzle exits.

(ii) the mass flux (2.757a, b) is zero (2.758c) and hence (2.757c) the pressure is continuous (2.758d); (iii) other flow variables like the tangential velocity (2.758e), mass density (2.758f) and enthalpy (2.758g) may be discontinuous. Thus a **vortex sheet** *(Figure 2.29) with discontinuous (2.758e) purely tangential velocity on both sides (2.758a, b) must be in pressure equilibrium (2.758d), and other flow quantities like the mass density (2.758f) or enthalpy (2.758g) may be discontinuous.* The vorticity (2.626c) is infinite across a vortex sheet, that is unstable to perturbations, and this **Kelvin – Helmholtz instability** applies both to incompressible (compressible) flow [Section II.8.9 (V.9.4)]. The vortex sheet is an idealization with zero thickness and discontinuous tangential velocity of a **shear layer** between two jets where the two flows mix in a region of increasing width with continuous tangential velocity. Case I (II) of purely tangential (normal) velocity leads [subsection 2.7.2 (2.7.3)] to a vortex sheet (normal shock wave).

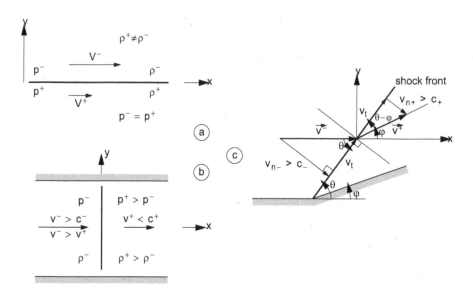

FIGURE 2.29 There are three types of discontinuities in a flow: (i) a tangential discontinuity like a vortex sheet (a) with equal pressure on the two sides, and unrestricted velocities and mass densities: (ii/iii) a normal (b) [oblique (c)] shock in a parallel-sided duct (concave corner) in which the normal upstream/downstream velocity is supersonic/subsonic relative to the local sound speed, and the pressure and density increase across the shock front.

TABLE 2.18
Classification of Flow Discontinuities

Discontinuity		Tangential	Oblique	Normal
Designation		Vortex Sheet	Oblique Shock	Normal Shock
velocity	Normal	$v_n = 0$	$[v_n] \neq 0$	$[v_n] \neq 0$
	Tangential	$[v_t] \neq 0$	$[v_t] = 0$	$v_t = 0$
Other flow variables	Pressure	$[p] = 0$	$[p] > 0$	$[p] > 0$
	Mass density	$[\rho] =, \neq 0$	$[\rho] > 0$	$[\rho] > 0$
Thermodynamic	Entropy	$[s] =, \neq 0$	$[s] > 0$	$[s] > 0$
functions of state	Enthalpy	$[h] =, \neq 0$	$[h] > 0$	$[h] > 0$
	Internal energy	$[u] =, \neq 0$	$[u] > 0$	$[u] > 0$

Note: Classification of flow discontinuities into (a) tangential, like a vortex sheet and non-tangential or shocks, either (b) normal shock or (c) oblique shock. For each case (a, b, c) is indicated whether there is continuity $[...] = 0$ or a jump $[...] \neq 0$ for: (i–ii) the normal and tangential velocity; (iii–iv) the other flow variables: pressure and mass density; (v–vii) thermodynamic functions of state: entropy, enthalpy and internal energy.

2.7.3 COMPRESSION ACROSS A NORMAL SHOCK FRONT

The opposite to (2.758a, b) case I is case II of a purely normal velocity on one side (2.759a) that: (i) by (2.757d) is equivalent to a purely normal velocity on the other side (2.759b):

Normal shock:

$$v_n^- \neq 0 = \bar{v}_t^- \quad \Leftrightarrow \quad v_n^+ \neq 0 = \bar{v}_t^+ : \quad j \neq 0;$$
$$v^- > v^+, \quad \rho^+ > \rho^-, \quad p^+ > p^- \quad ,h^+ > h^-; \qquad (2.759\text{a–g})$$

(ii) the mass flux (2.757a, b) is not zero (2.759c); (iii) if the normal velocity decreases (2.759d) then by (2.757a, c, g) the mass density (2.759e), pressure (2.759f) and enthalpy (2.759g) all increase corresponding to a compression. If the velocity would increase across a normal shock, then a succession of normal shocks would lead to an unbounded velocity while conserving energy, which is physically absurd. The case of a velocity increase opposite to (2.759d) is excluded, because it contradicts the second principle of thermodynamics stating that the entropy cannot decrease across a normal shock (subsections 2.7.6–2.7.9). Thus *across a **normal shock** (Figure 2.29): (i) the velocity is normal (2.759a, b) and decreases (2.759d); (ii) the conservation of the mass flux (2.759c) requires the mass density to increase (2.759e); (iii) the **compression** across the normal shock also implies an increase in pressure (2.759f) and enthalpy (2.759g).* The remaining case III concerns an oblique velocity (subsection 2.7.4).

2.7.4 DEVIATION OF VELOCITY TOWARDS AN OBLIQUE SHOCK FRONT

Case III is an oblique velocity on one side of the surface (2.760a) implying that: (i) the velocity (2.757d) is also oblique on the other side (2.760b):
Oblique shock:

$$v_n^- \neq 0 \neq \bar{v}_t^- \quad \Leftrightarrow \quad v_n^+ \neq 0 \neq \bar{v}_t^+ : \quad j \neq 0, \quad \bar{v}_t^- = \bar{v}_t^+, \quad v_n^- > v_n^+,$$
$$p^+ > p^-, \quad \rho^+ > \rho^-, \quad v^- > v^+, \quad h^+ > h^-; \qquad (2.760\text{a–i})$$

(ii) the mass flux (2.757a, b) is not zero (2.760c), and its conservation (2.757d) implies that the tangential velocity is continuous (2.760d); (iii) if the normal velocity decreases (2.760e) then by (2.757c) [(2.757a)] the pressure (2.760f) [mass density (2.760g)] increases; (iv) the continuity (decrease) of the tangential (2.760d) [normal (2.760e)] velocity leads to a decrease in total velocity (2.760h) and an increase in enthalpy (2.760i). Case III is not a vortex sheet because the tangential velocity is continuous (2.760d) and is rather a generalization of case II of a normal shock to allow for oblique velocity. Thus in an **oblique shock** *(Figure 2.29c) the tangential velocity is continuous (2.760d) and the normal velocity decreases (2.760e) so that: (i) the total velocity is smaller in modulus (2.760h) and its direction deviates more from the normal, that is towards the shock front; (ii) the angles $\theta^-(\theta^+)$ of the upstream $\bar{v}^-$ (downstream $\bar{v}^+$) velocity with the shock front are specified by (b):*

$$\theta \equiv \theta^- > \theta^+ = \theta^- - \varphi \equiv \theta - \varphi: \quad \tan\theta^- = \frac{v_n^-}{|\bar{v}_t|} > \frac{v_n^+}{|\bar{v}_t|} = \tan\theta^+, \qquad (2.761\text{a–d})$$

$$\tan \varphi = \tan\left(\theta^- - \theta^+\right) = \frac{\tan \theta^- - \tan \theta^+}{1 + \tan \theta^- \, \tan \theta^+} = \frac{\left|\bar{v}_t\right|\left(v_n^- - v_n^+\right)}{\left|\bar{v}_t\right|^2 + v_n^- \, v_n^+}, \qquad (2.761e\text{--}g)$$

corresponding to a deviation (2.531a–d) of the velocity towards the shock front and away from the normal (Figure 2.29c); (iii) across an oblique (as across a normal) shock increase the mass density (2.760g) [(2.759e)], pressure (2.760f) [(2.759f)] and enthalpy (2.760i) [(2.759g)]. The oblique shock will be reconsidered (Section 2.8) after a more detailed analysis of the normal shock (subsections 2.7.5–2.7.25).

2.7.5   CONSERVED AND DISCONTINUOUS QUANTITIES ACROSS A NORMAL SHOCK

A normal shock may be visualized as a front advancing normal to the walls in a parallel-sided duct (Figure 2.29b) or as a plane surface of discontinuity advancing in an unbounded medium (Figure 2.28). Since the velocity has only a normal component (2.528a–d) the index n is suppressed in the conditions of: (i) mass conservation (2.755a) $\equiv$ (2.762c) in terms of the mass density (2.762b) or specific volume (2.762a, d):

$$\vartheta \equiv \frac{1}{\rho}: \quad \rho^- v^- = \rho^+ v^+ \equiv j = \frac{v^-}{\vartheta^-} = \frac{v^+}{\vartheta^+}; \qquad (2.762a\text{--}d)$$

(ii) conservation of momentum (2.756a) $\equiv$ (2.763a) that specifies the pressure difference (2.763b–e):

$$p^- + \rho^-\left(v^-\right)^2 = p^+ + \rho^+\left(v^+\right)^2:$$

$$p^+ - p^- = \rho^-\left(v^-\right)^2 - \rho^+\left(v^+\right)^2 = j\left(v^- - v^+\right) = j^2\left(\frac{1}{\rho^-} - \frac{1}{\rho^+}\right) = j^2\left(\vartheta^- - \vartheta^+\right); \quad (2.763a\text{--}e)$$

(iii) conservation of the stagnation enthalpy (2.757e) $\equiv$ (2.764a) that specifies the enthalpy difference (2.694b–d):

$$2h^- + \left(v^-\right)^2 = 2h^+ + \left(v^+\right)^2 \equiv 2h_0:$$

$$2\left(h^+ - h^-\right) = \left(v^-\right)^2 - \left(v^+\right)^2 = j^2\left[\left(\rho^-\right)^{-2} - \left(\rho^+\right)^{-2}\right] = j^2\left[\left(\vartheta^-\right)^2 - \left(\vartheta^+\right)^2\right]. \qquad (2.764a\text{--}d)$$

The equations (2.762a–d; 2.763a–e; 2.764a–d) apply to any gas, and will be used to study the properties of normal shocks (subsections 2.7.5–2.7.9), before considering the particular case of ideal or perfect gases (subsections 2.7.10–2.7.26).

     The difference in normal velocity across the shock is given by (2.763c) $\equiv$ (2.765a) leading by (2.763c, d) to (2.765b, c) and hence to (2.765d, e):

$$v^- - v^+ = \frac{p^+ - p^-}{j} = (p^+ - p^-)\left[\frac{p^+ - p^-}{\frac{1}{\rho^-} - \frac{1}{\rho^+}}\right]^{-\frac{1}{2}}$$

$$= \left[(p^+ - p^-)\left(\frac{1}{\rho^-} - \frac{1}{\rho^+}\right)\right]^{\frac{1}{2}} = \sqrt{(p^+ - p^-)(\vartheta^- - \vartheta^+)}. \qquad (2.765a\text{--}e)$$

The difference of enthalpies (2.764b) $\equiv$ (2.766a) is given by (2.766b, c) using (2.762a–e; 2.763c):

$$2(h^+ - h^-) = j(v^- - v^+)\frac{v^- + v^+}{j} = (p^+ - p^-)\left(\frac{1}{\rho^-} + \frac{1}{\rho^+}\right)$$

$$= (p^+ - p^-)(\vartheta^+ + \vartheta^-). \qquad (2.766a\text{--}c)$$

In the jump in internal energy (2.767b–e) is used (2.767a; 2.766b):

$$h^{\pm} = u^{\pm} + \frac{p^{\pm}}{\rho^{\pm}}: \quad 2(u^+ - u^-) = 2(h^+ - h^-) - 2\left(\frac{p^+}{\rho^+} - \frac{p^-}{\rho^-}\right)$$

$$= (p^+ - p^-)\left(\frac{1}{\rho^-} + \frac{1}{\rho^+}\right) - 2\left(\frac{p^+}{\rho^+} - \frac{p^-}{\rho^-}\right) \qquad (2.767a\text{--}e)$$

$$= (p^+ + p^-)\left(\frac{1}{\rho^-} - \frac{1}{\rho^+}\right) = (p^+ + p^-)(\vartheta^- - \vartheta^+).$$

Thus have been *obtained the jumps across a normal shock of the pressure (2.763b–e), velocity (2.765a–e), enthalpy (2.766a–c) and internal energy (2.767a–e).*
From (2.763d) it follows that $p^+ - p^-$ and $1/\rho^- - 1/\rho^+$ have the same sign (2.768a):

$$(p^+ - p^-)\left(\frac{1}{\rho^-} - \frac{1}{\rho^+}\right) > 0: \quad p^+ > p^-, \quad \rho^+ > \rho^-, \quad \vartheta^+ < \vartheta^-, \quad (2.768a\text{--}d)$$

implying that the pressure and mass density both increase across the shock (2.768b, c), the specific volume decreases (2.768d), or all in reverse. From (2.762d) and (2.768d) follows that the velocity decreases across the normal shock (2.769a):

$$v^- > v^+, \quad h^+ > h^-, \quad u^+ > u^-, \qquad (2.769a\text{--}c)$$

and (2.764c; 2.769a) [(2.767e; 2.768d)] imply that the enthalpy (2.769b) [internal energy (2.769c)] increase. The inequalities (2.697b, c; 2.699a–c) hold together either with the same or reversed sign. The reversed signs would imply that a succession of shocks conserving the stagnation enthalpy could lead to an increasing unlimited velocity and decreasing mass density tending to zero, which seem unlikely to be true. It will be shown next that the signs in (2.768b, c; 2.769a–c) [the reverse signs] imply an increase (decrease) of entropy across the shock, which, by the second principle of thermodynamics (subsection 2.3.16) must (cannot) be the case. The proof will be made first for a weak shock (subsections 2.7.6–2.7.7) and then for a strong shock (subsections 2.7.8–2.7.9) for generic gases, including ideal gases. It is possible to proceed directly to the case of an ideal gas (subsections 2.7.10–2.7.15) for which the proof and relations between flow variables are simpler.

2.7.6   ENTROPY PRODUCTION IN A WEAK SHOCK

A **weak shock** corresponds to the lowest order non-vanishing term in the enthalpy jump, expressed in power series of its variables that are the jumps in entropy and pressure:

$$
\begin{aligned}
h^+ - h^- &= \left(\frac{\partial h}{\partial s^-}\right)_p \left(s^+ - s^-\right) + \left(\frac{\partial h}{\partial p^-}\right)_s \left(p^+ - p^-\right) \\
&+ \left(\frac{\partial^2 h}{\partial p^{-2}}\right)_s \frac{\left(p^+ - p^-\right)^2}{2!} + \left(\frac{\partial^3 h}{\partial p^{-3}}\right)_s \frac{\left(p^+ - p^-\right)^3}{3!},
\end{aligned}
\tag{2.770}
$$

where it will be found that: (i) the term of first order on entropy is sufficient, because it does not cancel; (ii) the first and second-order terms on pressure do cancel, so it is necessary to proceed to third order in (2.770). The enthalpy per unit mass is given by (2.615d) ≡ (2.771a) and thus its first-order partial derivative with regard to entropy (pressure) equals the temperature (2.771b) [specific volume (2.771c, d)]:

$$
dh = T\, ds + \frac{1}{\rho}\, dp: \quad \left(\frac{\partial h}{\partial s}\right)_p = T, \quad \left(\frac{\partial h}{\partial p}\right)_s = \frac{1}{\rho} = \vartheta.
\tag{2.771a–d}
$$

Substitution of (2.771b, d) in (2.770) leads to (2.540a–c):

$$
\begin{aligned}
h^+ - h^- &= T^- \left(s^+ - s^-\right) + \vartheta^- \left(p^+ - p^-\right) \\
&+ \left(\frac{\partial \vartheta^-}{\partial p^-}\right)_{s^-} \frac{\left(p^+ - p^-\right)^2}{2} + \left(\frac{\partial^2 \vartheta^-}{\partial p^{-2}}\right)_{s^-} \frac{\left(p^+ - p^-\right)^3}{6},
\end{aligned}
\tag{2.772}
$$

so the expansion (2.776) can be re-written (2.772).

The change in specific volume $\vartheta(s,p)$ can also be expanded in powers where: (i) the terms in $s^+ - s^-$ would lead in (2.540a–c) to terms like $(s^+ - s^-)(p^+ - p^-)$, that are of higher-order than $s^+ - s^-$ and hence of no interest; (ii) the expansion in powers of $p^+ - p^-$ must go to the second-order to involve the same derivatives:

$$\vartheta^+ - \vartheta^- = \left(\frac{\partial\vartheta^-}{\partial p^-}\right)_{s^-}(p^+ - p^-) + \left(\frac{\partial^2\vartheta^-}{\partial p^{-2}}\right)_{s^-}\frac{(p^+ - p^-)^2}{2}. \tag{2.773}$$

Substituting (2.773) in (2.766c) ≡ (2.774a) leads to (2.774b, c):

$$h^+ - h^- = \left(p^+ - p^-\right)\frac{\vartheta^+ + \vartheta^-}{2} = \left(p^+ - p^-\right)\vartheta^- + \frac{p^+ - p^-}{2}\left(\vartheta^+ - \vartheta^-\right)$$

$$= \left(p^+ - p^-\right)\vartheta^- + \frac{\left(p^+ - p^-\right)^2}{2}\left(\frac{\partial\vartheta^-}{\partial p^-}\right)_{s} + \frac{\left(p^+ - p^-\right)^3}{4}\left(\frac{\partial^2\vartheta^-}{\partial p^{-2}}\right)_{s^-}. \tag{2.774a–c}$$

The two expression for the enthalpy difference (2.772) ≡ (2.774c) lead to the condition of compatibility with the second principle of thermodynamics (subsection 2.7.7).

2.7.7 COMPATIBILITY WITH THE SECOND PRINCIPLE OF THERMODYNAMICS

Comparing (2.772) ≡ (2.774c) follows (2.775a, 2.467a–e):

$$T^-\left(s^+ - s^-\right) = \left(\frac{\partial^2\vartheta^-}{\partial p^{-2}}\right)_{s}\left(p^+ - p^-\right)^3\left(\frac{1}{4} - \frac{1}{6}\right) = \left(\frac{\partial^2\vartheta^-}{\partial p^{-2}}\right)_{s}\frac{\left(p^+ - p^-\right)^3}{12}, \tag{2.775a, b}$$

that specifies the entropy jump. If the condition (2.776a) is met, then entropy production (2.776b) implies an increase in pressure (2.776c) ≡ (2.768b) across the shock:

$$\left(\frac{\partial^2\vartheta}{\partial p^2}\right)_{s} > 0: \quad s^+ > s^- \quad \Rightarrow \quad p^+ > p^-, \tag{2.776a–c}$$

which in turn implies the remaining inequalities (2.768c,d; 2.769a–c). The condition (2.776a) is not a general thermodynamic property, but rather a property most gases have, for example, for a gas with constant adiabatic (2.777a) exponent (2.777b), the adiabatic law (2.478d; 2.473b) ≡ (2.777c) implies (2.777d, e):

$$s = const, \quad \gamma = const, \quad \vartheta = k\,p^{-1/\gamma}:$$

$$\left(\frac{\partial^2\vartheta}{\partial p^2}\right)_{s} = \frac{k}{\gamma}\left(1 + \frac{1}{\gamma}\right)p^{-2-1/\gamma} = \frac{\gamma+1}{\gamma^2}\frac{\vartheta}{p^2} > 0, \tag{2.777a–f}$$

that satisfies (2.776a) ≡ (2.777f). It has been proved that *the entropy jump at lowest order in a weak shock scales like the cube (2.775b) of the pressure jump. For a gas satisfying the adiabatic inequality (2.776a), the condition of entropy growth (2.776b) implies the inequalities (2.768b–d; 2.769a–c) for a weak shock.* Thus the conditions (2.768b–d; 2.769a–c) (the reverse conditions) are associated with an increase (decrease) in entropy across the shock front, and comply (are incompatible) with the second principle of thermodynamics (subsection 2.3.16). The conditions (2.768b–d, c; 2.769a–c) have been proved only for weak shocks of a gas satisfying (2.776a), and the result is next (subsections 2.7.8–2.7.9) extended to strong shocks, for which the entropy jump need not be small.

2.7.8   ADIABATIC VERSUS SHOCK PRESSURE – VOLUME RELATIONS

In Figure 2.30 are plotted two loci or the relations between pressure p and specific volume ϑ for: (i) an **adiabatic process** (Poisson 1816); (ii) a **normal shock** (Rankine

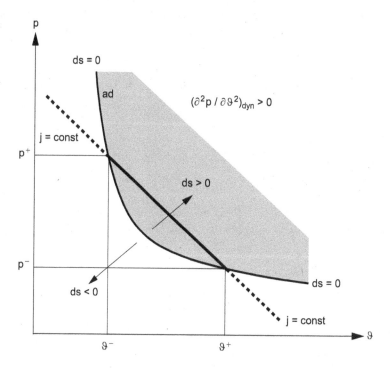

FIGURE 2.30   In a diagram of pressure p versus specific volume ϑ for any substance, hence not restricted to ideal or perfect gas, the: (i) adiabatic locus is a curve, thin solid line below shaded area of entropy increase; (ii) the locus of a normal shock is a straight line, thick solid or dotted line. Over the range of pressures $p^{\pm}$ and specific volumes $\vartheta^{\pm}$ across a normal shock its locus (thick solid line) lies above the adiabatic locus and in the shaded region of entropy increase and thus the process is irreversible with entropy production in agreement with the second principle of thermodynamics.

1870; Huginot 1884). The adiabatic pressure – specific volume relation (2.777c) ≡ (2.778a) or **adiabatic locus** leads to: (i) a slope specified (2.778b–e) by the adiabatic sound speed (2.604b; 2.762a):

$$p = k \vartheta^{-\gamma} : \quad \left(\frac{\partial p}{\partial \vartheta}\right)_s = -\gamma k \vartheta^{1-\gamma} = -\frac{\gamma p}{\vartheta} = -\frac{c^2}{\vartheta^2} = -\rho^2 c^2; \quad (2.778a\text{–}e)$$

(ii) a positive curvature implying (2.545a–d) that the curve $p(\vartheta)$ is concave upward:

$$\left(\frac{\partial^2 p}{\partial \vartheta^2}\right)_s = \gamma(\gamma+1)k \vartheta^{-\gamma-2} = \gamma(\gamma+1)\frac{p}{\vartheta^2} > 0. \quad (2.779a\text{–}c)$$

In contrast the pressure – specific volume relation for a normal shock (2.763e) ≡ (2.780a) or **shock locus** is a straight line (2.780b), hence with zero curvature (2.780c) and with slope specified by the constant mass flux (2.780d, e):

$$p + j^2 \vartheta = const : \quad dp + j^2 d\vartheta = 0, \quad (2.780a, b)$$

$$\left(\frac{\partial^2 p}{\partial \vartheta^2}\right)_j = 0, \quad \left(\frac{\partial p}{\partial \vartheta}\right)_j = -j^2 = -\rho^2 v^2. \quad (2.780c\text{–}e)$$

Considering the adiabatic (2.778a–e; 2.779a–c) [normal shock (2.780a–e)] loci both passing through the same upstream (p^-, ϑ^-) and downstream (p^+, ϑ^+) conditions, the straight shock line is a secant of the upward curved adiabatic line (Figure 2.30), and thus the slope in modulus is larger upstream (2.781a) and smaller downstream (2.781b) in the adiabatic case than in the shock case:

$$\left(\frac{\partial p}{\partial \vartheta}\right)_{s^-} > j^2 > \left(\frac{\partial p}{\partial \vartheta}\right)_{s^+} : \quad \left(\rho^- c^-\right)^2 < \left(\rho^- v^-\right)^2 = \left(\rho^+ v^+\right)^2 < \left(\rho^+ c^+\right)^2, \quad (2.781a\text{–}e)$$

substitution of (2.778e; 2.780e) in (2.781a, b) leads to (2.781c–e) proving that the velocity before (after) the shock is larger (2.782a) [smaller (2.782b)] than the local sound speed:

$$v^- > c^-, \quad v^+ < c^+ : \quad M^- \equiv \frac{v^-}{c^-} > 1 > \frac{v^+}{c^+} \equiv M^+, \quad (2.782a\text{–}f)$$

and thus the local Mach number is supersonic before (2.782c, d) [subsonic after (2.782e, f)] the normal shock. The relative position of the adiabatic and shock loci leads to the thermodynamic conditions for entropy increase across a strong normal shock (subsection 2.7.9).

2.7.9 THERMODYNAMIC CONDITIONS FOR A STRONG SHOCK

The adiabatic locus in the pressure – specific volume diagram (Figure 2.30) is the curve of constant entropy =0, and hence separates regions of positive entropy to one side and of negative entropy to the other side. If the entropy is positive to the right side of the adiabatic curve, then the range (2.783a) of normal shock pressures lies in the region of positive entropy (2.783b) and all the preceding inequalities are consistent with the second principle of thermodynamics:

$$p^+ > p > p^- : \quad ds > 0, \quad p < p^- \quad \text{or} \quad p > p^+, \quad ds < 0; \qquad \text{(2.783a–d)}$$

conversely (2.783c) outside the pressure range (2.783a) the entropy is negative (2.783d), and the reverse inequalities are impossible for the pressure – specific volume relation of a normal shock. Thus to prove all preceding inequalities (2.768b–d; 2.769a–c) it is sufficient to prove that $ds > 0$ to right of the adiabatic pressure – specific volume relation or locus. Any point will do, for example, a horizontal line to the right is chosen (2.784b) and entropy increase (2.784a, c) requires (2.784d):

$$ds > 0: \quad ds = \left(\frac{\partial s}{\partial \vartheta}\right)_p d\vartheta > 0 \quad \Rightarrow \quad \left(\frac{\partial s}{\partial \vartheta}\right)_p > 0. \qquad \text{(2.784a–d)}$$

The thermodynamic derivative (2.784d) is evaluated using jacobians (subsection 2.3.9) by (2.552a–c):

$$\left(\frac{\partial s}{\partial \vartheta}\right)_p = \frac{\partial(s,p)}{\partial(\vartheta,p)} = \frac{\partial(s,p)}{\partial(T,p)}\frac{\partial(T,p)}{\partial(\vartheta,p)} = \left(\frac{\partial s}{\partial T}\right)_p\left(\frac{\partial T}{\partial \vartheta}\right)_p$$

$$= \frac{c_p}{T}\left(\frac{\partial T}{\partial \vartheta}\right)_p = \frac{c_p}{T\,\vartheta\,K_p} > 0, \qquad \text{(2.785a–f)}$$

in terms of the specific heat at constant pressure (2.252e) and coefficient of thermal expansion (2.253a). This completes the proof of all preceding inequalities: *across a normal shock, a gas satisfying the inequality (2.785a–f): (i) increases the pressure (2.768b), mass density (2.768c), enthalpy (2.769b), internal energy (2.769c) and entropy (2.776b); (ii) decreases in velocity (2.769a) from locally supersonic before (2.782a, d) to locally subsonic after (2.782b, f).*

In the case of an ideal gas (2.473e) ≡ (2.786a) the coefficient of thermal expansion (2.489a) ≡ (2.786b) is given by (2.786c–e) implying (2.786f, g) that meets (2.786g) ≡ (2.785f):

$$p\vartheta = \bar{\bar{R}}T; \quad k_p \equiv \frac{1}{\vartheta}\left(\frac{\partial \vartheta}{\partial T}\right)_p = \frac{1}{\vartheta}\left[\frac{\partial}{\partial T}\left(\frac{\bar{\bar{R}}T}{p}\right)\right]_p = \frac{\bar{\bar{R}}}{p\vartheta} = \frac{1}{T}, \quad \left(\frac{\partial s}{\partial \vartheta}\right)_p = \frac{c_p}{\vartheta} > 0. \qquad \text{(2.786a–g)}$$

The same result can be obtained (2.787b–d) from the entropy of a perfect gas (2.477c) ≡ (2.787a), proving again (2.779f) ≡ (2.786f) = (2.817d):

$$s = c_V \log p + c_p \log \vartheta + const; \quad \left(\frac{\partial s}{\partial \vartheta}\right)_p = c_p \frac{d}{d\vartheta}(\log \vartheta) = \frac{c_p}{\vartheta} > 0. \qquad (2.787a\text{–}d)$$

The conditions for entropy growth in a weak (2.776a) [strong (2.785a)] shock are distinct and compatible: (i) the condition (2.776a) ≡ (2.788a) implies that the static or adiabatic locus is concave in agreement with (2.779a–c) ≡ (2.788b) exchanging the (p, ϑ) axis in Figure 2.30:

$$\left(\frac{\partial^2 \vartheta}{\partial p^2}\right)_s > 0 \leftrightarrow \left(\frac{\partial^2 p}{\partial \vartheta^2}\right)_s > 0; \qquad (2.788a, b)$$

(ii) the condition (2.785a–f) is one of several ways of stating (2.788b) that the dynamic pressure – specific volume straight line lies in the region in the region of entropy growth to the right of the static or adiabatic pressure – specific volume curve in Figure 2.30.

The thermodynamic inequalities (2.768b–d; 2.769a–c; 2.776b) across a weak (strong) normal shock [subsection(s) 2.7.6 (2.7.7–2.7.9)] have been proved for generic gases satisfying (2.776a) [(2.784d)] including perfect gases (2.777a–f) [2.786a–g ≡ (2.787a–d)]. The change of entropy across a strong normal (oblique) shock [Section 2.7 (2.8)] is given for a perfect gas (2.477d; 2.475a–c) by (2.556a–h):

$$s^+ - s^- = c_V \log\left(\frac{p^+}{p^-}\right) - c_p \log\left(\frac{\rho^+}{\rho^-}\right) = c_V \left[\log\left(\frac{p^+}{p^-}\right) - \gamma \log\left(\frac{\rho^+}{\rho^-}\right)\right]$$

$$= c_V \log\left[\frac{p^+}{p^-}\left(\frac{\rho^+}{\rho^-}\right)^{-\gamma}\right], \qquad (2.789a\text{–}c)$$

showing that entropy increase across a shock (2.789d) implies a pressure ratio larger than the adiabatic value (2.789e, f):

$$s^+ > s^- \Rightarrow \frac{p^+}{p^-} > \left(\frac{\rho^+}{\rho^-}\right)^\gamma = \left(\frac{\vartheta^-}{\vartheta^+}\right)^\gamma. \qquad (2.789d\text{–}f)$$

The geometrical interpretation (Figure 2.30) is that the shock locus (straight line) lies above the adiabatic locus (curved line), in the range $p^- < p < p^+$ and $\vartheta^- > \vartheta > \vartheta^+$, where the entropy increases from the adiabatic towards the shock locus. In order to obtain explicit relations between flow before and after the shock (subsections 2.7.10–2.7.14) the equation of state is needed, for example, for an ideal (perfect) gas [subsections 2.5.1–2.5.3 (2.5.4–2.5.7)]. As an example for an ideal gas, the pressure ratio across a normal shock is expressed as a function of the temperatures before and after the shock (subsection 2.7.10).

2.7.10   PRESSURE AND TEMPERATURE RATIOS FOR AN IDEAL GAS

In the case of an ideal gas the equation of state (2.786a) may be substituted in the enthalpy jump (2.766c) $\equiv$ (2.790a, b) leading to (2.790c, d):

$$2\left(h^{+}-h^{-}\right)=2\left[h\left(T^{+}\right)-h\left(T^{-}\right)\right]=\left(p^{+}-p^{-}\right)\left(\vartheta^{+}+\vartheta^{-}\right)$$

$$=\bar{\bar{R}}\left(p^{+}-p^{-}\right)\left(\frac{T^{+}}{p^{+}}+\frac{T^{-}}{p^{-}}\right)=\bar{\bar{R}}\left(T^{+}-T^{-}\right)+\bar{\bar{R}}T^{-}\frac{p^{+}}{p^{-}}-\bar{\bar{R}}T^{+}\frac{p^{-}}{p^{+}}, \qquad \text{(2.790a–d)}$$

that is a quadratic equation (2.790d) $\equiv$ (2.791b) for the pressure ratio involving the parameter (2.791a):

$$2b\left(T^{+},T^{-}\right)\equiv 1-\frac{T^{+}}{T^{-}}+2\frac{h\left(T^{+}\right)-h\left(T^{-}\right)}{\bar{\bar{R}}T^{-}}$$

$$: \quad \left(\frac{p^{+}}{p^{-}}\right)^{2}-2b\left(T^{+},T^{-}\right)\frac{p^{+}}{p^{-}}-\frac{T^{+}}{T^{-}}=0. \qquad \text{(2.791a, b)}$$

The roots (2.791c) of (2.791b) are:

$$\frac{p^{+}}{p^{-}}=b\left(T^{+},T^{-}\right)\pm\left|\left[b\left(T^{+}-T^{-}\right)\right]^{2}+\frac{T^{+}}{T^{-}}\right|^{1/2}, \qquad \text{(2.791c)}$$

and specify *the ratio of pressures (2.791c) before p⁻ and after p⁺ a normal shock in terms of the temperature before T⁻ and after T⁺ the shock for: (i) an ideal gas (2.791a); (ii) in particular (2.792b–d) for a perfect gas (2.623b) $\equiv$ (2.792a):*

$$h\left(T^{\pm}\right)=c_{p}T^{\pm}: \quad 2b\left(T^{+},T^{-}\right)=\left(\frac{2c_{p}}{\bar{\bar{R}}}-1\right)\left(\frac{T^{+}}{T^{-}}-1\right)$$

$$=\left(\frac{2\gamma}{\gamma-1}-1\right)\left(\frac{T^{+}}{T^{-}}-1\right)=\frac{\gamma+1}{\gamma-1}\left(\frac{T^{+}}{T^{-}}-1\right), \qquad \text{(2.792a–d)}$$

where was used (2.264e). Additional relations for normal shocks are obtained (subsections 2.7.11–2.7.15) for perfect gases, starting with the geometric (arithmetic) mean or average of the velocities before and after the shock [subsection 2.7.11 (2.7.12)].

2.7.11 GEOMETRIC/ARITHMETIC MEAN OF VELOCITIES

The enthalpy is given for a perfect gas by (2.793b–f):

$$n = \gamma: \quad h = c_p T = \frac{\gamma \overline{\overline{R}}}{\gamma - 1} T = \frac{c^2}{\gamma - 1} = \frac{\gamma}{\gamma - 1} \frac{p}{\rho} = \frac{\gamma}{\gamma - 1} p \vartheta, \qquad (2.793a\text{–}f)$$

and coincides with the polytropic (2.657e) in the adiabatic case (2.793a). *The conservation of the stagnation enthalpy (2.764a) across a normal shock implies (2.794a–d):*

$$2 h_0 = 2 h^\pm + \left(v^\pm\right)^2 = \left(v^\pm\right)^2 + \frac{2\left(c^\pm\right)^2}{\gamma - 1}$$

$$= \left(v^\pm\right)^2 + \frac{2\gamma}{\gamma - 1} \frac{p^\pm}{\rho^\pm} = \left(v^\pm\right)^2 + \frac{2\gamma}{\gamma - 1} p^\pm \vartheta^\pm. \qquad (2.794a\text{–}d)$$

*In the **critical (stagnation) case** of velocity equal to the sound speed (2.795a) [zero velocity (2.796a)] it follows that the critical c_* (stagnation c_0) sound speed is also conserved (2.795b–d) [(2.796b)]:*

$$v_* = c_*: \quad 2 h_0 = \frac{2 c_*^2}{\gamma - 1} + v_*^2 = \frac{\gamma + 1}{\gamma - 1} c_*^2 = \frac{\gamma + 1}{\gamma - 1} v_*^2, \qquad (2.795a\text{–}d)$$

$$v = 0: \quad h_0 = \frac{c_0^2}{\gamma - 1}, \quad c_0^2 = \frac{\gamma + 1}{2} c_*^2, \qquad (2.796a\text{–}c)$$

from the comparison of (2.795c) $\equiv$ (2.796b) follows the relation (2.796c) $\equiv$ (2.664c) between the stagnation and critical sound speeds.

The difference of velocities before and after the shock is given by (2.763c) $\equiv$ (2.797a) leading to (2.797b) by use of (2.762b, c):

$$v^- - v^+ = \frac{p^+ - p^-}{j} = \frac{p^+}{\rho^+ v^+} - \frac{p^-}{\rho^- v^-}. \qquad (2.797a, b)$$

From (2.794c) and (2.795d) follows (2.798a) $\equiv$ (2.798b):

$$\frac{\gamma + 1}{\gamma - 1} c_*^2 = \left(v^\pm\right)^2 + \frac{2\gamma}{\gamma - 1} \frac{p^\pm}{\rho^\pm} \quad \Leftrightarrow \quad \frac{p^\pm}{\rho^\pm} = \frac{\gamma + 1}{2\gamma} c_*^2 - \frac{\gamma - 1}{2\gamma} \left(v^\pm\right)^2. \qquad (2.798a, b)$$

Substitution of (2.798b) in (2.797b) leads to the identity (2.799a, 2.467a–e):

$$v^- - v^+ = \frac{\gamma - 1}{2\gamma} \left(v^- - v^+\right) + \frac{\gamma + 1}{2\gamma} c_*^2 \left(\frac{1}{v^+} - \frac{1}{v^-}\right) = \left(v^- - v^+\right) \left(\frac{\gamma - 1}{2\gamma} + \frac{\gamma + 1}{2\gamma} \frac{c_*^2}{v^- v^+}\right). \qquad (2.799a, b)$$

Since the velocities are distinct before and after the shock (2.800a) from (2.799b) follows (2.800b) ≡ (2.800c):

$$v^- \neq v^+: \quad 1 = \frac{\gamma-1}{2\gamma} + \frac{\gamma+1}{2\gamma}\frac{c_*^2}{v^- v^+} \quad \Rightarrow \quad \frac{c_*^2}{v^- v^+} = \frac{1-\dfrac{\gamma-1}{2\gamma}}{\dfrac{\gamma+1}{2\gamma}} = 1. \quad (2.800a\text{--}c)$$

It has been shown that *the product of the velocities of a perfect gas before v^- and after v^+ a normal shock is the square of the critical sound speed (2.800c) ≡ (2.801a):*

$$v^- v^+ = c_*^2: \quad v^- > v^+ \Rightarrow v^- > c_* > v^+, \qquad (2.801a\text{--}d)$$

$$M_*^- \equiv \frac{v^-}{c_*} > 1 > \frac{v^+}{c_*} \equiv M_*^+; \quad M_*^- M_*^+ = 1, \qquad (2.801e\text{--}g)$$

*since the velocity is larger before than after the shock (2.769a) ≡ (2.801b) it follows that for a perfect gas the velocity before (after) the shock is larger (2.801c) [smaller (2.801d)] than the critical sound speed, and hence: (i) the critical Mach number is supersonic (2.801e) before [subsonic (2.801f) after] the shock; (ii) the product of the two critical Mach numbers is unity (2.801g). Similar inequalities to (i) were obtained before for the local Mach number (2.782a–f) without the restriction to a perfect gas, that is for an arbitrary gas. The **geometric (arithmetic) mean or average** (2.802a) [(2.802c)] of the velocities before and after the shock equals the critical sound speed (2.800c) ≡ (2.801a) ≡ (2.802b) [the speed of advance of the shock front (2.802d)]:*

$$\overline{\overline{v^- v^+}} \equiv \sqrt{v^- v^+} = c_*, \quad \overline{v^- v^+} = \frac{v^- + v^+}{2} = \bar{v}, \qquad (2.802a\text{--}d)$$

The latter result is proved next (subsection 2.7.12) without restriction on the equation of state of the gas.

2.7.12   SPEED OF ADVANCE OF THE SHOCK FRONT

For an observer at rest: (i) the flow velocity is higher v^- before than after v^+ the shock; (ii) the shock front advances with a **shock velocity** $\bar{v}$; (iii) the flow velocity relative to the shock front (Figure 2.31) is given by the l.h.s. (r.h.s.) of (2.803a) before (after) the shock; (iv) since the flow velocity relative to the shock must be continuous follows (2.803b) ≡ (2.802d):

$$v^- - \bar{v} = v^+ + \bar{v} \Rightarrow \bar{v} = \frac{v^- + v^+}{2}. \qquad (2.803a, b)$$

It has been shown that *the speed of advance of a normal shock front is the arithmetic mean of the flow velocities before and after the shock (2.803b); in the case of a*

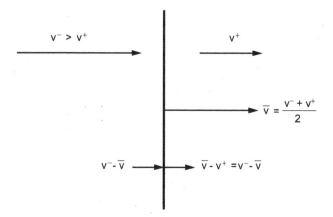

FIGURE 2.31 The velocity of advance $\bar{v}$ of a normal shock front must be equal relative to the upstream v^+ and downstream v^- velocities, and thus is the arithmetic mean of the two.

perfect gas (2.800c) ≡ (2.802a) ≡ (2.804a) the velocity of the shock front (2.802d) ≡ (2.803b) ≡ (2.804b–d) exceeds the critical sound speed (2.804e):

$$\frac{v^-}{c_*} = \frac{c_*}{v^+} : \quad \frac{\bar{v}}{c_*} = \frac{v^- + v^+}{2c_*} = \frac{1}{2}\left(\frac{v^-}{c_*} + \frac{c_*}{v^-}\right) = \frac{1}{2}\left(\frac{v^+}{c_*} + \frac{c_*}{v^+}\right) > 1. \quad (2.804\text{a–e})$$

The result (2.804e) is proved next in three distinct ways: algebraic, graphic and analytic.

2.7.13 ARITHMETIC LARGER THAN GEOMETRIC AVERAGE

First *the arithmetic mean of distinct values (2.805a) is larger than the geometric mean (2.805b–e):*

$$a \neq b: \quad \overline{ab} = \frac{a+b}{2} > \sqrt{ab} \equiv \overline{\overline{ab}}. \quad (2.805\text{a–e})$$

The proof follows from (2.573a–c):

$$0 < \left(\sqrt{a} - \sqrt{b}\right)^2 = a + b - 2\sqrt{ab} = 2\left(\overline{ab} - \overline{\overline{ab}}\right). \quad (2.806\text{a–c})$$

An alternative proof is made introducing the variable (2.807a) that is distinct from unity (2.807b) and showing that the function (2.804d) ≡ (2.804e) ≡ (2.807c, d) exceeds unity (2.807e):

$$0 < x \equiv \frac{v^\pm}{c_*} \neq 1: \quad \frac{\bar{v}}{c_*} = \frac{1}{2}\left(x + \frac{1}{x}\right) = j(x) > 1. \quad (2.807\text{a–e})$$

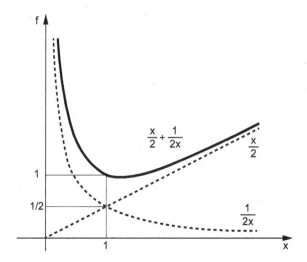

FIGURE 2.32 Geometric proof that the arithmetic mean of two unequal real positive quantities exceeds the geometric mean; this implies that the speed of advance of a normal shock front (Figure 2.31) exceeds the stagnation sound speed that is conserved across the shock.

The function (2.807d) ≡ (2.808a, b) has a minimum unity at point unity (2.808c) and exceeds unity at all other points (2.808d):

$$x > 0: \quad j(x) = \frac{1}{2}\left(x + \frac{1}{x}\right)\begin{cases} = 1 \; if \; x = 1, \\ > 1 \; if \; x \neq 1. \end{cases} \tag{2.808a–d}$$

The proof of (2.808a–d) can be made graphically (Figure 2.32) adding the functions $x/2$ and $1/(2x)$: (i) both take the value $1/2$ for $x = 1$ adding to 1; (ii) that is the minimum value so $j > 1$ for $x \neq 1$.

This can also be proved analytically considering the function (2.809a) ≡ (2.808b) ≡ (2.807d) and its first (2.809b) [second (2.809c)] order derivatives:

$$j(x) = \frac{1}{2}\left(x + \frac{1}{x}\right) \quad j'(x) = \frac{1}{2}\left(1 - \frac{1}{x^2}\right) \quad j''(x) = \frac{1}{x^3}. \tag{2.809a–c}$$

The first-order derivative (2.809b) vanishes (2.810a, b) at the point unity (2.810c) that is a minimum since the second-order derivative (2.808c) is positive at that point (2.810d–f):

$$0 = j'(x) = \frac{1}{2}\left(1 - \frac{1}{x^2}\right) \implies \bar{x} = 1: \quad j''(\bar{x}) = \frac{1}{\bar{x}^3} = 1 > 0; \quad j(1) = 1 \tag{2.810a–g}$$

also the minimum is unity (2.810g) ≡ (2.808c) so the function is larger (2.808d) for all other positive values of the variable (2.808a).

The shock locus for a perfect gas can be represented as a relation between the ratios of pressures and specific volumes before and after the shock (subsection 2.7.14) to compare with the adiabatic locus.

2.7.14 COMPARISON OF THE ADIABATIC AND SHOCK LOCI

The adiabatic locus or relation (2.811a) between the ratio of pressures and specific volumes for a perfect gas (2.777c) ≡ (2.811b) is not symmetric (2.811c):

$$s = const: \quad \frac{p^+}{p^-} = \left(\frac{\vartheta^+}{\vartheta^-}\right)^{-\gamma} \Leftrightarrow \frac{\vartheta^+}{\vartheta^-} = \left(\frac{p^+}{p^-}\right)^{-1/\gamma}, \quad (2.811a\text{–}c)$$

has (Figure 2.33) the coordinate axis as asymptotes (2.812c, d) and the closest point to the origin is the unit point (2.812a, b):

$$\frac{p^+}{p^-} = 1 = \frac{\vartheta^+}{\vartheta^-}: \quad \lim_{\frac{\vartheta^+}{\vartheta^-} \to \infty} \frac{p^+}{p^-} = 0 = \lim_{\frac{p^+}{p^-} \to \infty} \frac{\vartheta^+}{\vartheta^-}. \quad (2.812a\text{–}d)$$

For comparison (Figure 2.33) the shock locus (2.813b) can be obtained from the enthalpy difference (2.766c) ≡ (2.813a) for a perfect gas (2.793f):

$$\left(p^+ - p^-\right)\left(\vartheta^- + \vartheta^+\right) = 2\left(h^+ - h^-\right) = \frac{2\gamma}{\gamma - 1}\left(p^+ \vartheta^+ - p^- \vartheta^-\right), \quad (2.813a. b)$$

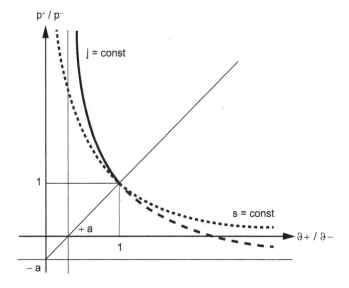

FIGURE 2.33 Comparison of the loci of the ratio of pressure p^+/p^- and specific volumes ϑ^+/ϑ^- for an adiabatic process (dotted line) and a normal shock (solid and dashed line). Since the pressure increases across a normal shock $p^+/p^- > 1$ only the solid line is relevant showing that the pressure ratio is larger than for an adiabatic process.

in terms of the ratio of pressures and specific volumes (2.747a, 2.467a–e):

$$j = const: \quad \left(\frac{p^+}{p^-}-1\right)\left(\frac{\vartheta^+}{\vartheta^-}+1\right) = \frac{2\gamma}{\gamma-1}\left(\frac{p^+}{p^-}\frac{\vartheta^+}{\vartheta^-}-1\right). \qquad (2.814a, b)$$

Solving for the ratio of pressures (2.815b) [specific volumes (2.815c)] there is symmetry with change of sign (2.815a):

$$\frac{p^+}{p^-} \to -\frac{\vartheta^+}{\vartheta^-} : \quad \frac{\vartheta^+}{\vartheta^-} = \frac{(\gamma-1)p^+ + (\gamma+1)p^-}{(\gamma-1)p^- + (\gamma+1)p^+},$$

$$\frac{p^+}{p^-} = \frac{(\gamma-1)\vartheta^+ - (\gamma+1)\vartheta^-}{(\gamma-1)\vartheta^- - (\gamma+1)\vartheta^+}, \qquad (2.815a\text{–}c)$$

and the asymptotes are parallel to the axis at equal distance (2.816a–c):

$$\lim_{\frac{p^+}{p^-}\to\infty} \frac{\vartheta^+}{\vartheta^-} = -\lim_{\frac{\vartheta^+}{\vartheta^-}\to\infty} \frac{p^+}{p^-} = \frac{\gamma-1}{\gamma+1} = \frac{1}{1+f} = \left\{\frac{1}{4},\frac{1}{6},\frac{1}{7}\right\} \equiv a, \qquad (2.816a\text{–}f)$$

specified by (2.816d, e), respectively (2.475a–c) for monoatomic/diatomic and polyatomic gases (2.816f).

The shock locus also passes (2.817b) through the unit point (2.812a, b) and thus lies above (2.817a) [below (2.817c)] the adiabatic locus before (after):

$$\left(\frac{p^+}{p^-}\right)_j - \left(\frac{p^+}{p^-}\right)_s \begin{cases} > 0 \text{ if } \vartheta^+ < \vartheta^-, \\ = 0 \text{ if } \vartheta^+ = \vartheta^-, \\ < 0 \text{ if } \vartheta^+ > \vartheta^-. \end{cases} \qquad (2.817a\text{–}c)$$

It has been shown that *the shock (2.814a, b) [adiabatic (2.811a–c)] loci for a perfect gas: (i) is symmetric (2.815a–c) [unsymmetric (2.811a–c]; (ii) has asymptotes (2.816a–f) [(2.812c, d)] parallel to (coincident with) the coordinate axis (Figure 2.33). Both pass (2.817b) through the unit point (2.812a, b), and the shock locus lies above (below) the adiabatic locus before (2.817a) [after (2.817c)]. Thus only the part of the shock locus (solid line in Figure 2.33) above the adiabatic locus (short dashes or dots) satisfies entropy growth in compliance with the second principle of thermodynamics (subsection 2.3.16), implying that the pressure is larger upstream (2.698b). The part of the shock locus (long dashes in Figure 2.33) below the adiabatic locus implying pressure reduction across a shock is unphysical, since it requires an entropy reduction inconsistent with the second principle of thermodynamics. The same conclusion can be reached noting that the flow conditions before the shock specify the flow after the shock (subsection 2.7.16).*

2.7.15 DOWNSTREAM IN TERMS OF UPSTREAM SHOCK CONDITIONS

The values of (v^-, ρ^-, p^-) before the normal shock of a perfect gas determine the values after, namely: (i) the velocity (2.801a) $\equiv$ (2.818a):

$$v^+ = \frac{(c_*)^2}{v^-} \quad \rho^+ = \rho^- \frac{v^-}{v^+} = \rho^- \frac{(v^-)^2}{v^- v^+} = \rho^- \left(\frac{v^-}{c_*}\right)^2 = \rho^- (M_*^-)^2; \quad (2.818a\text{-}e)$$

(ii) the mass density (2.818b–f) from (2.762b); (iii) the pressure difference (2.819a–e) from (2.763c; 2.762b; 2.801a, e):

$$p^+ - p^- = j(v^- - v^+) = \rho^- v^- (v^- - v^+) = \rho^- \left[(v^-)^2 - c_*^2\right] \quad (2.819a\text{-}c)$$

$$= \rho^- (v^-)^2 \left[1 - (M_*^-)^{-2}\right] = \rho^- c_*^2 \left[(M_*^-)^2 - 1\right]; \quad (2.819d, e)$$

(iv) the pressure ratio (2.820a–d) using (2.819c; 2.798b; 2.801e):

$$\frac{p^+}{p^-} = 1 + \frac{\rho^-}{p^-} \left[(v^-)^2 - c_*^2\right] = 1 + \frac{2\gamma\left[(v^-)^2 - c_*^2\right]}{(\gamma+1)c_*^2 - (\gamma-1)(v^-)^2} \quad (2.820a, b)$$

$$= \frac{(\gamma+1)(v^-)^2 - (\gamma-1)c_*^2}{(\gamma+1)c_*^2 - (\gamma-1)(v^-)^2} = \frac{(\gamma+1)(M_*^-)^2 - \gamma + 1}{\gamma+1 - (\gamma-1)(M_*^-)^2}. \quad (2.820c, d)$$

The general inequalities across the normal shock (2.768b–d; 2.769a–c; 2.782c–f) for a gas satisfying the adiabatic condition (2.785a–f), in the particular case of a perfect gas (2.786a–g) $\equiv$ (2.786a–d), can be also proved as follows: (i) from (2.818a) follow that the velocity before (after) the shock is larger (2.821a) [smaller (2.821b)] than the critical sound speed:

$$v^- > c_* > v^+, \quad \rho^+ > \rho^-, \quad p^+ > p^-, \quad s^+ > s^-, \quad (2.821a\text{-}e)$$

(ii) from (2.821a, b; 2.818b) follows (2.821c) that the mass density is larger after the shock; (iii) from (2.819c; 2.821a) follows that (2.821d) the pressure is larger after the shock; (iv) from (2.821d; 2.775b) it follows (2.821e) that the entropy increases across the shock as required by the second principle of thermodynamics. The latter condition would be violated by the opposite choice of signs in (2.821a–e), which is physically impossible.

From (2.796c) [(2.794b; 2.795a)] it follows that *the stagnation sound speed exceeds the critical (2.822a–c) [local (2.822d, e)] sound speeds:*

$$\gamma > 1: \quad c_* = \sqrt{\frac{2}{\gamma+1}}\, c_0 < c_0, \quad c = \sqrt{c_0^2 - \frac{\gamma-1}{2} v^2} < c_0, \qquad (2.822a\text{–}e)$$

and thus the stagnation Mach number (2.823c) is smaller than both the critical (2.822c) $\equiv$ (2.823a, b) [local (2.822d, e) $\equiv$ (2.823d, e)] Mach numbers:

$$M_* \equiv \frac{v}{c_*} > \frac{v}{c_0} \equiv M_0 < \frac{v}{c} \equiv M. \qquad (2.823a\text{–}e)$$

In particular for a normal shock of a perfect gas: (i) all downstream Mach numbers are subsonic (2.782b, e, f; 2.801f; 2.823a–e) $\equiv$ (2.824a–f):

$$1 > M^+ \equiv \frac{v^+}{c^+} > M_0^+ < M_*^+ \equiv \frac{v^+}{c_*} < 1; \qquad (2.824a\text{–}f)$$

(ii) the upstream local and critical Mach numbers are supersonic (2.782a, c, d; 2.801e) $\equiv$ (2.825a–d):

$$M^- \equiv \frac{v^-}{c^-} > 1 < M_*^- = \frac{v^-}{c_*} \qquad (2.825a\text{–}d)$$

Since the stagnation Mach number is smaller than the critical and local Mach numbers (2.824a–f), it follows that: (i) downstream it is always subsonic (2.824a–f); (ii) upstream it could kinematically be supersonic or subsonic (2.825a–d). The case of a subsonic upstream stagnation Mach number, with supersonic upstream local and critical Mach numbers (2.825a–d) leads to entropy decrease across the shock, and hence violates the second principle of thermodynamics, as shown in a subsequent example (subsections 2.7.19–2.7.24). Thus *the condition of supersonic upstream stagnation Mach number is not a kinematic condition but rather a thermodynamic constraint of entropy growth across the shock, to comply with the second principle of thermodynamics.* The jumps of flow variables across the normal shock of a perfect gas can be expressed in terms of: (i) the pressure ratio downstream to upstream (subsection 2.7.16); (ii/iii) the upstream (downstream) Mach number [subsection 2.7.17 (2.7.18)].

2.7.16   Variations across Shock in Terms of Pressure Ratios

From arbitrary normal shock in a perfect gas (2.479d) the ratio of temperature equals (2.793d–f) $\equiv$ (2.826b) the ratio of sound speed squared (2.826a) and is given (2.815b) [(2.815c)] by (2.826c) [(2.826d)] in terms of the pressure (specific volume) ratio:

$$\left(\frac{c^+}{c^-}\right)^2 = \frac{T^+}{T^-} = \frac{p^+ \, \vartheta^+}{p^- \, \vartheta^-} = \frac{(\gamma+1)\,p^+/p^- +(\gamma-1)\left(p^+/p^-\right)^2}{\gamma-1+(\gamma+1)\,p^+/p^-}$$

$$= \frac{(\gamma-1)\left(\vartheta^+/\vartheta^-\right)^2 -(\gamma+1)\,\vartheta^+/\vartheta^-}{\gamma-1-(\gamma+1)\,\vartheta^+/\vartheta^-}. \tag{2.826a–d}$$

The upstream p^- and downstream p^+ pressures and upstream specific volume ϑ^- specify: (i) the mass flux (2.827a–d) from (2.763c; 2.815b):

$$j^2 = \frac{p^+ - p^-}{\vartheta^- - \vartheta^+} = \frac{\left(p^+ - p^-\right)/\vartheta^-}{1-\vartheta^+/\vartheta^-} = \frac{p^+ - p^-}{\vartheta^-} \frac{(\gamma-1)\,p^- +(\gamma-1)\,p^+}{2\left(p^+ - p^-\right)}$$

$$= \frac{(\gamma-1)\,p^- +(\gamma+1)\,p^+}{2\,\vartheta^-}; \tag{2.827a–d}$$

(ii) the upstream velocity (2.828a, b) from (2.762c; 2.827d):

$$\left(v^-\right)^2 = \left(\vartheta^- \, j\right)^2 = \frac{\vartheta^-}{2}\left[(\gamma-1)\,p^- +(\gamma+1)\,p^+\right] = \frac{\left(c^-\right)^2}{2\gamma}\left[\gamma-1+(\gamma+1)\frac{p^+}{p^-}\right], \tag{2.828a–c}$$

using also the sound speed (2.492d) in (2.828c); (iii) the downstream velocity (2.593a–e) from (2.762d; 2.828c; 2.815b):

$$\left(v^+\right)^2 = \left(\vartheta^+ \, j\right)^2 = \left(\vartheta^- \, j\right)^2 \left(\vartheta^+/\vartheta^-\right)^2 = \frac{\vartheta^-}{2}\left[(\gamma-1)\,p^- +(\gamma+1)\,p^+\right]\left(\vartheta^+/\vartheta^-\right)^2$$

$$= \frac{\vartheta^-}{2}\frac{\left[(\gamma-1)\,p^+ +(\gamma+1)\,p^-\right]^2}{(\gamma-1)\,p^- +(\gamma+1)\,p^+}; \tag{2.829a–d}$$

(iv) the difference of velocities upstream and downstream (2.830a, b) from (2.763c; 2.827d):

$$v^- - v^+ = \frac{p^+ - p^-}{j} = \left(p^+ - p^-\right)\sqrt{2\vartheta^-}\left[(\gamma-1)\,p^- +(\gamma+1)\,p^+\right]^{-1/2}. \tag{2.830a, b}$$

Thus have been obtained: *(i) ratios of specific volumes (2.815b) $\equiv$ (2.831a), mass densities (2.831b) and velocities (2.831c) in terms of the ratio of pressures:*

$$\frac{(\gamma+1)\,p^- +(\gamma-1)\,p^+}{(\gamma-1)\,p^- +(\gamma+1)\,p^+} = \frac{\vartheta^+}{\vartheta^-} = \frac{\rho^-}{\rho^+} = \frac{v^+}{v^-}; \tag{2.831a–c}$$

(ii) the ratio of pressures in terms of the ratio of specific volumes (2.815c); (iii–iv) the ratio of temperatures and local sound speeds in terms of the ratios of pressures and specific volumes (2.826a–d); (v–viii) the mass flux (2.827a–d), upstream (2.828a–c) [downstream (2.829a–d)] velocities and their difference (2.830a, b) in terms of the upstream and downstream pressures and upstream specific volume. The changes in flow variables across a normal shock of a perfect gas can also be expressed in terms of the upstream (downstream) local Mach number [subsection 2.7.17–2.7.18)].

2.7.17   SHOCK RELATIONS IN TERMS OF UPSTREAM MACH NUMBER

The ratio of pressures is given by (2.820d) $\equiv$ (2.832a) in terms of the upstream critical Mach number (2.801e) and by (2.832b, c) in terms of the upstream local Mach number (2.782c):

$$\frac{p^+}{p^-} = \frac{(\gamma+1)(M_*^-)^2 - \gamma + 1}{\gamma + 1 - (\gamma-1)(M_*^-)^2} = \frac{2\gamma(M^-)^2 - \gamma + 1}{\gamma+1} = 1 + \frac{2\gamma}{\gamma+1}\left[(M^-)^2 - 1\right], \quad (2.832a\text{-}c)$$

using the relation (2.828c) $\equiv$ (2.833a, b):

$$\gamma - 1 + (\gamma+1)\frac{p^+}{p^-} = 2\gamma\left(\frac{v^-}{c^-}\right)^2 = 2\gamma(M^-)^2. \qquad (2.833a, b)$$

Substituting (2.832a) [(2.832b, c)] in (2.831a–c) specifies the ratio of velocities, mass densities and specific volumes (2.834a, b) in terms of the local (2.834d) [critical (2.834c)] upstream Mach number:

$$\frac{v^+}{v^-} = \frac{\rho^-}{\rho^+} = \frac{\vartheta^+}{\vartheta^-} = (M_*^-)^{-2} = \frac{\gamma - 1 + 2(M^-)^{-2}}{\gamma + 1}; \qquad (2.834a\text{-}d)$$

the result (2.834c) coincides with (2.676c) $\equiv$ (2.834c) and can also be obtained (2.835a–c) from (2.801e, a; 2.834c):

$$(M_*^-)^2 \equiv \left(\frac{v^-}{c_*}\right)^2 = \frac{(v^-)^2}{v^- v^+} = \frac{v^-}{v^+}. \qquad (2.835a\text{-}c)$$

The ratio of sound speeds squared, temperatures, internal energies and enthalpies is given by the product of (2.832c) and (2.834c) leading to (2.836a–e):

$$\left(\frac{c^+}{c^-}\right)^2 = \frac{T^+}{T^-} = \frac{u^+}{u^-} = \frac{h^+}{h^-} = \frac{\vartheta^+ \, p^+}{\vartheta^- \, p^-}$$

$$= \frac{\gamma + 1 - (\gamma-1)/(M_*^-)^2}{\gamma + 1 - (\gamma-1)(M_*^-)^2} = \frac{\left[2\gamma(M^-)^2 - \gamma + 1\right]\left[2(M^-)^{-2} + \gamma - 1\right]}{(\gamma+1)^2}. \qquad (2.836a\text{-}f)$$

The entropy difference across the shock (2.477c; 2.264c–e) ≡ (2.837a) is given by (2.837b, c) using (2.832a, b) and (2.834c):

$$\frac{\gamma-1}{R}\left(s^+ - s^-\right) = \log\left(\frac{p^+}{p^-}\right) + \gamma \log\left(\frac{9^+}{9^-}\right) = \log\left[\left(M_*^-\right)^{-2\gamma} \frac{(\gamma+1)\left(M_*^-\right)^2 - \gamma + 1}{\gamma + 1 - (\gamma-1)\left(M_*^-\right)^2}\right]$$

$$= \log\left\{\frac{2\gamma\left(M^-\right)^2 - \gamma + 1}{\gamma+1}\left[\frac{2\left(M^-\right)^{-2} + \gamma - 1}{\gamma+1}\right]^\gamma\right\}.$$

(2.837a–c)

Thus have been obtained *for the normal shock of a perfect gas, in terms of the upstream critical (2.801e) [local (2.782c] Mach number: (i) the ratio of pressures (2.832a) [(2.832b, c)]; (ii–iv) the ratio (2.834c) [(2.834d)] of velocities, mass densities and specific volumes (2.834a, b); (v–viii) the ratio (2.836e) [(2.836f)] of sound speeds squared, temperatures, internal energies and enthalpies (2.836a–d); (ix) the difference (2.837b) [(2.837c)] of entropies (2.837a).* The local upstream Mach number may be replaced by: (i) the stagnation (critical) upstream Mach number using (2.677a, b) [(2.677c, d)]; (ii) the corresponding downstream local, stagnation or critical Mach numbers (subsection 2.7.18).

2.7.18 RELATION BETWEEN UPSTREAM AND DOWNSTREAM MACH NUMBERS

Concerning the normal shock of a perfect gas, the relation between the upstream (downstream) Mach numbers before (after) the shock is: (i) from (2.801a, g), algebraic inverses (2.838a, b) for the critical Mach numbers:

$$M_*^- M_*^+ \equiv \frac{v^- v^+}{c_*^2} = 1: \quad M_0^- M_0^+ = \frac{v^- v^+}{c_0^2} = \frac{v^- v^+}{c_*^2}\left(\frac{c_*}{c_0}\right)^2 = \frac{2}{\gamma+1}; \quad (2.838a\text{–}e)$$

(ii) the product of the stagnation Mach numbers is given by (2.838c–e), since the stagnation and critical sound speeds are in a constant ratio (2.796c); (iii) substituting (2.834d) in (2.838a) specifies the relation (2.839a) between local Mach numbers:

$$\left[2 + (\gamma-1)\left(M^-\right)^2\right]\left[2 + (\gamma-1)(M^+)^2\right] = \left[(\gamma+1)M^- M^+\right]^2, \quad (2.839a)$$

that: (i) leads to the symmetric relation (2.603a, b):

$$2 = 2\gamma\left(M^- M^+\right)^2 - (\gamma-1)\left[\left(M^-\right)^2 + \left(M^+\right)^2\right], \quad (2.839b)$$

that can be solved for the upstream (2.840a) [downstream (2.840b)] local Mach numbers:

$$\left(M^{\pm}\right)^{2} = \frac{2+\left(\gamma-1\right)\left(M^{\mp}\right)^{2}}{2\gamma\left(M^{\mp}\right)^{2}-\gamma+1}. \tag{2.840a, b}$$

The relation (2.840a) can be used to express (2.832b, c; 2.834a, b, d; 2.836a, b, c, e; 2.837a, c) in terms of the local upstream Mach number. The theory of the normal shock (Section 2.7) is concluded by two numerical examples for a perfect gas with an upstream supersonic (subsonic) stagnation Mach number (subsection 2.7.19).

2.7.19 UPSTREAM SUPERSONIC (SUBSONIC) STAGNATION MACH NUMBER

Two examples of normal shock of a perfect gas are considered in standard atmospheric conditions at sea level corresponding to a stagnation temperature (2.716a) ≡ (2.841a), pressure (2.710g, 1) = (2.841b, c), mass density (2.720c) ≡ (2.841d) and stagnation sound speed (2.721d) ≡ (2.841e):

$$T_{0} = 288\,K, \quad p_{0} = 1.014\times10^{5}\,kg\,m^{-1}\,s^{-2} \equiv 1\,atm, \tag{2.841a–c}$$

$$\rho_{0} = 1.225\,kg\,m^{-3} \quad c_{0} = 340\,m\,s^{-1}. \tag{2.841d, e}$$

For a diatomic (2.475b) ≡ (2.842a) perfect gas with adiabatic exponent (2.842b–d) the critical sound speed (2.795c) is (2.842e–h)

$$f = 5, \quad \gamma = 1 + \frac{2}{f} = \frac{7}{5} = 1.40, \quad c_{*} = c_{0}\sqrt{\frac{2}{\gamma+1}} = \frac{340}{\sqrt{1.2}}\,m\,s^{-1} = 310\,m\,s^{-1} < c_{0}. \tag{2.842a–h}$$

Two upstream velocities are chosen: (i) one larger (2.843a) than the stagnation sound speed (2.841e) and hence (2.842e–h) also larger than the critical sound speed (2.843b, c)

$$v^{-} = 400\,m\,s^{-1} > c_{0} = 340\,m\,s^{-1} > c_{*} = 310\,m\,s^{-1}, \tag{2.843a–c}$$

$$c_{0} = 340\,m\,s^{-1} > v^{-} = 320\,m\,s^{-1} > c_{*} = 310\,m\,s^{-1}; \tag{2.844a–c}$$

(ii) the other (2.844b) smaller (larger) than the stagnation (2.844a) [critical (2.844c)] sound speed.

The two upstream flow velocities (2.843a; 2.844a) ≡ (2.845a) correspond (2.794b; 2.796a) ≡ (2.845b) to the local sound speeds (2.845c, d):

$$v^{-} = 400\left(320\right)m\,s^{-1}:$$

$$c^{-} = \left|c_{0}^{2} - \frac{\gamma-1}{2}\left(v^{-}\right)^{2}\right|^{1/2} = \left|340^{2} - 0.2\times\left[400\left(320\right)\right]^{2}\right|^{1/2} = 289\left(308\right)m\,s^{-1}. \tag{2.845a–d}$$

The local (2.846a, b) [critical (2.846c, d)] Mach numbers upstream before the shock are supersonic in all cases with the former larger (2.846e):

$$M^- \equiv \frac{v^-}{c^-} = \frac{400(320)}{289(308)} = 1.384(1.039), \tag{2.846a, b}$$

$$M_*^- \equiv \frac{v^-}{c_*} = \frac{400(320)}{310} = 1.290(1.032) < M,^- \tag{2.846c–e}$$

$$M_0^- \equiv \frac{v^-}{c_0} = \frac{400(320)}{340} = 1.176(0.941) < M_*^- < M^-. \tag{2.846f–j}$$

The upstream stagnation Mach number (2.846f, g) is supersonic (subsonic) for the larger (smaller) flow velocity (2.846h), and in both cases smaller than the critical Mach number (2.846i) and hence also smaller than the local Mach number (2.846j). In contrast the downstream local, stagnation and critical Mach numbers are all subsonic (2.590a–c) as can be confirmed next (subsection 2.7.20).

2.7.20   SUBSONIC DOWNSTREAM LOCAL, STAGNATION AND CRITICAL MACH NUMBERS

The relation (2.801a) $\equiv$ (2.847a) specifies (2.845a; 2.844c) the downstream flow velocity (2.847b):

$$v^+ = \frac{c_*^2}{v^-} = \frac{(310\,m\,s^{-1})^2}{400(320)\,m\,s^{-1}} = 240(300)\,ms^{-1}, \tag{2.847a–c}$$

that is (2.847c) smaller (larger) for the larger (smaller) upstream flow velocity (2.845a). The downstream sound speed is similar to (2.845b) replacing v^- by v^+ in (2.848a, b):

$$c^+ = \sqrt{c_0^2 - \frac{\gamma - 1}{2}(v^+)^2} = \left|340^2 - 0.2 \times \left[240(300)\right]^2\right|^{1/2}$$
$$= 323(312)\,m\,s^{-1} > c_* > c^-, \tag{2.848a–d}$$

is larger (2.848c) for the larger upstream velocity (2.845a) and exceeds (2.848d) the critical sound speed (2.842g). Thus downstream after the shock the Mach numbers are all subsonic, with the critical (stagnation) Mach number the largest (smallest), hence larger (smaller) than the local Mach number (2.849a–l):

$$1 > M_*^+ \equiv \frac{v^+}{c_*} = \frac{240(300)}{310} = 0.774(0.968), \tag{2.849a–d}$$

$$> M^+ \equiv \frac{v^+}{c^+} = \frac{240(300)}{323(312)} = 0.743(0.963), \qquad (2.849e\text{–}h)$$

$$> M_0^+ \equiv \frac{v^+}{c_0} = \frac{240(300)}{340} = 0.706(0.882). \qquad (2.849i\text{–}l)$$

The speed of advance of the shock front (2.803b) ≡ (2.850a) is the arithmetic mean (2.850b, c) of the upstream (2.845a) and downstream (2.847c) flow velocities

$$\bar{v} = \frac{v^+ + v^-}{2} = \frac{400(320) + 240(300)}{2}\, m\,s^{-1} = 320(310)\,m\,s^{-1}, \quad (2.850a\text{–}c)$$

and: (i) is larger than (2.851a–d) the critical sound speed (2.842g) in agreement with (2.804e), although it is almost coincident for the lower upstream flow velocity:

$$\frac{\bar{v}}{c_*} = \frac{320(310)}{310} = 1.032(1.000) \geq 1 > \frac{\bar{v}}{c_0} = \frac{320(310)}{340} = 0.941(0.912); \quad (2.851a\text{–}f)$$

(ii) is (2.851d–f) smaller than the stagnation sound speed (2.841e) for the both upstream flow velocities (2.851e,f). The two numerical examples are concluded by considering the upstream (downstream) flow conditions, that is before (after) the shock [subsections 2.7.21 (2.7.22)] and hence the changes in terms of ratios or differences (subsection 2.7.23).

2.7.21   UPSTREAM FLOW BEFORE THE SHOCK

The flow upstream of the shock is adiabatic corresponding (2.841e) to the adiabatic factor (2.621a–c) for the velocities (2.845a):

$$\psi^- \equiv 1 - \frac{\gamma - 1}{2}\left(\frac{v^-}{c_0}\right)^2 = 1 - 0.2 \times \left[\frac{400(320)}{340}\right]^2 = 0.723(0.823), \quad (2.852a\text{–}c)$$

that specifies: (i) the local sound speed (2.622) from (2.666b) in agreement with (2.845d); (ii) the temperature (2.623a–g) from (2.666c); (iii) the mass density (2.624a–b) from (2.666d); (iv) the pressure (2.625a-c) from (2.844c, 2.845c); (v/vi) the internal energy (2.857a–d) [enthalpy (2.858a–d)] per unit mass from (2.623a) [(2.623b)] with (2.854c) using (2.719h):

$$c^- = c_0 \sqrt{\psi^-} = 340 \times \sqrt{0.723(0.823)}\, m\,s^{-1} = 289(308)\,m\,s^{-1}, \quad (2.853a\text{–}c)$$

$$T^- = T_0 \psi^- = 288\,K \times 0.723(0.823) = 208(237)\,K, \qquad (2.854a\text{–}c)$$

$$\rho^- = \rho_0 \left(\psi^-\right)^{1/(\gamma-1)} = 1.225 \times \left[0.723(0.823)\right]^{2.5} kg\,m^{-3}$$
$$= 0.544(0.753)\,kg\,m^{-3},$$

(2.855a–c)

$$p^- = p_0 \frac{\rho^-}{\rho_0}\frac{T^-}{T_0} = \frac{0.544(0.753)}{1.225} \times \frac{208(237)}{288}atm = 0.324(0.506)\,atm,$$

(2.856a–c)

$$u^- = c_V\,T^- = 7.18 \times 10^2\,m^2\,s^{-2}\,K^{-1} \times 208(237)\,K$$
$$= 1.49(1.70) \times 10^5\,m^2\,s^{-2},$$

(2.857a–c)

$$h^- = c_p\,T^- = 1.005 \times 10^3\,m^2\,s^{-2}\,K^{-1} \times 208(237)\,K$$
$$= 2.09(2.38) \times 10^5\,m^2\,s^{-2}.$$

(2.858a–c)

The upstream flow conditions before the shock specify the downstream flow conditions after the shock (subsection 2.7.22).

2.7.22 DOWNSTREAM FLOW AFTER THE SHOCK

The upstream (2.845a) [downstream (2.847c)] velocities and upstream mass density (2.855c) specify by (2.762b) ≡ (2.859a) the downstream mass density (2.859b, c):

$$\rho^+ = \rho^- \frac{v^-}{v^+} = 0.544(0.753) \times \frac{400(320)}{240(300)}kg\,m^{-2} = 0.907803\,kg\,m^{-3}.$$

(2.859a–c)

The upstream (downstream) velocities (2.845a) [(2.847c)] and mass densities (2.855c) [(2.859c)] specify by (2.762b) ≡ (2.860a) the mass flux (2.860b–d) that is conserved:

$$j = \rho^{\pm}\,v^{\pm} = 0.907(0.803)\,kg\,m^{-3} \times 240(300)\,m\,s^{-1}$$
$$= 0.544(0.753)\,kg\,m^{-3} \times 400(320)\,m\,s^{-1} = 218(241)\,kg\,m^{-2}\,s^{-1}.$$

(2.860a–d)

The downstream pressure (2.861a) ≡ (2.763c) is specified by the upstream pressure (2.856c), mass flux (2.860d) and flow velocities before (2.845a) and after (2.847c) the shock leading to (2.861b–d):

$$p^+ = p^- + j\left(v^- - v^+\right) = 0.321(0.506)\,atm + 218(241)\,kg\,m^{-2}\,s^{-1}$$
$$\times \left[400(320) - 240(300)\right]m\,s^{-1}$$
$$= \left[0.321(0.506) + \frac{3.49(0.48)}{10.14}\right]atm = 0.665(0.553)\,atm.$$

(2.861a–d)

The downstream temperature (2.862a) $\equiv$ (2.666c) follows (2.862b, c) from the upstream temperature (2.854c) and the ratio of the squares of upstream (2.845b) [downstream (2.848b)] local sound speeds:

$$T^+ = T^- \left(\frac{c^+}{c^-}\right)^2 = 208(237)K \times \left[\frac{323(311)}{289(308)}\right]^2 = 260(243)K. \quad (2.862a\text{--}c)$$

The upstream internal energy (2.632a–c) [enthalpy (2.633a–d)] is specified (2.623c) [(2.623b)] by the upstream temperature (2.852c) with (2.719h):

$$u^+ = c_V T^+ = 7.18 \times 10^2 \, m^2 \, s^{-2} \, K^{-1} \times 260(243)K$$
$$= 1.87(1.74) \times 10^5 \, m^2 \, s^{-2}, \quad (2.863a\text{--}c)$$

$$h^+ = c_p T^+ = 1.005 \times 10^3 \, m^2 \, s^{-2} \, K^{-1} \times 260(243)K$$
$$= 2.61(2.44) \times 10^5 \, m^2 \, s^{-2}. \quad (2.864a\text{--}c)$$

The values of the upstream (downstream) flow variables before (after) the shock [subsection 2.7.21 (2.7.22)] specify the ratios and differences (subsection 2.7.23) in Table 2.19 and also the entropy production (subsection 2.7.24).

2.7.23   DIFFERENCES AND RATIOS ACROSS THE SHOCK FRONT

The differences (ratios) of downstream (to upstream) flow variables after (before) the shock are given: (i) by (2.855c; 2.859c) for the mass density (2.865a, b) [(2.865c, d)] and are larger for the larger upstream flow velocity:

$$\rho^+ - \rho^- = 0.907(0.803) - 0.544(0.753) \, kg \, m^{-3} = 0.363(0.050) \, kg \, m^{-3},$$
$$\frac{\rho^+}{\rho^-} = \frac{0.907(0.803)}{0.544(0.753)} = 1.667(1.066); \quad (2.865a\text{--}d)$$

(ii) by (2.845a; 2.847c) for the flow velocity (2.866a, b) [(2.866c, d)]:

$$v^- - v^+ = \left[400(320) - 240(300)\right] m \, s^{-1} = 160(20) m \, s^{-1},$$
$$\frac{v^-}{v^+} = \frac{400(320)}{240(300)} = 1.667(1.067); \quad (2.866a\text{--}d)$$

(iii) by (2.845d; 2.848d) for the local sound speeds (2.867a, b) [(2.867c, d)]:

$$c^+ - c^- = \left[323(312) - 289(308)\right] m \, s^{-1} = 34(4) m \, s^{-1},$$
$$\frac{c^+}{c^-} = \frac{323(312)}{289(308)} = 1.118(1.012); \quad (2.867a\text{--}d)$$

TABLE 2.19
Comparison of Normal Shock with Two Upstream Velocities

Variable			Value			
Symbol	Name	Unit	Upstream	Downstream	Difference	Ratio
v	Velocity	$m\ s^{-1}$	400 (320)	240 (300)	160 (20)	1.667 (1.067)
c	Sound speed	$m\ s^{-1}$	289 (308)	323 (312)	34 (4)	1.118 (1.012)
ρ	Mass density	$kg\ m^{-3}$	0.544 (0.753)	0.902 (0.803)	0.363 (0.050)	1.667 (1.066)
p	Pressure	atm	0.321 (0.508)	0.665 (0.553)	0.344 (0.047)	2.072 (1.093)
T	Temperature	K	208 (237)	260 (243)	52 (6)	1.250 (1.026)
u	Internal energy*	$10^4\ m^2\ s^{-2}$	14.9 (17.0)	18.7 (17.4)	3.80 (0.40)	1.255 (1.024)
h	Enthalpy*	$10^4\ m^2\ s^{-2}$	20.9 (23.8)	26.1 (24.4)	5.20 (0.60)	1.249 (1.025)
M	Local Mach number	-	1.384 (1.039)	0.743 (0.963)	−0.641 (−0.076)	0.542 (0.927)
M^*	Critical Mach number	-	1.290 (1.036)	0.774 (0.968)	−0.516 (−0.063)	0.600 (0.939)
M_0	Stagnation Mach number	-	1.176 (0.941)	0.706 (0.882)	−0.470 (−0.059)	0.600 (0.937)
c_0	Stagnation sound speed	$m\ s^{-1}$	340	340	0	1.000
T_0	Stagnation temperature	K	288	288	0	1.000
u_0	Stagnation internal energy	$10^5\ m^2\ s^{-2}$	2.067	2.067	0	1.000
h_0	Stagnation enthalpy	$10^5\ m^2\ s^{-2}$	2.894	2.894	0	1.000
p_0	Stagnation pressure	atm	1.000	0.963 (1.001)	−0.037 (+0.001)	0.963 (1.001)
ρ_0	Stagnation mass density	$kg\ m^{-3}$	1.225	1.180 (1.226)	−0.078 (+0.001)	0.963 (1.001)

* per unit mass: entropy difference: $s^+ - s^- = 280.0\,(-10.4)\,m^2\,s^{-2}\,K^{-1}$; shock propagation speed $\bar{v} = 320\,(310)\,m\,s^{-1}$ $\dfrac{\bar{v}}{c_*} = 1.032\,(1.000) > \dfrac{\bar{v}}{c_0} = 0.941\,(0.912)$

- mass flux: $j = \rho^+ v^+ = 218\,(241)\,kg\,m^{-2}\,s^{-1}$

Note: Comparison of the normal shock with upstream velocity larger (smaller) than the stagnation sound speed, ultimately leading to entropy production (reduction) that is physical (unphysical) because it complies with (violates) the second principle of thermodynamics. The comparison includes the values upstream and downstream of the normal shock and their difference and ratio for: (i–vii) seven flow variables: velocity, sound speed, mass density, pressure, temperature, internal energy and enthalpy; (viii–x) three Mach numbers: local, critical and stagnation; (xi–xvi) six more stagnation values: sound speed, temperature, internal energy, enthalpy, pressure and mass density.

(iv) by (2.854c; 2.862c) for the temperatures (2.868a, b) [(2.868c, 2.820c, d)]:

$$T^+ - T^- = \left[260(243) - 208(237)\right]K = 52(6)K, \qquad (2.868a, b)$$

$$\frac{T^+}{T^-} = \frac{280(243)}{208(237)} = 1.250(1.025); \qquad (2.868c, d)$$

(v) by (2.856c; 2.861d) for the pressure (2.869a, b) [(2.869c, d)]:

$$p^+ - p^- = \left[0.665(0.553) - 0.321(0.506)\right]atm = 0.344(0.047)atm,$$

$$\frac{p^+}{p^-} = \frac{0.665(0.553)}{0.321(0.506)} = 2.072(1.093); \qquad (2.869a{-}d)$$

(vi) by (2.857c; 2.863c) for the internal energy (2.870a, b) [(2.870c, d)]:

$$u^+ - u^- = \left[1.87(1.74) - 1.49(1.70)\right]\times10^5\, m^2\, s^{-2} = 3.80(0.40)\times10^4\, m^2\, s^{-2},$$

$$\frac{u^+}{u^-} = \frac{1.87(1.74)}{1.49(1.70)} = 1.255(1.024); \qquad (2.870a{-}d)$$

(vii) by (2.858c; 2.864c) for the enthalpy (2.871a, b) [(2.871c, d)]:

$$h^+ - h^- = \left[2.61(2.44) - 2.09(2.38)\right]\times10^5\, m^2\, s^{-2} = 5.20(0.60)\times10^4\, m^2\, s^{-2},$$

$$\frac{h^+}{h^-} = \frac{2.61(2.44)}{2.09(2.38)} = 1.249(1.025). \qquad (2.871a{-}d)$$

Since the entropy is specified only as a difference from a reference state, for example, stagnation, it is best considered as the jump between downstream and upstream values, specifying the entropy production for a strong (weak) shock (subsection 2.7.24).

2.7.24 ENTROPY PRODUCTION IN A STRONG (WEAK) SHOCK

The entropy difference (2.477d) ≡ (2.872a) across a shock is given (2.869d; 2.865d; 2.719h) by (2.872b, c):

$$s^+ - s^- = c_V \log\left(\frac{p^+}{p^-}\right) - c_p \log\left(\frac{\rho^+}{\rho^-}\right)$$

$$= \left\{718\log\left[2.072(1.093)\right] - 1005\log\left[1.667(1.066)\right]\right\}m^2\, s^{-2}\, K^{-1} \qquad (2.872a{-}c)$$

$$= 280.0(-10.4)m^2\, s^{-2}\, K^{-1}.$$

Thus the larger (smaller) upstream velocity in (2.845a) is physically possible (unphysical) since it corresponds to an entropy increase (decrease) across the shock, that is consistent with (contradicts) the second principle of thermodynamics. The flow is isentropic both upstream and downstream of the shock, and the physical (unphysical) case of entropy increase (decrease) corresponds to an upstream velocity larger (2.843a–c) [smaller (2.844a–c)] than the stagnation sound speed, that is super-sonic (subsonic) stagnation Mach number. Thus the flow upstream of the normal shock must be supersonic relative to the: (i) critical and local sound speeds due to the kinematic shock relations; (ii) stagnation sound speed for entropy growth complying with the second principle of thermodynamics. In the case of a weak shock the entropy change (2.775b) for a perfect gas (2.777f) is given by (2.873a, b) using (2.473d):

$$\Delta s = \frac{\left(p^+ - p^-\right)^3}{12T^-}\frac{\gamma+1}{\gamma^2}\frac{\vartheta^-}{\left(p^-\right)^2} = \frac{(\gamma+1)\overline{R}}{12\gamma^2}\left(\frac{p^+}{p^-}-1\right)^3. \qquad \text{(2.873a, b)}$$

Using (2.869d; 2.842d; 2.718e) in (2.873b) leads to the entropy difference in the weak shock approximation (2.874a, b):

$$\Delta s = \frac{2.4\times287}{12\times1.4^2}\times1.072^3\,m^2\,s^{-2}\,K^{-1} = 36.1\,m^2\,s^{-2}\,K^{-1}; \qquad \text{(2.874a, b)}$$

the approximate value (2.874b) is significantly smaller than the exact value (2.872c) for the physical shock, showing that the weak shock estimate can be quite inaccurate for a strong shock. The entropy change across the shock is compatible with the con-servation of some but not all stagnation flow variables (subsection 2.7.25).

2.7.25 CONSERVATION OR VARIATION OF STAGNATION FLOW VARIABLES

Since the stagnation sound speed (2.844a) ≡ (2.875a, b) is conserved across the shock for a perfect gas, the same applies to: (i) the stagnation temperature (2.841a) ≡ (2.875c, d); (ii/iii) the stagnation internal energy (2.623a) ≡ (b) [enthalpy (2.623b) ≡ (2.647a–d)] per unit mass using (2.719h):

$$c_0 = 340\,m\,s^{-1} = c_0^{\pm}, \quad T_0 = 288\,K = T_0^{\pm}, \qquad \text{(2.875a–d)}$$

$$u_0 = c_{V_0}T_0 = 718\,m^2\,s^{-2}\,K^{-1}\times288\,K = 2.067\times10^5\,m^2\,s^{-2} = u_0^{\pm}, \qquad \text{(2.875e–h)}$$

$$h_0 = c_{p_0}T_0 = 1005\,m^2\,s^{-2}\,K^{-1}\times288\,K = 2.894\times10^5\,m^2\,s^{-2} = h_0^{\pm}. \qquad \text{(2.875i–l)}$$

The downstream Mach factor (2.876a) is given (2.841d; 2.842d; 2.847c) by (2.876b, c):

$$\psi^+ = 1 - \frac{\gamma-1}{2}\left(\frac{v^+}{c_0}\right)^2 = 1 - 0.2\times\left[\frac{240(300)}{340}\right]^2 = 0.900(0.844), \qquad \text{(2.876a–c)}$$

and determines $(2.666e) \equiv (2.877a)$ the upstream stagnation pressure $(2.877b, c)$ using $(2.842d; 2.876c; 2.861d)$:

$$p_0^+ = p^+ \left(\psi^+\right)^{\gamma/(\gamma-1)} = 0.665(0.553) \times \left[0.900(0.844)\right]^{-3.5} atm$$

$$= 0.963(1.001) atm.$$

$$(2.877a\text{--}c)$$

The case of a physical (unphysical) shock with increase (decrease) in entropy leads to a decrease (increase) in stagnation pressure relative to the atmospheric pressure upstream $(2.841b) = (2.878a)$. In both cases the stagnation pressure is close to the atmospheric pressure and their ratio $(2.878b)$ applies also to the ratio of stagnation mass densities $(2.878c)$ because the stagnation temperature $(2.875c, d)$ is conserved:

$$p_0^- = 1.000\, atm : \quad \frac{p_0^+}{p_0^-} = 0.963(1.001) = \frac{\rho_0^+}{\rho_0^-}. \qquad (2.878a\text{--}c)$$

Thus the upstream stagnation mass density $(2.841d) \equiv (2.879a)$ specifies the downstream stagnation mass density $(2.879b, c)$:

$$\rho_0^- = 1.225\, kg\, m^{-3} : \quad \rho_0^+ = \left[0.963(1.001)\right] \times 1.225\, kg\, m^{-3} \qquad (2.879a, b)$$

$$= 1.180(1.226)\, kg\, m^{-3}. \qquad (2.879c)$$

The difference of downstream to upstream stagnation pressure $(2.880a, b)$ [mass density $(2.880c, 2.820c, d)$] are negative:

$$p_0^+ - p_0^- = \left[0.963(1.001) - 1.000\right] atm = -0.037(+0.001) atm, \quad (2.880a, b)$$

$$\rho_0^+ - \rho_0^- = \left[1.179(1.226) - 1.225\right] kg\, m^{-3} = -0.046(+0.001) kg\, m^{-3}, \quad (2.880c, d)$$

because the shock is an irreversible thermodynamic process causing a loss of **head**, that is the designation of stagnation pressure in engineering; an increase in head is unphysical, and indicates a decrease in entropy, violating the second principle of thermodynamics. The upstream and downstream flow variables, and their differences and ratios, for the normal shock of a perfect gas in the International Standard Atmosphere (I.S.A.) at sea level with the physical $(2.843a\text{--}c)$ [unphysical $(2.844a\text{--}c)$] upstream velocity above (below) the stagnation sound speed are indicated in Table 2.19 showing which are conserved or not. The sequence of calculations used (subsections 2.7.19–2.7.25) starts with the simplest shock relation $(2.818a)$ in order to obtain all flow variables for the normal shock of a perfect gas. A generalization of the normal shock (Section 2.7) is the oblique shock (Section 2.8) for which the upstream and downstream velocities are not normal to the shock front.

2.8 OBLIQUE SHOCK (BUSEMANN 1931) AND ADIABATIC TURN (PRANDTL, MEYER 1908)

The oblique shock wave (subsections 2.7.4 and 2.8.2–2.8.17) adds a continuous tangential velocity to the normal shock (subsections 2.7.3 and 2.7.5–2.7.25). The normal velocity satisfies the same relations for a normal shock and oblique shock (subsection 2.8.2), and the continuous tangential velocity (subsection 2.8.4) changes the stagnation enthalpy and limits the angle of incidence to the Mach angle (subsection 2.8.1). The upstream flow conditions including incidence within the Mach angle (subsection 2.8.5) specify the downstream velocity components (subsection 2.8.3) and show that the locus of the downstream velocity (Busemann 1931) or shock polar (subsection 2.8.6) is the strophoid or folio curve of Descartes (1638). The remarkable points of the folio are the normal shock and the adiabatic sound wave (subsection 2.8.7) as the extremes between which lie strong and weak oblique shocks (subsection 2.8.8). The relations across the oblique shocks (subsections 2.8.9 and 2.8.11) include the angle of deflection of the velocity towards the shock front (subsections 2.8.10 and 2.8.12) and simplify in the opposite limits of: (i) a strong shock wave with large upstream Mach number (subsections 2.8.13–2.8.15); (ii) a weak oblique shock with small angle of deflection of the velocity (subsections 2.8.16–2.8.18) leading (subsections 2.8.19–2.8.21) to a gradual adiabatic turn. A linear sound wave is the adiabatic limit of a weak shock wave, that specifies in the limit of small relative pressure change the lowest-order non-linear approximation (subsections 2.8.22–2.8.23). A numerical example is given of an oblique shock at an angle of incidence larger than and close (not close) to the Mach angle (subsection 2.8.24), including the values of the flow variables upstream and downstream (subsections 2.8.25–2.8.29) in both cases.

2.8.1 MACH (1883) CONE FOR A SUPERSONIC VELOCITY

A body pulsating or moving in a fluid generates pressure perturbations; if the perturbations are small relative to the ambient pressure they correspond to **acoustic waves** propagating in all directions at the **sound speed** c. Thus for a static pulsating body (Figure 2.34) the **acoustic wavefronts** are concentric spheres with radius (2.881a). If the body moves at subsonic speed $<c$, for example, $v = c/2$ in Figure 2.34, the acoustic wavefronts move ahead of it in all directions. If the body moves at the sound speed it travels together with the pressure perturbations it generates, and the spherical acoustic wavefronts have a common tangent (Figure 2.34c) that is a **normal shock**. If the body moves at a supersonic speed covering in a time t a distance (2.881b) larger than sound waves (2.881a) the envelope of the spherical acoustic wavefronts is the **Mach cone** (Figure 2.34d) with aperture $\tilde{\theta}$ at the vertex (2.881c) specified by the inverse of the Mach number (2.881d–f):

$$r = ct, \quad x = vt: \quad \sin \tilde{\theta} = \frac{r}{x} = \frac{ct}{vt} = \frac{c}{v} = \frac{1}{M}. \tag{2.881a–f}$$

The pressure perturbations lie within the **Mach angle** (2.881f) from the body, and far away form an **oblique shock wave**.

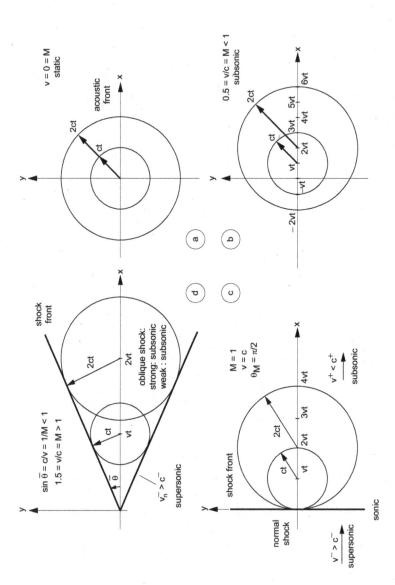

FIGURE 2.34   A sound source at rest (a) emits concentric spherical waves. If the sound source moves subsonically (b) the wave fronts move ahead of it. If the sound source moves at sound speed (c) the wave fronts coincide at the location of a normal shock. If the sound source moves supersonically (d) it leaves the wave fronts behind in a region whose envelope is the Mach cone, with aperture specified by the Mach number. In the case (d) of a supersonic sound source the separation between the flow affected inside (unaffected outside) the Mach cone corresponds in the far-field to an oblique shock wave.

Thus *the Mach angle (2.819a):*

$$\tilde{\theta} = \arcsin\left(\frac{1}{M}\right) \begin{cases} 0 \le \tilde{\theta} < \dfrac{\pi}{2} \text{ if } M > 1 : supersonic, \\[2mm] \tilde{\theta} = \dfrac{\pi}{2} \text{ if } M = 1 : sonic, \\[2mm] \tilde{\theta} = i\arcsinh\left(\dfrac{1}{M}\right) \text{if } M < 1 : subsonic, \end{cases} \qquad (2.882a\text{--}d)$$

(i) is real (2.882b) for a body in supersonic motion (Figure 2.34d) when all pressure perturbations are confined within the Mach cone of aperture $\tilde{\theta}$ in the downstream direction, corresponding to an oblique shock wave, outside which the fluid is at rest since no acoustic pressure perturbations reach there; (ii) corresponds to a plane (2.882c) at sonic velocity (Figure 2.34c) separating perturbed fluid downstream from unperturbed fluid upstream, and corresponding to a normal shock; (iii) imaginary (2.882d) for subsonic motion, since there is no Mach cone, and acoustic pressure perturbations can reach all space (Figures 2.34a, b). The curvature of the Mach cone can be neglected for a small area far downstream leading to an oblique shock wave with a flat shockfront (subsection 2.8.2).

2.8.2 CONTINUOUS TANGENTIAL VELOCITY FOR AN OBLIQUE SHOCK

In the case of an oblique shock there is a continuous non-zero tangential velocity (2.760d) that does not affect the conservation of: (i) mass (2.762b) ≡ (2.883a); (ii) normal component of momentum (2.763c) ≡ (2.883b):

$$j = \rho^- v_n^- = \rho^+ v_n^+ \quad p^- + j v_n^- = p^+ + j v_n^+ \quad 2h^- + \left(v_n^-\right)^2 = 2h^+ + \left(v_n^+\right)^2; \quad (2.883a\text{--}c)$$

(iii) stagnation enthalpy (2.764a) ≡ (2.883c) because the tangential velocity is continuous and drops in both sides of (2.883c). The stagnation enthalpy per unit mass for a perfect gas (2.796b,c) ≡ (2.884a, b):

$$2h_0 = \frac{2c_0^2}{\gamma-1} = \frac{\gamma+1}{\gamma-1}c_*^2 = 2h^\pm + \left(v^\pm\right)^2 \qquad (2.884a\text{--}c)$$

$$= \frac{2\left(c^\pm\right)^2}{\gamma-1} + \left(v_n^\pm\right)^2 + \left(v_t\right)^2 = \frac{2\gamma}{\gamma-1}\frac{p^\pm}{\rho^\pm} + \left(v_n^\pm\right)^2 + \left(v_t\right)^2, \qquad (2.884d, e)$$

equals the enthalpy plus the kinetic energy per unit mass (2.794a–c) ≡ (2.884c–e) where the total velocity involves the normal and tangential velocities; the stagnation enthalpy is related to the stagnation (2.884a) and critical (2.884b) sound speeds, and all three are conserved across the shock front. Thus *the relations for an oblique shock wave are: (i)* *the same as for a normal shock (2.762b; 2.763c; 2.764a) for the velocity components normal to the shock (2.883a–c); (ii) the tangential velocity appears only in the Bernoulli equation (2.884a–e), where the stagnation enthalpy h_0 and critical sound speeds c_* (c_0) are the same on both sides of the shock.* The normal velocities across an oblique shock satisfy a relation obtained as for a normal shock of a perfect gas (subsection 2.7.11) modified (subsection 2.8.3) to take into account a non-zero tangential velocity.

2.8.3   Jump of Normal Velocities across an Oblique Shock

The difference of normal velocities is calculated from (2.883a, b) as for a normal shock (2.765a) leading to (2.885a, b):

$$v_n^- - v_n^+ = \frac{p^+ - p^-}{j} = \frac{p^+}{\rho^+ v_n^+} - \frac{p^-}{\rho^- v_n^-}, \tag{2.885a, b}$$

with the difference that the Bernoulli equation (2.798b) is replaced by (2.884b, e) ≡ (2.886b) involving the tangential velocity:

$$\frac{p^\pm}{\rho^\pm} = \frac{\gamma+1}{2\gamma} c_*^2 - \frac{\gamma-1}{2\gamma}\left[\left(v_n^\pm\right)^2 + \left(v_t\right)^2 \right]. \tag{2.886}$$

Substitution of (2.886) in (2.885b) leads to (2.887):

$$2\gamma\left(v_n^- - v_n^+\right) = \left[(\gamma+1)c_*^2 - (\gamma-1)\left(v_t\right)^2 \right]\left(\frac{1}{v_n^+} - \frac{1}{v_n^-} \right) - (\gamma-1)\left(v_n^+ - v_n^-\right), \tag{2.887}$$

that is equivalent to:

$$v_n^- - v_n^+ = \left[c_*^2 - \frac{\gamma-1}{\gamma+1}\left(v_t\right)^2 \right]\frac{v_n^- - v_n^+}{v_n^- v_n^+}. \tag{2.888}$$

A shock exists if the normal velocities are distinct (2.889a) and then (2.888) implies (2.889b):

$$v_n^- \neq v_n^+ : \quad c_*^2 - v_n^- v_n^+ = \frac{\gamma-1}{\gamma+1}\left(v_t\right)^2 = a\left(v_t\right)^2, \tag{2.889a–c}$$

where appears (2.889c) the constant (2.816d–f). Thus *the distinct normal velocities before and after the oblique shock of a perfect gas (2.889a) are related by (2.889b) that: (i) reduces to (2.801a) for a normal shock; (ii) shows that for an oblique shock the normal velocities are reduced due to the tangential velocity, involving (2.889c) the constant factor (2.816d–f) that appears in the shock locus or pressure – specific volume relation (2.815a–c) for a normal shock (Figure 2.33).* In the case of a normal shock, say in a parallel-sided nozzle (Figure 2.29b) a natural cartesian reference frame has x-axis along the normal velocity and y-axis along the orthogonal shock front. In the case of an oblique shock (subsections 2.8.5–2.8.15) the choice of cartesian axis (subsection 2.8.4) is made for a shock wave "attached" to a concave or convex corner (Figure 2.29c).

2.8.4 SUPERSONIC FLOW PAST CONVEX AND CONCAVE CORNERS

Considering a corner with walls making an angle φ, if it is: (i) zero (2.890b) it is a **flat wall** with internal angle π like a duct (Figure 2.29b) with a normal shock and velocity normal to the shock and decreasing across it; (ii) if it is positive (2.890a) in a **concave corner** (Figure 2.29c ≡ 2.35a), with internal angle less than π, with an oblique shock associated with continuous tangential velocity, decreasing normal velocity and entropy increase; (iii) if it is negative (2.890c) in a **convex corner** (Figure 2.35b) with internal angle more than π, the continuous tangential velocity and increasing normal velocity would lead to an entropy decrease that is not possible, so there must be an infinite sequence of infinitesimally weak shocks corresponding to an **adiabatic turn**.

$$\varphi \begin{cases} > 0 : concave\ corner - oblique\ shock, \\ = 0 : flat\ wall - normal\ shock, \\ < 0 : convex\ corner - adiabatic\ turn. \end{cases} \qquad (2.890a\text{–}c)$$

The cases (ii) [(iii)] of an oblique shock (adiabatic turn) are compared [Figure 2.35a(b)] before a more detailed analysis [subsections 2.8.5–2.8.16 and 2.8.21–2.8.29 (2.8.17–2.8.21)].

In both cases of concave (convex) corner [Figure 2.35] the upstream velocity parallel to the x-axis along the first wall $\vec{v}^-$ has normal v_n^- and tangential v_t components. The tangential velocity v_t is continuous and the upstream velocity $\vec{v}^+$ parallel to the second wall, so that in the case of a concave corner (Figure 2.35) the normal velocity v_n^+ is smaller upstream (2.891a); the conservation of the mass flux (2.883a) leads to a compression (2.891b) and the normal momentum (2.883b) to an increase in pressure (2.891c); this implies an increase in entropy (2.891d) for a weak shock (2.775b), that would be larger for a strong shock:

$$v_n^+ < v_n^- \Rightarrow \rho^+ > \rho^-, \quad p^+ > p^-, \Rightarrow s^+ > s^-. \qquad (2.891a\text{–}d)$$

Thus a supersonic flow incident in a concave corner (Figure 2.35a) forms an oblique shock attached to the apex, that is an irreversible process increasing entropy in agreement with the second principle of thermodynamics (subsection 2.3.16).

If a shock could exist in a convex corner (Figure 2.35b) then the continuity of tangential velocities and the total velocity upstream (downstream) parallel to the first (second) wall would lead to an increase of normal velocity, that is the reverse of (2.891a); thus all signs in (2.454a–c) would be reversed, implying a decrease in entropy, contradicting the second principle of thermodynamics. Thus a single oblique shock or a finite succession of finite shocks cannot turn a supersonic flow around a

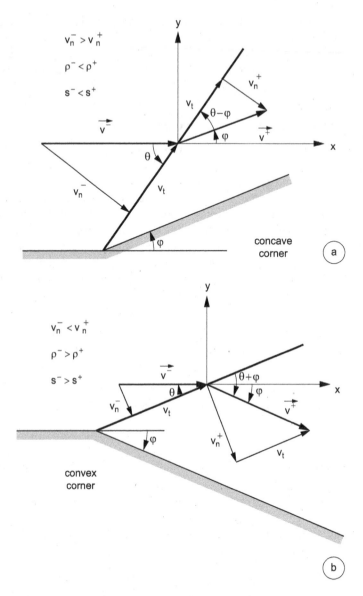

FIGURE 2.35 Comparison of an oblique shock wave (a) [adiabatic turn (b)] in a concave (convex) corner: (i) the tangential velocity is conserved in both cases; (ii) the normal velocity decreases (increases); (iii–iv) the pressure and mass density increase (decrease); (v) the process is irreversible (adiabatic) since the entropy increases (is conserved).

convex corner. The only possibility is infinitesimally weak adiabatic shocks in an infinite succession forming an adiabatic turn (2.890c). The normal shock (2.890b) was considered before (Section 2.7), the oblique shock (2.890a) is considered next (subsections 2.8.5–2.8.16) followed (subsections 2.8.17–2.8.21) by the adiabatic turn.

2.8.5 INCIDENT VELOCITY WITHIN THE MACH CONE

Consider a supersonic flow with velocity v^- parallel to a wall (Figure 2.35) and let it be deflected at a corner by an angle φ into a subsonic flow of velocity v^+. The transition will occur across an **oblique shock wave** attached to the corner at an angle θ, that is also the angle of the upstream velocity $\bar{v}^-$ with the shock front (2.892a, b) implying that the downstream velocity makes an angle $\theta - \varphi$ with the shock front (2.892c, d):

$$\{v_t, v_n^-\} = v^- \{\cos\theta, \sin\theta\}, \quad \{v_t, v_n^+\} = v^+ \{\cos(\theta - \varphi), \sin(\theta - \varphi)\}. \quad (2.892\text{a–d})$$

A cartesian reference frame is chosen with x-axis along the upstream velocity parallel to the wall before the corner and orthogonal y-axis into the flow, implying (2.893a, b) [(2.893c, d)] for the upstream (downstream) velocity:

$$v_x^- = v^- \neq 0 = v_y^-, \quad v_x^+ = v^+ \cos\varphi \quad v_y^+ = v^+ \sin\varphi. \quad (2.893\text{a–d})$$

The normal and tangential components of the velocity are given upstream by (2.892a, b) and downstream by (2.892c, d; 2.893a–d) $\equiv$ (2.894a–d):

$$v_t = v^+ \left(\cos\theta \cos\varphi + \sin\theta \sin\varphi\right) = v_x^+ \cos\theta + v_y^+ \sin\theta, \quad (2.894\text{a, b})$$

$$v_n^+ = v^+ \left(\sin\theta \cos\varphi - \cos\theta \sin\varphi\right) = v_x^+ \sin\theta - v_y^+ \cos\theta; \quad (2.894\text{c, d})$$

the relations (2.894b, d) $\equiv$ (2.895) correspond to a rotation by the angle θ of the oblique shock front with the x-axis

$$\begin{bmatrix} v_t \\ v_n^+ \end{bmatrix} = \begin{bmatrix} \cos\theta & \sin\theta \\ \sin\theta & -\cos\theta \end{bmatrix} \begin{bmatrix} v_n^+ \\ v_y^+ \end{bmatrix}. \quad (2.895)$$

The condition that the upstream normal velocity be supersonic (2.896a) leads (2.892b) to (2.896b):

$$1 < M_n^- \equiv \frac{v_n^-}{c^-} = \frac{v^-}{c^-} \sin\theta \quad \Rightarrow \quad \frac{\pi}{2} \geq \theta \geq \arcsin\left(\frac{c^-}{v^-}\right) = \arcsin\left(\frac{1}{M^-}\right) = \tilde{\theta}^-, \quad (2.896\text{a–e})$$

and implies (2.896c–e) that *the upstream velocity in an oblique shock must make an angle with the shock front larger than the Mach angle (2.882a), and thus must lie within the Mach cone (2.882b)*. This condition also limits the range of angles of the downstream velocity with the shock front, as shown by the deflection of streamlines across the oblique shock front (Section 2.8.6).

2.8.6   OBLIQUE SHOCK POLAR (BUSEMANN 1931) AND FOLIUM (DESCARTES 1638)

The **shock polar** for the downstream velocity (Figure 2.36a) uses as axis the cartesian components of the velocity, so that the distance from the origin is the modulus of the velocity and the angle with the x-axis the angle with the upstream flow. The downstream velocity polar for an oblique shock wave is thus a relation between the cartesian components of the downstream velocity involving only "known" upstream flow variables and constant (conserved) quantities like the adiabatic exponent (critical sound speed). To obtain the shock polar of a perfect gas the starting point may be the ratio of normal velocities for a normal shock (2.834a) ≡ (2.897) that also applies to an oblique shock:

$$\left(\gamma+1\right)\frac{v_n^+}{v_n^-} = \gamma - 1 + 2\left(\frac{c^-}{v_n^-}\right)^2. \tag{2.897}$$

Substituting (2.892b; 2.894d) in (2.897) leads to (2.898):

$$\left(\gamma+1\right)\left(v_x^+ - v_y^+ \cot\theta\right) = \left(\gamma-1\right)v^- + 2\frac{\left(c^-\right)^2}{v^-}\csc^2\theta. \tag{2.898}$$

The continuity of the tangential velocity (2.892a) ≡ (2.892c) ≡ (2.894b) leads to (2.899):

$$v^- \cos\theta = v_t = v_x^+ \cos\theta + v_y^+ \sin\theta, \tag{2.899}$$

that specifies *the angle of the upstream velocity with the shock front (2.899)* ≡ *(2.900a):*

$$\cot\theta = \frac{v_y^+}{v^- - v_x^+}: \quad \csc^2\theta \equiv 1 + \cot^2\theta = 1 + \frac{\left(v_y^+\right)^2}{\left(v^- - v_x^+\right)^2}. \tag{2.900a–c}$$

and implies (2.900b, c).

     Substitution of (2.900a, c) in (2.898) eliminates the incidence angle θ and leads to (2.901):

$$\left(v_y^+\right)^2 = \left(v^- - v_x^+\right)^2 \frac{\left(\gamma+1\right)v_x^+ - \left(\gamma-1\right)v^- - 2\left(c^-\right)^2/v^-}{\left(\gamma+1\right)\left(v^- - v_x^+\right) + 2\left(c^-\right)^2/v^-}, \tag{2.901}$$

that expresses *the transverse in terms of the longitudinal downstream velocity, and upstream conditions (velocity and sound speed), and thus describes the **refraction of***

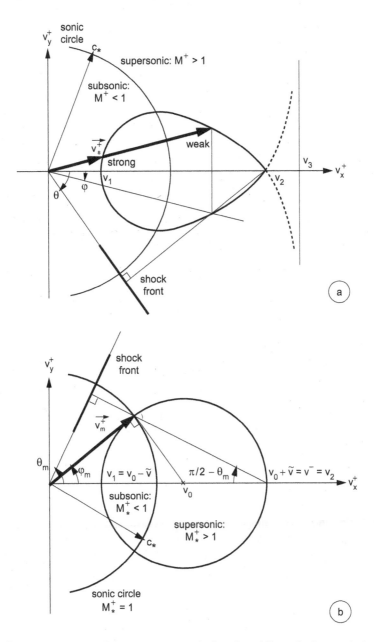

FIGURE 2.36 The locus of the downstream velocity of an oblique shock wave is the closed part of the folium of Descartes, that is the solid line in (a), and becomes a circle (b) in the case of large upstream Mach number. In both cases: (i) there is maximum deflection angle φ_{max} obtained by taking a tangent from the origin; (ii) below the maximum deflection $0 < \varphi < \varphi_{max}$ there are two possibilities, namely a strong (weak) oblique shock with smaller (larger) downstream velocity for the same upstream velocity, and hence closer to (farther from) the origin; (iii) in the limit of no deflection $\varphi = 0$ the strong (weak) shock becomes an irreversible normal shock (degenerates into an adiabatic sound wave).

the streamlines by the oblique shock of a perfect gas. The upstream sound speed c^- may be replaced (2.794b; 2.795c) ≡ (2.902) by the critical sound speed:

$$(\gamma+1)c_*^2 = 2(c^-)^2 + (\gamma-1)(v^-)^2, \tag{2.902}$$

in (2.901) leading to:

$$(v_y^+)^2 = (v^- - v_x^+)^2 \frac{v_x^+ v^- - c_*^2}{c_*^2 - v^- v_x^+ + 2(v^-)^2/(\gamma+1)}. \tag{2.903}$$

This is the equation of the strophoid or folium (Descartes 1638) that has 3 special points (subsection 2.8.7), including the normal shock and adiabatic sound wave.

2.8.7 NORMAL SHOCK AND SOUND WAVE

*The **oblique shock polar** (2.903) that is the polar for the downstream velocity in the oblique shock of a perfect gas is (Figure 2.36) a **strophoid** or **folium** (Descartes 1638):*

$$(v_y^+)^2 = (v_x^+ - v_2)^2 \frac{v_x^+ - v_1}{v_3 - v_x^+}, \tag{2.904}$$

with three special points: (i/ii) two (2.905b, c) are intersections with the v_x - axis (2.905a); (iii) the third (2.905e) is an asymptote parallel to the v_y – axis (2.905d):

$$v_y^+ = 0: \quad v_x^+ = \left\{ v_1 \equiv \frac{c_*^2}{v^-}, v_2 \equiv v^- \right\}; \quad v_y^+ = \infty: \quad v_x^+ = v_3 = \frac{c_*^2}{v^-} + \frac{2v^-}{(\gamma+1)}. \tag{2.905a–e}$$

Since the flow upstream is supersonic in terms of critical Mach number (2.906a) the **minimum point** (2.905b) ≡ (2.906b): (i) is the closest to the origin on the v_x^+ - axis (2.906c–f) and is a single root of (2.904):

$$v^- > c_*: \quad v_1 \equiv \frac{c_*^2}{v^-} = \left(\frac{c_*}{v^-}\right)^2 v^- = \frac{v^-}{(M^-)^2} < v^- < v_2; \tag{2.906a–f}$$

$$v_y^+ = 0: \quad v_x^+ = v^+ = v_1, \quad v^- v^+ = v^- v_1 = c_*^2, \quad v_t = 0, \tag{2.906g–k}$$

(ii) since the transverse velocity is zero (2.906g) follows (2.906g, h), and the relation (2.906i, j) ≡ (2.801c) by (2.889c) implies zero tangential velocity (2.906k). This corresponds to a normal shock, that is the strongest shock possible.

The **maximum point** on the v_x - axis (2.905c) $\equiv$ (2.907a) is: (i) a double root of (2.904) and corresponds (2.905a) $\equiv$ (2.907a) to (2.907b, c) to a continuous velocity, so there is no shock:

$$v_y^+ = 0: \quad v^+ = v_x^+ = v^-: \quad \theta = \frac{\pi}{2}, \quad v_t = 0, \tag{2.907a–e}$$

$$v_n^+ = v^+ = v^- = v_n^- = c_* = c^- = c^+. \tag{2.907f–k}$$

(ii) from (2.907c) and (2.907a) it follows that the velocity is normal (2.907d) and hence the tangential velocity is zero (2.907e); (iii) the velocities are normal and equal (2.907f–i) to the critical sound speed (2.889b), that also equals (2.465a, 2.445a, b) the local sound speeds (2.907j, k). This corresponds to an adiabatic **acoustic wave**, that is the weakest possible shock, in the limit of constant entropy.

The third point (2.905e) $\equiv$ (2.908a) is an **asymptote** (2.905d) that lies (2.902) beyond (2.908b–e) the farthest point on the v_x^+ - axis:

$$v_3 = \frac{c_*^2}{v^-} + \frac{2v^-}{\gamma+1} = \frac{2}{\gamma+1}\frac{\left(c^-\right)^2}{v^-} + \frac{\gamma-1}{\gamma+1}v^- + \frac{2v^-}{\gamma+1} \tag{2.908a, b}$$

$$= \frac{2}{\gamma+1}\frac{\left(c^-\right)^2}{v^-} + v^- = \frac{2}{\gamma+1}\frac{\left(c^-\right)^2}{v^-} + v_2 > v_2. \tag{2.908c–e}$$

The entropy is zero at the maximum point (2.907a–k) on the v_x^+ - axis that corresponds to an adiabatic acoustic wave, and beyond the entropy would be negative, implying that the branches $v_2 < v_x^+ < v_3$ contradict the second principle of thermodynamics, and are unphysical (dotted lines in Figure 2.36a). Thus *the physical range for oblique shocks $v_1 \le v \le v_2$ lies between the extremes of the normal shock (2.906a–k) [acoustic wave (2.907a–k)]*. For every angle of deflection intersecting the oblique shock polar there are two possibilities, namely an oblique strong (weak) shock closer to the normal shock (acoustic wave), up to a maximum deflection angle when the weak and strong shocks coincide (subsection 2.8.8).

2.8.8   WEAK/STRONG OBLIQUE SHOCKS AND MAXIMUM DEFLECTION

A straight line from the origin to the shock polar (Figure 2.36a) determines the **deflection angle** *φ of the downstream velocity relative to the upstream velocity (Figure 2.35a); there is a* **maximum deflection angle** *φ_{max} obtained taking a tangent to the shock polar through the origin (Figure 2.36b). The maximum transverse velocity occurs for a smaller angle. From (2.900a) it follows that the* **angle of incidence** *of the upstream velocity with the shock front (Figure 2.35a) corresponds to the perpendicular through the origin to the line joining the shock polar to the sound point (Figure 2.36a). For intermediate deflection angles (2.909a) there are two oblique* *shock waves, namely the **strong (weak) shock** closer to the normal shock v_1 (sound wave v_2):*

$$0 < \varphi < \varphi_{max}: \quad v_1 < \left(v_x^+\right)_s < \left(v_x^+\right)_w < v_2, \quad \left(v_y^+\right)_s < \left(v_y^+\right)_w, \quad \theta_s < \theta_w, \qquad \text{(2.909a–d)}$$

where the strong shock has: (i/ii) the smaller downstream velocity both longitudinal (2.909b) and transversal (2.909c); (iii) the smaller angle with the shock front (2.909d). The upstream velocity is always supersonic in both cases, and the downstream velocity is: (i) always subsonic for the strong shock; (ii) may be supersonic for the weak shock. The downstream flow is subsonic (supersonic) for the part of the shock polar lying within (2.910b) [outside (2.910d)] the sonic line (2.910c):

$$c_* = \sqrt{v_1 v_2}: \quad v^+ \equiv \left|\left(v_x^+\right)^2 + \left(v_y^+\right)^2\right|^{1/2} \begin{cases} < c_*: \text{subsonic downstream,} \\ = c_*: \text{sonic line,} \\ > c_*: \text{supersonic downstream,} \end{cases} \quad \text{(2.910a–d)}$$

the latter is (Figure 2.36a) the circle (2.910a) of radius equal to the geometric mean of the normal shock and sound wave points (2.905b, c), and thus cuts the shock polar between them. The qualitative properties of weak and strong oblique shocks of a perfect gas (subsections 2.8.5–2.8.8) are quantified next by relations across the oblique shock front (subsections 2.8.9–2.8.11).

2.8.9 Pressure, Mass Density, Velocity and Entropy Jumps

Replacing the Mach number M^- upstream of a normal shock by (Figure 2.35a) the normal component of the Mach number of an oblique shock (2.911a) extends from a normal to an oblique shock the relations (2.832c) → (2.911b), (2.834a, b, c) → (2.912a–e), (2.836a, b, c, f) → (2.913a–d), (2.837c) → (2.914):

$$M^- \to M^- \sin\theta: \quad \frac{p^+ - p^-}{p^-} = \frac{2\gamma}{\gamma+1}\left[\left(M^- \sin\theta\right)^2 - 1\right], \qquad \text{(2.911a, b)}$$

$$\frac{v_n^+ - v_n^-}{v_n^-} = \frac{\rho^- - \rho^+}{\rho^+} = \frac{\vartheta^+ - \vartheta^-}{\vartheta^-} = \frac{2}{\gamma+1}\left[\left(M^- \sin\theta\right)^{-2} - 1\right], \qquad \text{(2.912a–c)}$$

$$\left(\frac{c^+}{c^-}\right)^2 - 1 = \frac{T^+ - T^-}{T^-} = \frac{u^+ - u^-}{u^-} = \frac{h^+ - h^-}{h^-} \qquad \text{(2.913a–c)}$$

$$= \frac{2(\gamma-1)}{(\gamma+1)^2}\left[1 + \gamma\left(M^- \sin\theta\right)^2\right]\left[1 - \left(M^- \sin\theta\right)^{-2}\right], \qquad \text{(2.913d)}$$

$$\frac{\gamma-1}{R}\left(s^+ - s^-\right) = \log\left\{\frac{2\gamma\left(M^- \sin\theta\right)^2 + 1 - \gamma}{1+\gamma}\left[\frac{\gamma-1}{\gamma+1} + \frac{2}{\gamma+1}\left(M^- \sin\theta\right)^{-2}\right]^{-\gamma}\right\}. \quad \text{(2.914)}$$

*The relative changes across the oblique shock of a perfect gas for the pressure
(2.911b), velocity, mass density and specific volume v2.912a–c), sound speed, temperature, internal energy and enthalpy (2.913a–d) and entropy (2.914) are measures
of the shock strength, and increase monotonically along the shock polar from the
sound wave point (2.915c–k) on the Mach cone (2.882a) ≡ (2.915a, b):*

$$\theta_m = \tilde{\theta} \equiv \arcsin\left(\frac{1}{M^-}\right): \tag{2.915a, b}$$

$$\frac{p^+}{p^-} = \frac{v^+}{v^-} = \frac{\rho^+}{\rho^-} = \frac{\vartheta^+}{\vartheta^-} = \frac{c^+}{c^-} = \frac{T^+}{T^-} = \frac{u^+}{u^-} = \frac{h^+}{h^-} = \frac{s^+}{s^-} = 1, \tag{2.915c–k}$$

to the normal shock $\theta = 0 = \varphi$ that is the strongest shock. The relation between
upstream and downstream Mach number can also be generalized from the normal (to
the oblique) shock [Section 2.7.18 (2.8.10], and in the case of the oblique shock the
deflection angle is also specified by the upstream flow conditions (subsection 2.8.10).

2.8.10 DEFLECTION OF THE VELOCITY ACROSS AN OBLIQUE SHOCK

The angles of incidence (2.916a) and of deflection (2.916b) are related (Figure 2.35)
by (2.916c, d):

$$\tan\theta = \frac{v_n^-}{v_t}, \quad \tan(\theta-\varphi) = \frac{v_n^+}{v_t}: \quad \frac{\tan(\theta-\varphi)}{\tan\theta} = \frac{v_n^+}{v_n^-} = \frac{\rho^-}{\rho^+}. \tag{2.916a–d}$$

Substituting (II.5.48a) ≡ (2.761f) [(2.912b)] in the r.h.s. (l.h.s.) of (2.916d) leads to
(2.917a,b):

$$\frac{1-\tan\varphi\cot\theta}{1+\tan\varphi\tan\theta} = \frac{\rho^-}{\rho^+} = \frac{\gamma-1+2(M^-\sin\theta)^{-2}}{\gamma+1}, \tag{2.917a, b}$$

that is equivalent to (2.918a,b):

$$\cot\varphi\cot\theta = \frac{\rho^- + \rho^+\cot^2\theta}{\rho^+ - \rho^-} = \frac{1}{2}\frac{\gamma+1}{\sin^2\theta-(M^-)^{-2}} - 1. \tag{2.918a, b}$$

Thus *the angle of deflection of the velocity (Figure 2.35a) across an oblique shock of
a perfect gas is given by (2.918b) as a function of the upstream Mach number and
angle of incidence. The deflection angle (Figure 2.37): (i/ii) is zero for a normal
shock (2.919b) and (2.919c) also for an acoustic wave (2.915a); (iii) hence the maximum deflection (2.919a) occurs for an intermediate angle of incidence (2.919d):*

$$\varphi_m \equiv \varphi(\theta_m): \quad \varphi\left(\frac{\pi}{2}\right) = 0 = \varphi(\tilde{\theta}), \quad \tilde{\theta} < \theta_m < \frac{\pi}{2}. \tag{2.919a–d}$$

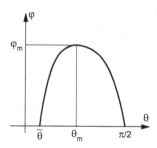

FIGURE 2.37   The angle of deflection φ of the velocity (Figure 2.36a) in an oblique shock wave (Figure 2.35a) depends on the angle of incidence θ, and goes through a maximum φ_m for an angle of incidence $\theta_m > \tilde{\theta}$ within the Mach cone and less than $\pi/2 = 90°$.

Concerning (i) for normal incidence $\theta = \pi/2$ then $\cot\theta = 0$, and since the r.h.s. of (2.918b) is not zero, then $\cot\varphi = \infty$ implying zero deflection $\varphi = 0$; concerning (ii) at the Mach angle (2.915a) upstream $M^- \sin\tilde{\theta} = 1$ the r.h.s. of (2.918b) is infinite, and since $\cot\tilde{\theta}$ is finite then $\cot\varphi = \infty$ implying again zero deflection. The upstream Mach number and angle of incidence specify all flow variables for an oblique shock, including the angle of deflection of velocity (subsection 2.8.9) and also the downstream Mach number (subsection 2.8.11).

2.8.11   RELATION BETWEEN UPSTREAM AND DOWNSTREAM MACH NUMBERS

The transformation from a normal to an oblique shock corresponds to substituting the upstream M^- (downstream M^+) local Mach number by their normal components (2.920a) [(2.920b)]:

$$M^- \rightarrow M_n^- = M^- \sin\theta, \quad M^+ \rightarrow M_n^+ = M^+ \sin(\theta - \varphi); \qquad (2.920a, b)$$

for example, the normal shock relation (2.840a) with the substitutions (2.920a, b) specifies the normal downstream local Mach number (2.921a, b):

$$\left(M_n^+\right)^2 = \left(M^+\right)^2 \sin^2(\theta - \varphi) = \frac{2 + (\gamma - 1)\left(M^- \sin\theta\right)^2}{1 - \gamma + 2\gamma\left(M^- \sin\theta\right)^2}. \qquad (2.921a, b)$$

The tangential downstream Mach number (2.922a) is given (2.892c, a) by (2.922b–d):

$$\left(M_t^+\right)^2 \equiv \left(M^+\right)^2 \cos^2(\theta - \varphi) = \left(\frac{v_t}{c^+}\right)^2 = \left(\frac{v^-}{c^+}\right)^2 \cos^2\theta \qquad (2.922a–c)$$

$$= \left(M^-\right)^2 \left(\frac{c^-}{c^+}\right)^2 \cos^2 \theta, \qquad (2.922d)$$

and substitution of (2.913a) or (2.836a; 2.911a) leads to (2.922e):

$$\left(M_t^+\right)^2 = \frac{\left[(\gamma+1)M^-\cos\theta\right]^2}{\left[1-\gamma+2\gamma\left(M^-\sin\theta\right)^2\right]\left[\gamma-1+2\left(M^-\sin\theta\right)^{-2}\right]}. \qquad (2.922e)$$

Adding (2.921b) and (2.922e) specifies the total downstream local Mach number (2.923):

$$\left[1-\gamma+2\gamma\left(M^-\right)^2\sin^2\theta\right]\left(M^+\right)^2 = 2+(\gamma-1)\left(M^-\sin\theta\right)^2 + \frac{(\gamma+1)^2\left(M^-\cos\theta\right)^2}{\gamma-1+\left(M^-\sin\theta\right)^{-2}}, \qquad (2.923)$$

that for a normal shock $\theta = \pi/2$ simplifies to (2.840a). Thus *the downstream normal (2.921a,b), tangential (2.922a–e) and total (2.923) local Mach numbers are specified by the upstream local Mach number and angle-of-incidence, that also appear in the deflection angle (2.918b) for the oblique shock of a perfect gas.* The deflection angle is calculated (Table 2.20) next (subsection 2.8.12) for a wide range of upstream Mach numbers and angles of incidence.

TABLE 2.20
Deflection Angle of the Velocity across an Oblique Shock

Angle of Incidence: φ	$M^- = 1.2$	1.5	2	2.5	3	5	10	30
				Upstream Mach Number				
$\tilde{\theta}$ = arcsin (1/M^-)	56.443°	41.810°	30.000°	23.578°	19.471°	11.537°	5.739°	1.910°
$\theta_m = 10°$	-	-	-	-	-	-	5.533°	8.007°
20°	-	-	-	-	0.773°	10.665°	15.051°	16.364°
30°	-	-	0.000°	7.994°	12.774°	20.174°	23.413°	24.382°
40°	-	-	10.623°	17.675°	21.846°	28.275°	31.090°	31.933°
50°	-	6.450°	18.130°	24.854°	28.860°	35.087°	37.876°	38.662°
60°	1.605°	11.157°	22.411°	29.180°	33.320°	39.915°	42.885°	43.785°
70°	3.882°	11.840°	22.115°	28.886°	33.277°	40.680°	44.180°	45.260°
80°	3.006°	7.820°	14.807°	20.103°	23.927°	31.254°	35.174°	36.450°

Note: Deflection angle of the velocity across an oblique shock of a diatomic perfect gas as a function of: (i) the upstream local Mach number M^- with eight values $M^- = 1.2, 1.5, 2, 2.5, 3, 5, 10, 30$ up to orbital re-entry speed, specifying the Mach angle; (ii) eight values of the angle of incidence $\theta = 10°, 20°, 30°, 40°, 50°, 60°, 70°, 80°$ of which are possible only those $\theta^- > \tilde{\theta}$ for incidence within the Mach cone.

2.8.12 RELATION BETWEEN THE ANGLES OF INCIDENCE AND DEFLECTION

Table 2.20 indicates the angle of deflection of the velocity across an oblique shock of a perfect gas for all combinations of a wide range of: (i) eight values of the upstream Mach number (2.924a) ranging from low supersonic to re-entry in the Earth's atmosphere (2.691a–g) but not considering effects (2.692a–d) beyond a perfect gas; (ii) eight angles of incidence (2.924b) spaced 10° between near normal $\theta = 80°$ and almost tangential $=10°$:

$$M^- = 1.2, 1.5, 2, 2.5, 3, 5, 10, 30, \quad \theta = 10, 20, 30, 40, 50, 60, 70, 80. \qquad \text{(2.924a, b)}$$

The maximum deflection $\varphi = \varphi_m$ in Figure 2.36b corresponds to the coincidence of weak (strong) shocks $\varphi \geq \varphi_m$ ($\varphi \leq \varphi_m$). The possible values of the angles of incidence are above (2.896c–e) the Mach angles (2.882a, b) $\equiv$ (2.925) for the local Mach numbers (2.924a):

$$\tilde{\theta} = \arcsin\left(\frac{1}{M^-}\right)$$

$$= \{56.443°, 41.810°, 30.000°, 23.578°, 19.471°, 11.537°, 5.739°, 1.910°\}. \qquad \text{(2.925)}$$

Thus Table 2.20 omits for each upstream local Mach number M^- the values of the angle of incidence $\theta^- < \tilde{\theta}$ below (2.925). For each upstream local Mach number M^- there is a maximum value φ_m of the angle of deflection of the velocity across the oblique shock and the corresponding angle of incidence θ_m. The preceding results for an oblique shock or arbitrary strength (subsections 2.8.2–2.8.12) simplify in the limit of a shock [subsections 2.8.13–2.8.15 (2.8.16–2.8.18)] with large upstream Mach number (small angle of deflection of the velocity across the shock).

2.8.13 SHOCK WITH LARGE UPSTREAM MACH NUMBER

The oblique shock polar can be written in the form (2.901) $\equiv$ (2.926b) in terms of the upstream local Mach number (2.926a):

$$M^- \equiv \frac{v^-}{c^-}: \quad \left(v_y^+\right)^2 = \left(v^- - v_x^+\right)^2 \frac{(\gamma+1)v_x^+ - (\gamma-1)v^- - 2\dfrac{v^-}{\left(M^-\right)^2}}{(\gamma+1)\left(v^- - v_x^+\right) + 2\dfrac{v^-}{\left(M^-\right)^2}}. \qquad \text{(2.926a, b)}$$

The maximum deflection of the velocity across the oblique shock front of a perfect gas (2.918b) occurs for large upstream Mach number (2.927a) when the shock polar (2.926b) simplifies to (2.927b):

$$\left(M^-\right)^2 \gg 1 \quad \left(v_y^+\right)^2 = \left(v^- - v_x^+\right)\left(v_x^+ - \frac{\gamma-1}{\gamma+1}v^-\right). \qquad \text{(2.927a, b)}$$

Rewriting the r.h.s. of (2.927a,b) in the form (2.928a,b):

$$\left(v_y^+\right)^2 = -\left(v_x^+\right)^2 + \frac{2\gamma}{\gamma+1}v^-v_x^+ - \frac{\gamma-1}{\gamma+1}\left(v^-\right)^2 = -\left(v_x^+ - \frac{\gamma}{\gamma+1}v^-\right)^2 + \left(\frac{v^-}{\gamma+1}\right)^2, \quad (2.928a, b)$$

it follows that for large upstream Mach number (2.927a) ≡ (2.929a) the shock polar (2.927b) ≡ (2.928a,b) ≡ (2.929b) becomes a circle (Figure 2.36b) with centre at (2.929c) and radius (2.929d):

$$\left(M^-\right)^2 \gg 1: \quad \left(v_y^+\right)^2 + \left(v^{+x} - v_0\right)^2 = \tilde{v}^2, \quad v_0 \equiv \frac{\gamma}{\gamma+1}v^- > \frac{v^-}{\gamma+1} \equiv \tilde{v}. \quad (2.929a\text{–}d)$$

The circular oblique shock polar in the limit of high upstream Mach number lies in the range (2.495a–d):

$$v_1 \equiv \frac{\gamma-1}{\gamma+1}v^- = v_0 - \tilde{v} \le v_x^+ \le v_0 + \tilde{v} = v^- \equiv v_2, \quad (2.930a\text{–}f)$$

with the normal shock (sound wave) at the lower (upper) end, because the downstream velocity is smaller than (equal to) the upstream velocity.

2.8.14   CIRCULAR POLAR FOR A HIGH MACH NUMBER SHOCK

The maximum deflection angle is given (Figure 2.38) by (2.931a):

$$\varphi_m = \arcsin\left(\frac{\tilde{v}}{v_0}\right) = \arcsin\left(\frac{1}{\gamma}\right) = \arcsin\left(\frac{f}{f+2}\right) = \arcsin\left(\frac{3}{5}, \frac{5}{7}, \frac{3}{4}\right) \quad (2.931a\text{–}e)$$
$$= \{36.870°, 45.585°, 48.590°\},$$

leading (2.929c, d) to (2.931b), and (2.931c–e), respectively, for a monoatomic/ diatomic/polyatomic perfect gas (2.475a–c). The sonic line (2.910a) ≡ (2.932a) is (2.930b, d) a circle with centre at the origin and radius (2.932b–g):

$$c_* = \sqrt{v_1 v_2} = \left|\left(v_0\right)^2 - \left(\tilde{v}\right)^2\right|^{1/2} = \frac{v^-}{\gamma+1}\sqrt{\gamma^2-1} = v^-\sqrt{\frac{\gamma-1}{\gamma+1}} = \frac{v^-}{\sqrt{1+f}} \quad (2.932a\text{–}i)$$
$$= v^-\left\{\frac{1}{2}, \frac{1}{\sqrt{6}}, \frac{1}{\sqrt{7}}\right\} = v^-\{0.50000, 0.40825, 0.37796\} = bv^- = v^-\sqrt{a},$$

involving (2.932h, i) the constant (2.889c) ≡ (2.816a–f). For a large upstream Mach number (2.929a) the sonic circle (2.932b) ≡ (2.933a) intersects the circular polar at the point of maximum deflection (Figure 2.36b), whose coordinates correspond to the downstream velocity (2.933b–d):

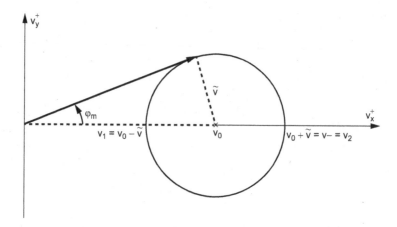

FIGURE 2.38   In the case of an oblique shock (Figure 2.35a) with a large upstream Mach number (Figure 2.36b) the locus of the downstream velocity is a circle specified by its centre v_0 and radius $\tilde{v}$.

$$(v_0)^2 = (c_*)^2 + (\tilde{v})^2 : \quad \vec{v}_m^+ = c_* \{\cos\varphi_m, \ \sin\varphi_m\}$$

$$= v^- \sqrt{\frac{\gamma-1}{\gamma+1}} \{\cos\varphi_m, \sin\varphi_m\} = v^- \, b \{\cos\varphi_m, \ \sin\varphi_m\}. \qquad (2.933\text{a–d})$$

The shock front lies on the perpendicular through the origin to the line joining the sound wave point $(v_2, 0)$ to the shock polar at φ_m, implying (Figure 2.36b) that is the angle of incidence corresponding to maximum deflection (2.931a–e) is (2.934a, b):

$$\cos\theta_m = \sin\left(\frac{\pi}{2} - \theta_m\right) = \frac{c_* \sin\varphi_m}{v_2 - c_* \cos\varphi_m} = \frac{b \sin\varphi_m}{1 - b \cos\varphi_m} = \frac{b}{\csc\varphi_m - b \cot\varphi_m}. \quad (2.934\text{a–d})$$

Substituting (2.931b) ≡ (2.935a) and (2.935b–d) together with (2.932h) and (2.935e):

$$\sin\varphi_m = \frac{1}{\gamma}: \quad \cos\varphi_m = \sqrt{1 - \sin^2\varphi_m} = \sqrt{1 - \frac{1}{\gamma^2}} = \frac{\sqrt{\gamma^2 - 1}}{\gamma}, \quad b = \sqrt{\frac{\gamma-1}{\gamma+1}}, \quad (2.935\text{a–e})$$

in (2.934c) leads to (2.501a, b):

$$\cos\theta_m = \frac{\dfrac{1}{\gamma}\sqrt{\dfrac{\gamma-1}{\gamma+1}}}{1 - \sqrt{\dfrac{\gamma-1}{\gamma+1}}\dfrac{\sqrt{\gamma^2-1}}{\gamma}} = \frac{\sqrt{\dfrac{\gamma-1}{\gamma+1}}}{\gamma - (\gamma-1)} \sqrt{\dfrac{\gamma-1}{\gamma+1}} = b. \qquad (2.936\text{a–d})$$

Thus the incidence angle for maximum deflection (2.931a–c) in the strong (2.937a) oblique shock of a, respectively, for monoatomic/diatomic/polyatomic perfect gas is given (2.936d; 2.932h) by (2.502a–e):

$$\theta_m = \arccos\sqrt{\frac{\gamma-1}{\gamma+1}} = \arccos\left(\sqrt{\frac{1}{1+f}}\right) = \arccos\left(\frac{1}{2}, \frac{1}{\sqrt{6}}, \frac{1}{\sqrt{7}}\right) \qquad (2.937a\text{--}d)$$

$$= 60.000°, 65.905°, 67.792°.$$

It has been shown that *in the limit (2.926a; 2.927a) of large upstream Mach number (Figure 2.36b): (i) the oblique shock polar is a circle (2.928b) ≡ (2.929b) with centre at (2.929c) and radius (2.929d) passing through normal shock and sound wave; (ii) the sonic circle (2.932a–i) has centre at the origin and radius such that it passes through the point of greatest deflection (2.933a–c); (iii) the maximum deflection angle (2.937a–d) corresponds (2.934a–d) to the angles of incidence (2.937a–d), respectively, for monoatomic, diatomic and polyatomic perfect gases.* The limit of large upstream Mach number (subsections 2.8.13–2.8.14) also simplifies the relations between flow variables across an oblique shock (subsection 2.8.15).

2.8.15 RELATIONS ACROSS A HIGH MACH NUMBER SHOCK

The ratios of pressures (2.911b) ≡ (2.938), normal velocities, mass densities and specific volumes (2.912a–c) ≡ (2.939a–c) and of sound speeds, temperatures, internal energies and enthalpies (2.913a–d) ≡ (2.940a, b):

$$\frac{p^+}{p^-} = \frac{1-\gamma+2\gamma\left(M^-\right)^2\sin^2\theta}{1+\gamma}, \qquad (2.938)$$

$$\frac{v_n^-}{v_n^+} = \frac{\rho^+}{\rho^-} = \frac{\vartheta^-}{\vartheta^+} = \frac{\gamma+1}{\gamma-1+2\left(M^-\sin\theta\right)^{-2}}, \qquad (2.939a\text{--}c)$$

$$\left(\frac{c^+}{c^-}\right)^2 = \frac{T^+}{T^-} = \frac{u^+}{u^-} = \frac{h^+}{h^-} = \frac{\left[1-\gamma+2\gamma\left(M^-\sin\theta\right)^2\right]\left[\gamma-1+2\left(M^-\sin\theta\right)^{-2}\right]}{\left(\gamma+1\right)^2}, \qquad (2.940a\text{--}d)$$

across an oblique shock of a perfect gas, simplify, respectively: (i) to (2.832b)/(2.834a, b, d)/(2.836a, b, c, f) for a normal shock θ = π/2 with arbitrary upstream Mach number; (ii) to (2.941b)/(2.942a–f)/(2.943a–f) for an oblique shock with large upstream normal Mach number (2.941a):

$$\left(M^-\sin\theta\right)^2 \gg 1: \quad \frac{p^+}{p^-} = \frac{2\gamma}{\gamma+1}\left(M^-\right)^2\sin^2\theta, \qquad (2.941a, b)$$

$$\frac{v_n^-}{v_n^+} = \frac{\rho^+}{\rho^-} = \frac{\vartheta^-}{\vartheta^+} = \frac{\gamma+1}{\gamma-1} = \frac{1}{a} = f+1 = \{4, 6, 7\}, \qquad (2.942a\text{--}f)$$

$$\left(\frac{c^+}{c^-}\right)^2 = \frac{T^+}{T^-} = \frac{u^+}{u^-} = \frac{h^+}{h^-} = \frac{2\gamma(\gamma-1)}{(\gamma+1)^2}\left(M^-\right)^2\sin^2\theta = \frac{\gamma-1}{\gamma+1}\frac{p^+}{p^-} = a\frac{p^+}{p^-}, \quad (2.943a\text{–}f)$$

where (2.941b) was used in (2.943e, f).

Thus *across an oblique shock of a perfect gas with large local upstream normal Mach number (2.941a): (i) the ratio of velocities, mass densities and specific volumes is a constant (2.942a–f); (ii) the ratio of sound speeds squared, temperatures, internal energies and enthalpies (2.943a–f) scales on the ratio of pressures (2.941b) that is large (2.941a). Also the local downstream normal (2.921a,b)/tangential (2.922a–e)/total (2.923) Mach numbers simplify, respectively, to (2.944b)/(2.944c)/ (2.944d) in the same approximation (2.944a):*

$$\left(M^-\sin\theta\right)^2 \gg 1: \quad \left(M_n^+\right)^2 = \frac{\gamma-1}{2\gamma}, \quad \left(M_t^+\right)^2 = \frac{(\gamma+1)^2}{2\gamma(\gamma-1)}\cot^2\theta, \quad (2.944a\text{–}c)$$

$$\left(M^+\right)^2 = \frac{1}{2\gamma}\left[\gamma-1+\frac{(\gamma+1)^2}{\gamma-1}\cot^2\theta\right], \quad (2.944d)$$

where appear the constant factors (2.945a–d):

$$\frac{\gamma-1}{2\gamma} = \frac{1}{f+2} = \left\{\frac{1}{5},\frac{1}{7},\frac{1}{8}\right\}, \quad \frac{(\gamma+1)^2}{2\gamma(\gamma-1)} = \frac{(f+1)^2}{f+2} = \left\{\frac{16}{5},\frac{36}{7},\frac{49}{8}\right\}, \quad (2.945a\text{–}d)$$

respectively, for monoatomic/diatomic/polyatomic perfect gases as in (2.942f). The oblique shock equations simplify assuming a small deviation of the velocity (subsection 2.8.15), that can occur (Figure 2.36a) in two cases: (i) close to the downstream velocity v_2 corresponding to acoustic waves, that are adiabatic, that is for an incidence angle close to the Mach angle (subsections 2.8.16–2.8.18); (ii) close to the downstream velocity v_1 corresponding to a normal shock, for example, a shock turning gradually over a smooth curved wall, leading the Prandtl – Meyer (1908) function (subsections 2.8.19–2.8.21).

2.8.16   OBLIQUE SHOCK WITH SMALL DEFLECTION OF VELOCITY

The relation between the angles of incidence θ and deflection φ for the oblique shock of a perfect gas (2.918b) simplifies to (2.946b–d) for small deflections (2.946a):

$$\varphi^2 \ll 1: \quad \frac{\gamma+1}{2}\frac{\left(M^-\right)^2}{\left(M^-\sin\theta\right)^2-1} = 1+\cot\theta\cot\varphi \sim 1+\frac{\cot\theta}{\varphi} \sim \frac{\cot\theta}{\varphi}, \quad (2.946a\text{–}d)$$

and (2.946a, d) $\equiv$ (2.947a, b) can be re-written:

$$\varphi^2 \ll 1: \quad \left(M^-\right)^2 \sin^2\theta - 1 = \varphi\frac{\gamma+1}{2}\left(M^-\right)^2 \tan\theta. \qquad (2.947a, b)$$

Thus *in the oblique shock of a perfect gas the deflection angle φ is related to the incidence angle by (2.946b) ≡ (2.918b), that simplifies to (2.946c, d) ≡ (2.947b) for small deflection angle (2.946a) ≡ (2.947a); if in addition the upstream Mach number is large (2.948a, b) the incidence and deflection angles are in a constant ratio (2.948c)*

$$\varphi^2 \ll 1 \ll \left(M^-\theta\right)^2: \quad \frac{\varphi}{\theta} = \frac{2}{\gamma+1} = \frac{f}{f+1} = \left\{\frac{3}{4},\frac{5}{6},\frac{6}{7}\right\} = \{0.750, 0.833, 0.857\}, \quad (2.948a\text{--}f)$$

respectively, for monoatomic/diatomic/polyatomic perfect gases (2.948d–f). The deflection angle is small in the nearly adiabatic case of incidence angle close to the Mach angle, that is considered next (subsection 2.8.17).

2.8.17   INCIDENCE ANGLE CLOSE TO THE MACH ANGLE

The Mach angle (2.881f) ≡ (2.882a) corresponds to (2.549a–c):

$$\tan\tilde\theta = \frac{\sin\tilde\theta}{\cos\tilde\theta} = \frac{\sin\tilde\theta}{\sqrt{1-\sin^2\tilde\theta}} = \frac{1}{\sqrt{\left(M^-\right)^2-1}}. \qquad (2.949a\text{--}c)$$

The small difference (2.950a) between the incidence angle θ and Mach angle (2.882a) satisfies (2.950b–f):

$$\left(\theta-\tilde\theta\right)^2 \ll 1: \quad \left(M^-\right)^2 \sin^2\theta = \left(M^-\right)^2 \sin^2\left[\tilde\theta + \left(\theta-\tilde\theta\right)\right]$$
$$= \left(M^-\right)^2 \left[\sin\tilde\theta\cos\left(\theta-\tilde\theta\right) + \cos\tilde\theta\sin\left(\theta-\tilde\theta\right)\right]^2$$
$$= \left(M^-\right)^2 \sin^2\tilde\theta\left[1 + \left(\theta-\tilde\theta\right)\cot\tilde\theta\right]^2$$
$$= 1 + 2\left(\theta-\tilde\theta\right)\cot\tilde\theta = 1 + 2\left(\theta-\tilde\theta\right)\sqrt{\left(M^-\right)^2-1}. \; (2.950a\text{--}f)$$

Substituting (2.950a, f) in (2.947a,b) specifies *the relation (2.951c) between the small angle of deflection (2.951a) ≡ (2.947a) and the small deviation between the incidence and Mach angle (2.951b) ≡ (2.950a):*

$$\varphi^2 \ll 1 \gg \left(\theta-\tilde\theta\right)^2: \quad \theta-\tilde\theta = \varphi\frac{\gamma+1}{4}\left(M^-\right)^2 \tan^2\tilde\theta = \varphi\frac{\gamma+1}{4}\frac{\left(M^-\right)^2}{\left(M^-\right)^2-1}, \quad (2.951a\text{--}d)$$

with (2.949c) leading to (2.951d).
The relative pressure change (2.911b) is given (2.950f; 2.951c) by (2.952a,b):

$$\frac{p^+ - p^-}{p^-} = \frac{4\gamma}{\gamma + 1}\left(\theta - \tilde{\theta}\right)\sqrt{\left(M^-\right)^2 - 1} = \varphi\gamma\frac{\left(M^-\right)^2}{\sqrt{\left(M^-\right)^2 - 1}}. \qquad (2.952a, b)$$

The entropy change for the weak shock of a perfect gas (2.873b) scales on the cube of the small deflection angle:

$$s^+ - s^- = \bar{R}\gamma\frac{\gamma + 1}{12}\varphi^3\left(M^-\right)^3\left[1 - \frac{1}{\left(M^-\right)^2}\right]^{-\frac{3}{2}}. \qquad (2.953)$$

The relations between flow variables across the oblique shock of a perfect gas for small deflection of the velocity are obtained next (subsection 2.8.18).

2.8.18   RELATIONS BETWEEN FLOW VARIABLES AND DEFLECTION ANGLE

The ratio between the flow velocity after (2.892c, d) and before (2.892a, b) the oblique shock is given by (2.954a–c):

$$\left(\frac{v^+}{v^-}\right)^2 = \frac{\left(v_n^+\right)^2 + \left(v_t\right)^2}{\left(v_n^-\right)^2 + \left(v_t\right)^2} = \frac{1 + \tan^2\left(\theta - \varphi\right)}{1 + \tan^2\theta} = \frac{\cos^2\theta}{\cos^2\left(\theta - \varphi\right)}. \qquad (2.954a\text{–}c)$$

For small deviation between the incidence and Mach angles (2.950f) the numerator of (2.954c) is evaluated by (2.955a–d):

$$\left(\frac{v_t}{v^-}\right)^2 = \cos^2\theta = 1 - \sin^2\theta = 1 - \frac{1}{\left(M^-\right)^2} - 2\frac{\theta - \tilde{\theta}}{\left(M^-\right)^2}\sqrt{\left(M^-\right)^2 - 1}$$

$$= \frac{\left(M^-\right)^2 - 1}{\left(M^-\right)^2}\left[1 - 2\frac{\theta - \tilde{\theta}}{\sqrt{\left(M^-\right)^2 - 1}}\right]. \qquad (2.955a\text{–}d)$$

An expression similar to (2.955d) holds for the denominator of (2.954c) replacing θ by $\theta - \varphi$, so that the ratio of velocities after and before the shock is given by (2.956b) for small deviations (2.956a):

$$\left(\theta - \tilde{\theta}\right)^2 \ll 1: \quad \left(\frac{v^+}{v^-}\right)^2 = \left[1 - 2\frac{\theta - \tilde{\theta}}{\sqrt{\left(M^-\right)^2 - 1}}\right]\left[1 - 2\frac{\theta - \varphi - \tilde{\theta}}{\sqrt{\left(M^-\right)^2 - 1}}\right]^{-1} \qquad (2.956a\text{–}c)$$

$$= \left[1 - 2 \frac{\theta - \tilde{\theta}}{\sqrt{\left(M^-\right)^2 - 1}} \right] \left[1 + 2 \frac{\theta - \varphi - \tilde{\theta}}{\sqrt{\left(M^-\right)^2 - 1}} \right], \qquad (2.956d)$$

implying (2.956a, c) ≡ (2.957a, b) and hence (2.957c):

$$\left(\theta - \tilde{\theta}\right)^2 \lll : \quad \left(\frac{v^+}{v^-}\right)^2 = 1 - \frac{2\varphi}{\sqrt{\left(M^-\right)^2 - 1}} \quad \Leftrightarrow \quad \frac{v^+}{v^-} = 1 - \frac{\varphi}{\sqrt{\left(M^-\right)^2 - 1}}. \qquad (2.957a\text{--}c)$$

Thus have been obtained *the ratio of velocities (2.957c) and pressures (2.952a,b), and the incidence angle (2.946a, b; 2.948a–f) and its deviation (2.951a–d) from the Mach angle (2.881f; 2.882a–d) as a function of the upstream Mach number and velocity deflection φ; the latter always appears linearly, except in the entropy change (2.953) where it appears to the cube.* The small deflection of the velocity across the oblique shock of a perfect gas also occurs close to the smaller down-stream velocity v_1 in the Figure 2.36a corresponding to a nearly normal shock turn-ing gradually (Figure 2.39a) along a smooth gently curved concave wall (subsections 2.8.19–2.8.21).

2.8.19 GRADUAL ROTATION OF A FRONT ALONG A SMOOTH CONCAVE WALL

The relative change of velocity for the nearly normal oblique shock of a perfect gas with small deflection angle (2.957c) ≡ (2.958a) can be put in the differential form (2.958b):

$$\frac{v^+ - v^-}{v^+} = -\frac{\varphi}{\sqrt{M^2 - 1}} \quad \Leftrightarrow \quad -d\varphi = \sqrt{M^2 - 1}\, \frac{dv}{v}. \qquad (2.958a, b)$$

The local Mach number (2.959a) leads by logarithmic differentiation (2.959b) to (2.959c):

$$M \equiv \frac{v}{c}: \quad \log M = \log v - \log c \Rightarrow \frac{dM}{M} = \frac{dv}{v} - \frac{dc}{c}, \qquad (2.959a\text{--}c)$$

and likewise for the logarithmic differentiation (2.960c, d) of the local sound speed (2.661a–d) ≡ (2.960a, b):

$$2c_0^2 = 2c^2 + \left(\gamma - 1\right)v^2 = c^2\left[2 + \left(\gamma - 1\right)M^2\right]:$$

$$\log 2 + 2\log c_0 = 2\log c + \log\left[2 + \left(\gamma - 1\right)M^2\right] \Rightarrow \frac{dc}{c} = -\frac{\left(\gamma - 1\right)M\, dM}{2 + \left(\gamma - 1\right)M^2}. \qquad (2.960a\text{--}d)$$

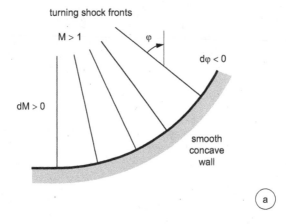

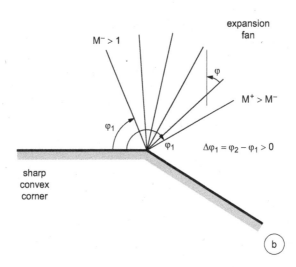

FIGURE 2.39 A supersonic flow in a sharp convex corner (Figure 2.35b) corresponds to an expansion fan (Figure 2.39b); the case of a strong oblique shock at a sharp concave corner (Figure 2.35a) leads to a gradual turn of a weak oblique shock along a smooth curved concave corner (Figure 2.39a). In both cases (a) and (b) given the initial Mach number M^- and angle φ^- the final Mach number M^+ specifies the final angle φ^+ or vice-versa.

Substitution of (2.959c) [(2.960d)] in (2.958b) determines the gradual change of angle of deflection (2.961a) [(2.961b)] that simplifies to (2.961c):

$$-d\varphi = \sqrt{M^2-1}\left(\frac{dM}{M}+\frac{dc}{c}\right) = \sqrt{M^2-1}\left[\frac{dM}{M}-\frac{(\gamma-1)M\,dM}{2+(\gamma-1)M^2}\right]$$

$$= \sqrt{M^2-1}\left(1+\frac{\gamma-1}{2}M^2\right)^{-1}\frac{dM}{M}.$$

$$(2.961a\text{–}c)$$

The integration of (2.961c) uses the exact differential (II.7.124a) ≡ (2.962a,b):

$$d\left\{\arctan\left(\sqrt{M^2-1}\right)\right\} = \frac{1}{M^2-1+1}d\left(\sqrt{M^2-1}\right) = \frac{dM}{M\sqrt{M^2-1}}, \qquad (2.962a, b)$$

added to both sides of (2.961c), leading to (2.963a–c):

$$-d\varphi + d\left[\arctan\left(\sqrt{M^2-1}\right)\right] = \frac{dM}{M\sqrt{M^2-1}}\left[1+\frac{2\left(M^2-1\right)}{2+(\gamma-1)M^2}\right]$$

$$= \frac{(\gamma+1)M}{2+(\gamma-1)M^2}\frac{dM}{\sqrt{M^2-1}} = \sqrt{\frac{\gamma+1}{\gamma-1}}\,d\left\{\arctan\left[\sqrt{\frac{\gamma-1}{\gamma+1}\left(M^2-1\right)}\right]\right\}, \qquad (2.963a–c)$$

where is used (2.963c) ≡ (2.964a, b):

$$d\left\{\arctan\left[\sqrt{\frac{\gamma-1}{\gamma+1}\left(M^2-1\right)}\right]\right\} = \left[1+\frac{\gamma-1}{\gamma+1}\left(M^2-1\right)\right]^{-1}d\left\{\sqrt{\frac{\gamma-1}{\gamma+1}\left(M^2-1\right)}\right\}$$

$$= \sqrt{\frac{\gamma-1}{\gamma+1}}\frac{\gamma+1}{2+(\gamma-1)M^2}\frac{M\,dM}{\sqrt{M^2-1}}. \qquad (2.964a, b)$$

From (2.963a–c) follows the Prandtl – Meyer (1908) function (subsection 2.8.20).

2.8.20 SUPERSONIC EXPANSION FAN IN A CONVEX CORNER (PRANDTL – MEYER 1908)

It has been shown (2.963c) ≡ (2.965) that *for a perfect gas in an adiabatic turn over a smooth wall (Figure 2.39) the Mach number is related to the deflection angle by the* **Prandtl – Meyer (1908) function**:

$$\varphi(M) = \arctan\left[\sqrt{M^2-1}\right] - a^{-1/2}\arctan\left[\sqrt{a\left(M^2-1\right)}\right], \qquad (2.965)$$

where a is the constant (2.816b–f). As the Mach number varies from 1 to ∞ in (2.966a) the deflection angle reaches a maximum (2.966b–g):

$$1 \le M \le \infty \quad 0 \le -\varphi \le -\varphi_{max} = -\varphi(\infty) = \frac{\pi}{2}\left(\frac{1}{\sqrt{a}}-1\right) = \frac{\pi}{2}\left(\sqrt{\frac{\gamma+1}{\gamma-1}}-1\right)$$

$$= \frac{\pi}{2}\left(\sqrt{1+f}-1\right) = \frac{\pi}{2}\left\{1, \sqrt{6}-1, \sqrt{7}-1\right\} \qquad (2.966a–i)$$

$$= \left\{90.000°, 130.454°, 148.118°\right\},$$

respectively (2.966h, i) for monoatomic/diatomic/polyatomic perfect gases (2.475a–c).
The Prandtl – Meyer function (2.965) ≡ (2.967a–c):

$$\varphi(M) = \varphi_1(M) - \frac{\varphi_2(M)}{\sqrt{a}} \quad \varphi_{1,2}(M) = \arctan\left[\sqrt{(1,a)(M^2-1)}\right], \quad (2.967a\text{–}c)$$

appears in Table 2.21 over a wide range (2.968a–d) of Mach numbers: (i) from 1.2 to 2 with 0.2 intervals (2.968a); (ii) from 2 to 5 with 0.5 intervals (2.968b); (iii) from 5 to 10 with unit intervals (2.968c); (iv) from 10 to 30 with intervals (2.968d) of 5:

$$M = \{1.2, 1.4, 1.6, 1.8, 2.0; 2.5, 3.0, 3.5, 4.0, 4.5; \atop 5.0, 6.0, 7.0, 8.0, 9.0; 10.0, 15.0, 20.0, 25.0, 30.0\}. \qquad (2.968a\text{–}d)$$

thus covering the full range (2.691a–g) with variable resolution, finer for lower Mach number. The Table 2.21 shows that in the adiabatic turn of a perfect gas (Figure 2.39) the modulus of the turn angle: (i) increases with the Mach number up to the limit

---

TABLE 2.21
Adiabatic Turn for a Perfect Gas Flow

Mach Number M	Partial Angle φ_1	Total Angle: φ		
		f = 3	f = 5	f = 6
		a = 1/4	a = 1/6	a = 1.7
1.2	33.56	−3.14	−3.56	−3.68
1.4	44.16	−8.04	−9.25	−9.60
1.6	51.32	−12.64	−14.86	−15.54
1.8	56.25	−17.37	−20.73	−21.79
2.0	60.00	−21.79	−26.38	−27.87
2.5	66.42	−31.37	−39.13	−41.77
3.0	70.53	−38.94	−49.76	−53.59
3.5	73.40	−44.99	−58.53	−63.47
4.0	75.52	−49.86	−65.79	−71.75
4.5	77.16	−53.83	−71.83	−78.70
5.0	78.46	−57.12	−76.93	−84.59
6.0	80.41	−62.23	−84.95	−93.96
7.0	81.79	−66.01	−90.96	−112.21
8.0	82.82	−69.89	−96.24	−106.52
10.0	84.26	−73.01	−102.32	−114.46
15.0	86.18	−79.60	−111.51	−125.41
20.0	87.13	−81.44	−116.20	−131.03
25.0	87.71	−83.13	−119.03	−134.41
30.0	88.09	−84.27	−120.92	−136.69
∞	90.00	−90.00	−130.45	−148.12

Note: Adiabatic turn for a monoatomic/diatomic/polyatomic perfect gas, respectively, with f = 3/5/6 degrees-of-freedom of the molecules, indicating for each of 22 Mach numbers the partial φ_1 and total φ angles.

(2.966i) for infinite Mach number; (ii) is larger for larger number of degrees of free-dom of a molecule. Table 2.21 may be used to solve direct (inverse) problem of find-ing (subsection 2.8.21) for given initial conditions, the adiabatic turn angle for a given final Mach number (vice-versa).

2.8.21   DIRECT AND INVERSE PROBLEMS OF ADIABATIC TURN

The adiabatic expansion is specified by the Prandtl – Meyer (1908) function (2.965; 2.816b–f) both for: (i) the gradual rotation of a weak shock along a smooth concave surface (Figure 2.39a) with φ a continuous function of M; (ii) for an expansion fan near a convex corner (Figure 2.39b) involving the angles φ∓ corresponding to two Mach numbers M∓. In both cases can be solved two problems for the same initial conditions as regards Mach number (2.969a) and angle of rotation (2.969b): (i) in the direct problem the final Mach number is given (2.969c) and the final angle (2.969d) or rotation (2.969e) is sought:

$$\left\{ M^-, \varphi^- \right\}, \quad \varphi^- \equiv \varphi\left(M^-\right): \quad M^+ \leftrightarrow \varphi^+ \equiv \varphi\left(M^+\right), \quad \Delta\varphi = \varphi^+ - \varphi^-; \quad (2.969\text{a–e})$$

(ii) vice-versa in the inverse problem from (2.969e) ≡ (2.969d) to (2.969c). As an example consider an initial Mach number (2.970a) and corresponding (Table 2.21) angle (2.970c, d) for monoatomic/diatomic/polyatomic molecules (2.970b):

$$M^- = 2.00, \quad f = \left\{3,5,6\right\}: \quad \varphi^- \equiv \varphi\left(M^-\right) = \left\{-21.79, -26.38, -27.87\right\}. \quad (2.970\text{a–c})$$

For a final Mach number (2.971a) the corresponding angle is (2.971b, c):

$$M^+ = 5.00: \quad \varphi^+ = \varphi\left(M^+\right) = \left\{-57.12, -76.93, -84.59\right\}, \quad (2.971\text{a–c})$$

$$\Delta\varphi = \varphi^+ - \varphi^- = \left\{-35.33, -50.55, -56.72\right\} \quad (2.971\text{d,e})$$

leading to rotation by (2.971d, e).

If for the same initial conditions (2.970a–e), an angle of turn (2.972a) is sought, the final angle is (2.972b–d) corresponding to the final Mach number (2.972e):

$$\Delta\varphi = -60.00°: \quad \varphi^- + \Delta\varphi = \left\{-81.79, -86.38, -87.87\right\} = \varphi^+ = \varphi\left(M^+\right), \quad (2.972\text{a–d})$$

$$M^+ = \left\{21.04, 6.24, 5.35\right\} \quad (2.972\text{e})$$

In (2.970a–c; 2.971b) the angles of rotation are read directly from Table 2.21 using the Mach numbers to solve (2.971c–e) the **direct problem**: *given the initial angle (2.970b, c) and initial (2.970a) and final (2.971a) Mach numbers find the angle of adiabatic turn (2.971e). The **inverse problem** for the same initial Mach number (2.970a) and*

angle (2.970b, c) is to specify the rotation (2.972a) and obtain the final Mach number
(2.972e), by interpolation from Table 2.21. In the case (2.972e) the values were
obtained by linear interpolation (2.965) between adjacent values (2.973a, b):

$$\varphi_1 = \{\varphi(20), \varphi(6.0), \varphi(5.0)\} = \{-81.44, -84.95, -84.59\}, \qquad (2.973a)$$

$$\varphi_2 = \{\varphi(25), \varphi(7.0), \varphi(6.0)\} = \{-83.13, -90.96, -93.96\}: \qquad (2.973b)$$

$$M^+ = \{20,6,5\} + \{5,1,1\} \frac{\varphi^+ - \varphi_1}{\varphi_2 - \varphi_1} = \{20,6,5\} + \{5,1,1\} \times \frac{\{0.35, 1.43, 3.28\}}{\{1.69, 6.01, 9.37\}} \quad (2.973c\text{–}e)$$

$$= \{21.04, 6.24, 5.35\},$$

in agreement with (2.973e) = (2.972e). Since a full Table 2.21 is available more
elaborate methods than linear interpolation could be used. Another case of weak
nearly adiabatic shocks is the lowest order corrections to acoustic waves (subsections
2.8.22–2.8.23).

2.8.22   NON-LINEAR SECOND-ORDER APPROXIMATIONS FOR WEAK SHOCKS

A linear acoustic wave is adiabatic, and the lowest order non-linear correction for
weak normal shock *corresponds to a small relative pressure change (2.974a), for*
example, for the upstream (2.974b) [downstream (2.974e)] Mach number:

$$\varepsilon = \frac{p^+ - p^-}{p^-}: \quad M^- = 1 + \frac{\gamma+1}{4\gamma}\varepsilon - \frac{(\gamma+1)^2}{32\gamma^2}\varepsilon^2 \qquad (2.974a, b)$$

$$M^+ = 1 - \frac{\gamma+1}{4\gamma}\varepsilon + \frac{(\gamma+1)(7\gamma-1)}{32\gamma^2}\varepsilon^2 \qquad (2.974c)$$

that are: (i) unity to order zero, for a linear sound wave; (ii) to first order are supersonic (subsonic) by the same small amount; (iii) the second-order correction is distinct and opposite to the first-order. The result (2.974b) is derived from (2.832c) ≡
(2.975a) in two stages: (i) the substitution of (2.974a) leading to (2.975b):

$$\left(M^-\right)^2 = 1 + \frac{\gamma+1}{2\gamma}\left(\frac{p^+}{p^-} - 1\right) = 1 + \frac{\gamma+1}{2\gamma}\varepsilon: \qquad (2.975a, b)$$

(ii) use of the binomial series (I.25.37a–c) ≡ (2.976a) leading to (2.976b) ≡ (2.974b):

$$M^- = \sqrt{1 + \frac{\gamma+1}{2\gamma}\varepsilon} = 1 + \frac{1}{2} \times \frac{\gamma+1}{2\gamma}\varepsilon + \binom{1/2}{2}\left(\frac{\gamma+1}{2\gamma}\varepsilon\right)^2 = 1 + \frac{\gamma+1}{4\gamma}\varepsilon - \frac{(\gamma+1)^2}{32\gamma^2}\varepsilon^2. \quad (2.976a, b)$$

In the case (2.974c): (i) the start is (2.840a) $\equiv$ (2.977a) and using (2.975b) leads to (2.977b–d):

$$\left(M^{+}\right)^{2} = \frac{2+(\gamma-1)\left(M^{-}\right)^{2}}{2\gamma\left(M^{-}\right)^{2}-(\gamma-1)} = \frac{1+\dfrac{\gamma-1}{2\gamma}\varepsilon}{1+\varepsilon} = \frac{1-\dfrac{\gamma-1}{2\gamma}+\dfrac{\gamma-1}{2\gamma}(1+\varepsilon)}{1+\varepsilon}$$

$$= \frac{\gamma-1}{2\gamma}+\frac{\gamma+1}{2\gamma}\frac{1}{1+\varepsilon};$$

(2.977a–d)

(ii) using the geometric series (I.21.62a, c) $\equiv$ (2.978a) leads to (2.978b):

$$\left(M^{+}\right)^{2} = \frac{\gamma-1}{2\gamma}+\frac{\gamma+1}{2\gamma}\left(1-\varepsilon-\varepsilon^{2}\right) = 1-\frac{\gamma+1}{2\gamma}\varepsilon\left(1-\varepsilon\right);$$

(2.978a, b)

(iii) the binomial series then leads to (2.543a–f):

$$M^{+} = \sqrt{1-\frac{\gamma+1}{2\gamma}\varepsilon\left(1-\varepsilon\right)} = 1-\frac{\gamma+1}{4\gamma}\varepsilon\left(1-\varepsilon\right)-\frac{1}{8}\left[\frac{\gamma+1}{2\gamma}\varepsilon\left(1-\varepsilon\right)\right]^{2}$$

$$= 1-\frac{\gamma+1}{4\gamma}\varepsilon+\frac{(\gamma+1)(7\gamma-1)}{32\gamma^{2}}\varepsilon^{2},$$

(2.979a–c)

in agreement with (2.979c) = (2.974c). The ratio of velocities, mass densities and specific volumes (2.831a–c) leads to (2.544a–d) to second-order in (2.974a):

$$\frac{v^{+}}{v^{-}} = \frac{\rho^{-}}{\rho^{+}} = \frac{\vartheta^{+}}{\vartheta^{-}} = \frac{\gamma+1+(\gamma-1)(1+\varepsilon)}{\gamma-1+(\gamma+1)(1+\varepsilon)} = \frac{2\gamma+(\gamma-1)\varepsilon}{2\gamma+(\gamma+1)\varepsilon}$$

$$= \left(1+\frac{\gamma-1}{2\gamma}\varepsilon\right)\left[1-\frac{\gamma+1}{2\gamma}\varepsilon+\left(\frac{\gamma+1}{2\gamma}\varepsilon\right)^{2}\right] = 1-\frac{\varepsilon}{\gamma}+\frac{\gamma+1}{2\gamma^{2}}\varepsilon^{2}.$$

(2.980a–f)

Other relations between flow variables across the weak shock of a perfect gas are obtained in a similar way to second-order (subsection 2.8.22), for example, the speed of advance of the shock front exceeds the sound speed to first-order in the relative pressure change (2.974a) that acts as a non-linearity parameter.

2.8.23 SPEED OF ADVANCE OF THE SHOCK RELATIVE TO THE SOUND SPEED

For the shock of a perfect gas with small relative pressure difference (2.974a) the ratio of temperatures (2.981a–c) [sound speeds (2. 982a–c)] follows from (2.980f) [(2.981c)]:

$$\frac{T^{+}}{T^{-}} = \frac{p^{+}}{p^{-}}\frac{\rho^{-}}{\rho^{+}} = (1+\varepsilon)\left(1-\frac{\varepsilon}{\gamma}+\frac{1+\gamma}{2\gamma^{2}}\varepsilon^{2}\right) = 1+\left(1-\frac{1}{\gamma}\right)\varepsilon+\frac{1-\gamma}{2\gamma^{2}}\varepsilon^{2}, \quad (2.981a\text{–}c)$$

$$\frac{c^+}{c^-} = \sqrt{\frac{T^+}{T^-}} = 1 + \left(1 - \frac{1}{\gamma}\right)\frac{\varepsilon}{2} + \frac{1-\gamma}{4\gamma^2}\varepsilon^2 - \frac{1}{8}\left(1-\frac{1}{\gamma}\right)^2\varepsilon^2$$

$$= 1 + \left(1 - \frac{1}{\gamma}\right)\frac{\varepsilon}{2} + \frac{1-\gamma^2}{8\gamma^2}\varepsilon^2. \tag{2.982a–c}$$

The difference of velocities (2.830b) leads to (2. 983a–c):

$$v^- - v^+ = \frac{\varepsilon\,p^+}{\sqrt{p^-\rho^-/2}}\left|\gamma - 1 + (\gamma+1)(1+\varepsilon)\right|^{-1/2} = \frac{\varepsilon\,p^+}{\sqrt{p^-\rho^-\gamma}}\left|1+\frac{\gamma+1}{2\gamma}\varepsilon\right|^{-1/2}$$

$$= \frac{\varepsilon\,p^+}{\sqrt{p^-\rho^-\gamma}}\left[1 - \frac{\gamma+1}{4\gamma}\varepsilon + \frac{3(\gamma+1)^2}{32\gamma^2}\varepsilon^2\right]. \tag{2.983a–c}$$

Comparing with (2.980f) and (2.983c) specifies the upstream (2.984a–h) [downstream (2.985a–e)] velocities to first-order:

$$v^- = \frac{v^- - v^+}{1 - v^+/v^-} = \frac{\gamma}{\varepsilon}(v^- - v^+) = p^+\sqrt{\frac{\gamma}{p^-\rho^-}}\left(1 - \frac{\gamma+1}{4\gamma}\varepsilon\right)$$

$$= \sqrt{\frac{\gamma p^-}{p^-}}\,\frac{p^+}{p^-}\left(1 - \frac{\gamma+1}{4\gamma}\varepsilon\right) = c^-\sqrt{1+\varepsilon}\left(1 - \frac{\gamma+1}{4\gamma}\varepsilon\right)$$

$$= c^-\left[1 + \varepsilon - \frac{\gamma+1}{4\gamma}\varepsilon\right] = c^-\left(1 + \frac{3-\gamma}{4}\varepsilon\right) > c^-, \tag{2.984a–h}$$

$$v^+ = \frac{v^+}{v^-}v^- = \left(1 - \frac{\varepsilon}{\gamma}\right)c^-\left(1 + \frac{3-\gamma}{4}\varepsilon\right)$$

$$= c^-\left(1 - \frac{\varepsilon}{\gamma} + \frac{3-\gamma}{4}\varepsilon\right) = \left(1 - \frac{4+\gamma^2-3\gamma}{4\gamma}\varepsilon\right) < c^-. \tag{2.985a–e}$$

In the upstream (2.984a–h) [downstream (2.985a–e)] velocity, the upstream may be replaced by the downstream local sound speed (2.986a–e) using (2.982c);

$$\{v^-, v^+\} = c^+\frac{c^-}{c^+}\left\{1 + \frac{3-\gamma}{4}\varepsilon, 1 - \frac{4+\gamma^2-3\gamma}{4\gamma}\right\}$$

$$= c^+\left[1 - \left(1 - \frac{1}{\gamma}\right)\frac{\varepsilon}{2}\right]\left\{1 + \frac{3-\gamma}{4}\varepsilon, 1 - \frac{4+\gamma^2-3\gamma}{4\gamma}\varepsilon\right\}$$

$$= c^+\left[1 - \varepsilon\left[\left(\frac{1}{2} - \frac{1}{2\gamma}\right) - \left\{\frac{3-\gamma}{4}, \frac{3\gamma-4-\gamma^2}{4\gamma}\right\}\right]\right] \tag{2.986a–e}$$

$$= c^+\left\{1 + \frac{\gamma+2-\gamma^2}{4\gamma}\varepsilon, 1 - \frac{2+\gamma^2-\gamma}{4\gamma}\varepsilon\right\} = \{>,<\}c^+.$$

The velocity is supersonic (2.984g) [subsonic (2.985e)] upstream (downstream) relative to the local sound speed in agreement with (2.752c,d) [2.752e,f].

The arithmetic mean of the velocities specifies (2.803b) the speed of advance of the shock front (2.987a):

$$
\bar{v} = \frac{v^+ + v^-}{2} = c^- \left[1 + \left(\frac{3-\gamma}{\varepsilon} - \frac{4+\gamma^2-3\gamma}{8\gamma} \right) \varepsilon \right]
$$

$$
= c^- \left(1 - \frac{2+3\gamma-\gamma^2}{4\gamma} \varepsilon \right) < c^-
$$

$$
= c^+ \left(1 + \left(\frac{\gamma+2-\gamma^2}{8\gamma} - \frac{2+\gamma^2-\gamma}{8\gamma} \right) \varepsilon \right)
$$

$$
= c^+ \left(1 - \frac{\gamma-4}{4} \varepsilon \right) < c^+,
$$

(2.987a–g)

that is smaller than the local sound speeds both upstream (2.987b-d) and downstream (2.987e-g).

It has been shown that *the limit of small relative pressure difference (2.974a) of the normal shock of a perfect gas leads to: (i) first-order to the upstream (2.984a–h) and downstream (2.985a–e) velocity and hence (2.986a–e) speed of advance (2.987a–g) of the "weak shock" front; (ii) second-order for the ratios of velocities, mass densities and specific volumes (2.980a–f), temperatures (2.981a–c) and sound speeds (2.982a–c), the difference of velocities (2.983a–c) and upstream (2.974b) [downstream (2.974c)] Mach numbers.* The account on oblique shocks (Section 2.8) is concluded with the numerical example of the same upstream Mach number and two distinct incidences larger than the Mach angle (subsection 2.8.24), specifying the upstream (downstream) flow variables [subsection 2.8.25 (2.8.26)] and their changes across the oblique shock front (subsections 2.8.27–2.8.29).

2.8.24 OBLIQUE SHOCK WITH TWO UPSTREAM INCIDENCE ANGLES

The upstream Mach number (2.988a) corresponds (2.882a) to the angle of the Mach cone or Mach angle (2.988b–e):

$$
M^- = 2.000: \quad \tilde{\theta} = \arcsin\left(\frac{1}{M^-}\right) = \arcsin\left(\frac{1}{2}\right) = \frac{\pi}{6} = 30°. \quad (2.988a\text{–}e)
$$

Two angles of incidence (2.989a, b) both larger (2.989c) than the Mach angle (2.988e) are considered, so that the upstream velocity lies within the Mach cone, and the normal upstream local Mach number (2.989d–g) is supersonic (2.989h):

$$
\theta \equiv \theta^- = \frac{\pi}{4}\left(\frac{\pi}{3}\right) = 45°(60°) > \tilde{\theta}:
$$

$$
M_n^- = M^- \sin\theta = 2\sin\left[\frac{\pi}{4}\left(\frac{\pi}{3}\right)\right] = \sqrt{2}(\sqrt{3}) = 1.414(1.732) > 1.
$$

(2.989a–h)

The upstream stagnation Mach number (2.677a) is given for a diatomic perfect gas (2.990a) by (2.990b–d) and (2.676a) the upstream critical Mach number by (2.990e–g):

$$\gamma = 1.4: \quad M_0^- = \left| \frac{1}{\left(M^-\right)^2} + \frac{\gamma-1}{2} \right|^{-1/2} = \left| \frac{1}{4} + 0.2 \right|^{-1/2} = 1.491, \quad (2.990\text{a–d})$$

$$M_*^- = \sqrt{\frac{\gamma+1}{2}} \, M_0^- = 1.491 \times \sqrt{1.2} = 1.633. \quad (2.990\text{e–g})$$

For the International Standard Atmosphere (ISA) at sea level the stagnation sound speed (2.721d) ≡ (2.991a) leads to the upstream velocity (2.991b–d):

$$c_0 = 340 \, m\,s^{-1}: \quad v^- = c_0 \, M_0^- = 1.491 \times 340 \, m\,s^{-1} = 507 \, m\,s^{-1}, \quad (2.991\text{a–d})$$

whose normal (2.892b) ≡ (2.992a–c) [tangential (2.892a) ≡ (2.557a–d)] component:

$$v_n^- = v^- \sin\theta = 507 \times \frac{1}{\sqrt{2}}\left(\frac{\sqrt{3}}{2}\right) m\,s^{-1} = 359(439) \, m\,s^{-1} > v_n^+, \quad (2.992\text{a–d})$$

$$v_t^- = v^- \cos\theta = 507 \times \frac{1}{\sqrt{2}}\left(\frac{1}{2}\right) m\,s^{-1} = 359(253) \, m\,s^{-1} = v_t^+, \quad (2.993\text{a–d})$$

is discontinuous (2.992d) [continuous (2.993d)] across the shock front.

The critical sound speed (2.664c) ≡ (2.994a–c) is constant across the shock and smaller (2.994d) than the stagnation sound speed (2.991a):

$$c_* = c_0 \sqrt{\frac{2}{\gamma+1}} = \frac{340}{\sqrt{1.2}} \, m\,s^{-1} = 310 \, m\,s^{-1} < c_0. \quad (2.994\text{a–d})$$

The local sound speed is not continuous and is given upstream by (2.661d) ≡ (2.684i) ≡ (d)

$$c^- = \left| c_0^2 - \frac{\gamma-1}{2}\left(v^-\right)^2 \right|^{1/2} = \sqrt{340^2 - 0.2 \times 507^2} \, m\,s^{-1} = 253 \, m\,s^{-1} < c_* < c_0, \quad (2.995\text{a–e})$$

and is the smallest (2.995d, e) of the three sound speeds. Since the stagnation (2.991a) is larger (2.994d) ≡ (2.995e) than the critical (2.994c) and in turn larger (2.995d) than the local (2.995c) sound speed, the reverse applies to the Mach numbers: the local is larger than the critical and larger than the stagnation Mach number for the total (2.996a–k), normal (2.997a–k) and tangential (2.560a–c) Mach numbers:

$$M^- \equiv \frac{v^-}{c^-} = \frac{507}{253,5} = 2.000$$

$$> M_*^- \equiv \frac{v^-}{c_*} = \frac{507}{310} = 1.635$$

$$> M_0^- \equiv \frac{v^-}{c_0} = \frac{507}{340} = 1.491 > 1,$$

(2.996a–l)

$$M_n^- \equiv \frac{v_n^-}{c^-} = \frac{359(439)}{253} = 1.419(1.735)$$

$$> M_{*n}^- \equiv \frac{v_n^-}{c_*} = \frac{359(439)}{310} = 1.158(1.416)$$

(2.997a–l)

$$> M_{0n}^- \equiv \frac{v_n^-}{c_0} = \frac{359(439)}{340} = 1.056(1.291) > 1,$$

$$M_t^+ \equiv \frac{v_t}{c^+} = \frac{359(253)}{285(309)} = 1.260(0.819)$$

$$> M_{*t} \equiv \frac{v_t}{c_*} = \frac{359(253)}{310} = 1.158(0.816)$$

(2.998a–k)

$$> M_{0t} \equiv \frac{v_t}{c_0} = \frac{359(253)}{340} = 1.056(0.744),$$

using the total (2.991d), normal (2.992c) and tangential (2.993c) upstream velocities. The normal (2.997l), and hence total (2.996l) upstream Mach numbers are supersonic, but the tangential Mach numbers need not be (2.988a–k), in particular for larger incidences closer to normal. The preceding results specify all upstream flow variables (subsection 2.8.25).

2.8.25 UPSTREAM FLOW VARIABLES AND ANGLE OF DEFLECTION

From (2.990d) follows the upstream Mach factor (2.561a–d):

$$\psi^- = 1 - \frac{\gamma-1}{2}\left(\frac{v^-}{c_0}\right)^2 = 1 - 0.2\left(M_0^-\right)^2 = 1 - 0.2 \times 1.491^2 = 0.555, \quad (2.999a–d)$$

that specifies the pressure (2.666e) ≡ (2.1000b–d) [mass density (2.666d) ≡ (2.1000b–d)] relative to the stagnation values (2.1000a) ≡ (2.716j) [(2.929a) ≡ (2.720c)] for ISA at sea level:

$$p_0 = 1\,atm: \quad p^- = p_0^-\left(\psi^-\right)^{\gamma/(\gamma-1)} = 0.555^{3.5}\,atm = 0.127\,atm, \quad (2.1000a–d)$$

$$\rho_0 = 1.225 \, kg \, m^{-3} : \quad \rho^- = \rho_0 \left(\psi^- \right)^{1/(\gamma - 1)} = 1.275 \times 0.555^{2.5} \, kg \, m^{-3}$$

$$= 0.281 \, kg \, m^{-3}. \qquad (2.1001a\text{–}d)$$

The upstream temperature (2.666c) $\equiv$ (2.1002b–d) follows from the stagnation temperature (2.716a) $\equiv$ (2.1002a):

$$T_0 = 288 \, K : \quad T^- = T_0 \psi^- = 288 \times 0.555 \, K = 160 \, K, \qquad (2.1002a\text{–}d)$$

and specifies the internal energy (2.623a) $\equiv$ (2.1003a–c) [enthalpy (2.623b) $\equiv$ (2.1004a–c)] per unit mass using the specific heat at constant volume (2.719g) [pressure (2.719h)]:

$$u^- = c_V \, T^- = 718 \, m^2 \, s^{-2} \, K^{-1} \times 160 \, K = 1.149 \times 10^5 \, m^2 \, s^{-2}, \qquad (2.1003a\text{–}c)$$

$$h^- = c_p \, T^- = 1005 \, m^2 \, s^{-2} \, K^{-1} \times 160 \, K = 1.608 \times 10^5 \, m^2 \, s^{-2}. \qquad (2.1004a\text{–}c)$$

The downstream normal velocity (2.889b) $\equiv$ (2.1005a–c) is specified by the upstream normal velocity (2.992d), the stagnation sound speed (2.994d) and the adiabatic exponent (2.990a), and the tangential velocity (2.993d) that is continuous:

$$v_n^+ = \frac{c_*^2}{v_n^-} - \frac{\gamma - 1}{\gamma + 1} \frac{v_t^2}{v_n^-} = \frac{310^2 - \left(\dfrac{1}{6} \right) \times \left[359(253) \right]^2}{359(439)} \, m \, s^{-1} = 208(195) \, m \, s^{-1}. \qquad (2.1005a\text{–}c)$$

The continuity of the tangential velocity (2.993c, d) specifies the downstream incidence angle (2.567a–c)

$$\tan \theta^+ = \frac{v_n^+}{v_t} = \frac{208(195)}{359(253)} = 0.579(0.771), \quad \theta^+ = 30.1°(37.6°). \qquad (2.1006a\text{–}d)$$

The difference of the upstream (2.989b) and downstream (2.1006d) incidence angles is the angle of deflection of the velocity across the oblique shock (2.568a–c):

$$\varphi = \theta^- - \theta^+ = 60.0(45.0)° - 30.1(37.6)° = 29.9(7.4)°, \qquad (2.1007a\text{–}c)$$

that is larger for larger incidence. The total downstream velocity follows (2.1008a–c) [(2.1009a–c)] from its normal (2.1005c) [tangential (2.993c)] component and downstream incidence angle (2.1006d):

$$v^+ = v_n^+ \csc \theta^+ = \frac{208}{0.502} \left(\frac{195}{0.610} \right) m \, s^{-1} = 415(319) \, m \, s^{-1}, \qquad (2.1008a\text{–}c)$$

$$v^+ = v_t \sec\theta^+ = \frac{359}{0.865}\left(\frac{253}{0.792}\right)m\,s^{-1} = 415(319)\,m\,s^{-1}. \qquad (2.1009a\text{–}c)$$

The preceding results specify the remaining downstream flow variables (subsection 2.8.26).

2.8.26 DOWNSTREAM ANGLE OF INCIDENCE AND FLOW VARIABLES

The local downstream sound speed (2.661d) ≡ (2.684i) ≡ (2.1010a–c):

$$c^+ = \left|c_0^2 - \frac{\gamma-1}{2}(v^+)^2\right|^{\frac{1}{2}} = \sqrt{340^2 - 0.2\times\left[415(319)\right]^2} \qquad (2.1010a\text{–}e)$$

$$= 285(309)\,m\,s^{-1} < c_* = 310\,m\,s^{-1} < c_0 = 340\,m\,s^{-1},$$

is smaller (2.1010d) [(2.1010e)] than the critical (2.994e) [stagnation (2.991a)] sound speeds, and the reverse applies to the downstream total (2.572a–c), normal (2.573a–c) and tangential (2.574) Mach numbers:

$$M^+ \equiv \frac{v^+}{c^+} = \frac{415(319)}{285(309)} = 1.456(1.032)$$

$$> M_*^+ \equiv \frac{v^+}{c_*} = \frac{415(319)}{310} = 1.339(1.029) \qquad (2.1011a\text{–}l)$$

$$> M_0^+ \equiv \frac{v^+}{c_0} = \frac{415(319)}{340} = 1.221(0.938)(1 > f),$$

$$M_n^+ \equiv \frac{v_n^+}{c^+} = \frac{208(195)}{285(309)} = 0.730(0.631)$$

$$> M_{*n}^+ \equiv \frac{v_n^+}{c_*} = \frac{208(195)}{310} = 0.670(0.629) \qquad (2.1012a\text{–}k)$$

$$> M_{0n}^+ \equiv \frac{v_n^+}{c_0} = \frac{208(195)}{340} = 0.612(0.574) < 1,$$

$$M_t^+ \equiv \frac{v_t}{c^+} = \frac{359(253)}{285(309)} = 1.260(0.819)$$

$$> M_{*t}^+ \equiv \frac{v_t}{c_*} = \frac{359(253)}{310} = 1.158(0.816) \qquad (2.1013a\text{–}g)$$

$$> M_{0t}^+ \equiv \frac{v_t}{c_0} = \frac{359(253)}{340} = 1.056(0.744),$$

using the total (2.1008c), normal (2.1005c) and tangential (2.993c) downstream velocities. The continuity of the tangential velocity (2.993a–d) implies: (i) the continuity of the tangential critical (2.998d–g) ≡ (2.1013d, e) [stagnation (2.998h–k) ≡ (2.1013f, g)] Mach numbers because the critical (2.994c) [stagnation (2.991a)] sound speeds are continuous; (ii) the discontinuity of the local tangential Mach number (2.998a–c) ≠ (2.1013a–c) because the local sound speeds are discontinuous (2.995a–e) ≠ (2.1010a–e).

The upstream (2.1011h–k) Mach factor is (2.1014a–c):

$$\psi^+ = 1 - \frac{\gamma-1}{2}\left(M_0^+\right)^2 = 1 - 0.2 \times \left[1.221(0.938)\right]^2 = 0.702(0.824). \quad (2.1014a\text{-}c)$$

Since the stagnation sound speed (2.991a), and hence the stagnation temperature (2.1002a) are conserved, adiabatic relations may be used for: (i) the upstream temperature (2.666c) ≡ (2.1015a–c):

$$T^+ = T_0 \psi^+ = 0.702(0.824) \times 288\,K = 202(237)\,K; \quad (2.1015a\text{-}c)$$

(ii/iii) the internal energy (2.623a) ≡ (2.1016a–c) [enthalpy (2.623b) ≡ (2.1017a–c)] per unit mass:

$$u^+ = c_V\,T^+ = 718\,m^2\,s^{-2}\,K^{-1} \times 202(237)\,K = 1.450(1.702) \times 10^5\,m^2\,s^{-2}, \quad (2.1016a\text{-}c)$$

$$h^+ = c_p\,T^+ = 1005\,m^2\,s^{-2}\,K^{-1} \times 202(237)\,K = 2.030(2.382) \times 10^5\,m^2\,s^{-2}. \quad (2.1017a\text{-}c)$$

The adiabatic relations for the pressure (2.666e) [mass density (2.666d)] cannot be used downstream of the shock because the upstream stagnation values (2.1000a) [(2.1001a)] are not conserved. The conservation of the normal mass flux (2.883a) ≡ (2.1018a) for an oblique shock specifies (2.992c; 2.1001d; 2.1005c) the downstream mass density (2.1018b, c):

$$\rho^+ = \frac{\rho^-\,v_n^-}{v_n^+} = \frac{359(439)}{208(195)} \times 0.281\,kg\,m^{-3} = 0.485(0.637)\,kg\,m^{-3}. \quad (2.1018a\text{-}c)$$

Using the equation of state for a perfect gas (2.473g) and the upstream (2.1002d) [downstream (2.1015c)] temperature and mass density (2.1001d) [(2.1018c)] relates the upstream (2.1000d) to the downstream (2.580a–c) pressure:

$$p^+ = p^- \frac{\rho^+\,T^+}{\rho^-\,T^-} = \frac{0.485(0.633)}{0.281} \times \frac{202(237)}{160} \times 0.127\,atm$$

$$= 0.277(0.424)\,atm. \quad (2.1019a\text{-}c)$$

The changes, such as differences or ratios of flow variables across the oblique shock front follow (subsection 2.8.27) from the upstream (downstream) values [subsections 2.8.24–2.8.25 (2.8.26)].

2.8.27 CHANGES ACROSS THE OBLIQUE SHOCK FRONT

Table 2.22 indicates all upstream and downstream flow variables and their ratios, that are larger for the larger upstream angle of incidence (2.989a, b) with the same upstream Mach number (2.988a), for example, for: (i) the Mach factor (2.1020a, b) using (2.999d; 2.1014c); (ii) temperature (2.1020c, d) using (2.1002d; 2.1015c); (iii) pressure (2.1021a, b) using (2.1000d; 2.1019c); (iv) mass density (2.1021c, d) using (2.1001d; 2.1018c):

$$\frac{\psi^+}{\psi^-} = \frac{0.702(0.824)}{0.555} = 1.265(1.485) \sim 1.263(1.481) = \frac{202(237)}{160} = \frac{T^+}{T^-}; \quad (2.1020\text{a–d})$$

$$\frac{p^+}{p^-} = \frac{0.277(0.424)}{0.127} = 2.181(3.339) > 1.726(2.253) = \frac{0.485(0.633)}{0.281} = \frac{\rho^+}{\rho^-}; \quad (2.1021\text{a–d})$$

(v–vi) total (2.1022a–c) [normal (2.1022d–f)] velocity using (2.991d; 2.1008c) [(2.992c; 2.1005c)]; (vii) angle of incidence (2.1023a, b) using (2.989c; 2.1006d); (viii) local sound speed (2.1023c, d) using (2.995c; 2.1010c):

$$\frac{v^+}{v^-} = \frac{415(319)}{507} = 0.819(0.629) > 0.579(0.444) = \frac{208(195)}{359(439)} = \frac{v_n^+}{v_n^-}; \quad (2.1022\text{a–f})$$

$$\frac{\theta^+}{\theta^-} = \frac{30.1(37.6)}{45.0(60.0)} = 0.669(0.627): \quad \frac{c^+}{c^-} = \frac{285(309)}{253} = 1.126(1.221); \quad (2.1023\text{a–d})$$

(ix–xi) local (2.1024a, b), critical (2.1024c, 2.584a–c) and stagnation (2.1024e, 2.585a–c) Mach numbers:

$$\frac{M^+}{M^-} = \frac{1.456(1.032)}{2.000} = 0.728(0.516), \quad (2.1024\text{a, b})$$

$$\frac{M_*^+}{M_*^-} = \frac{1.334(1.029)}{1.635} = 0.816(0.629), \quad (2.1024\text{c, d})$$

$$\frac{M_0^+}{M_0^-} = \frac{1.22(0.938)}{1.491} = 0.819(0.629,) \quad (2.1024\text{e, f})$$

using, respectively (2.988a; 2.1011c), (2.996g; 2.1011g) and (2.996k; 2.1011k), and their normal and tangential components (Table 2.22). The changes across the shock front imply that it is supersonic relative to the stagnation sound speed and there is entropy production (subsection 2.8.28).

TABLE 2.22
Comparison of Two Oblique Shock Waves with Different Incidence Angles

Variable			Value		
Symbol	Name	Unit	Upstream (U)	Downstream (D)	Ratio (D/U)
M	Local Mach number	-	2.000	1.456 (1.032)	0.726 (0.516)
θ	Angle of incidence*	°	45.0 (60.0)	30.1 (37.6)	0.637 (0.625)
M_0	Stagnation Mach number	-	1.491	1.221 (0.938)	0.819 (0.629)
$M*$	Critical Mach number	-	1.635	1.339 (1.029)	0.816 (0.629)
v	Velocity	$m\ s^{-1}$	508	415 (319)	0.819 (0.629)
v_n	Normal velocity	$m\ s^{-1}$	359 (439)	208 (195)	0.579 (0.444)
v_t	Tangential velocity	$m\ s^{-1}$	359 (253)	359 (253)	1.000 (1.000)
c_0	Stagnation sound speed	$m\ s^{-1}$	340	340	1.000
$c*$	Critical sound speed	$m\ s^{-1}$	310	310	1.000
c	Local sound speed	$m\ s^{-1}$	253	285 (309)	1.126 (1.221)
M_n	Normal local Mach number	-	1.414 (1.735)	0.730 (0.631)	0.516 (0.364)
$M*_n$	Normal critical Mach number	-	1.158 (1.416)	0.670 (0.629)	0.579 (0.444)
M_{0n}	Normal stagnation Mach number	-	1.056 (1.291)	0.612 (0.574)	0.580 (0.445)
M_t	Tangential local Mach number	-	1.419 (1.000)	1.260 (0.819)	0.888 (0.819)
$M*_t$	Tangential critical Mach number	-	1.158 (0.816)	1.158 (0.816)	1.000 (1.000)
M_{0t}	Tangential stagnation Mach number	-	1.058 (0.744)	1.058 (0.744)	1.000 (1.000)
ψ	Mach factor	-	0.555	0.702 (0.824)	1.265 (1.485)
p	Pressure	atm	0.127	0.277 (0.424)	2.181 (3.339)
ρ	Mass density	$kg\ m^{-3}$	0.281	0.485 (0.633)	1.710 (2.258)
T	Temperature	K	160	202 (237)	1.263 (1.481)
u	Internal energy	$10^5\ m^2\ s^{-2}$	1.149	1.450 (1.702)	1.263 (1.481)
h	Enthalpy	$10^5\ m^2\ s^{-2}$	1.608	2.030 (2.382)	1.263 (1.481)
T_0	Stagnation temperature	K	288	288	1.000
u_0	Stagnation internal energy	$10^5\ m^2\ s^{-2}$	2.079	2.079	1.000
h_0	Stagnation enthalpy	$10^5\ m^2\ s^{-2}$	2.915	2.915	1.000
p_0	Stagnation pressure	atm	1.000	0.955 (0.835)	0.955 (0.835)
ρ_0	Stagnation mass density	$kg\ m^{-3}$	1.225	1.170 (1.023)	0.955 (0.835)

* Deflection angle of velocity: $= \theta^- - \theta^+ = 29.9°$ (7.4°); Normal speed of the shock front: $\bar{v}_n = \dfrac{v^- + v^+}{2} = 284\,(317)\,m\,s^{-1}$; Entropy increase across the shock front: $s^+ - s^- = 11.34$ (42.12) $m^2\,s^{-2}\,K^{-1}$

Note: Comparison of two oblique shock waves with the same upstream Mach number and different angles of incidence, indicating the upstream and downstream values and their ratio for: (i, ii) stagnation and critical Mach numbers; (iii–v) total, normal and tangential velocities; (vi–viii) stagnation, critical and local sound speeds; (ix–xvi) normal and tangential local, critical and stagnation Mach numbers; (xvii–xix) Mach factor used to calculate pressure and mass density; (xx–xxii) the preceding specify the temperature, internal energy and enthalpy; (xxiii–xxv) their stagnation values; (xxvi–xxvii) the stagnation pressure and mass density.

2.8.28 ENTROPY PRODUCTION AND SUPERSONIC SHOCK FRONT

The entropy change across the shock (2.477d) ≡ (2.1025a) is given (2.1021b, d; 2.719g, h) by (2.1025b, c):

$$s^+ - s^- = c_V \log\left(\frac{p^+}{p^-}\right) - c_p \log\left(\frac{\rho^+}{\rho^-}\right)$$
$$= \{718\log[2.181(3.339)] - 1005\log[1.726(2.253)]\} m^2 s^{-2} K^{-1} \quad (2.1025a\text{–}d)$$
$$= 11.34(42.12) m^2 s^{-2} K^{-1} > 0,$$

is positive, and significantly larger for the shock with larger angle of incidence, that is stronger because it is closer to a normal shock, for the same upstream Mach number. The velocity of advance along the normal direction of the front of the oblique shock is (2.803b) ≡ (2.1026a–c) the arithmetic mean of the upstream (2.992c) and downstream (2.1005c) normal velocities:

$$\bar{v}_n = \frac{1}{2}(v_n^- + v_n^+) = \frac{359(439) + 208(195)}{2} m s^{-1} = 284(317) m s^{-1}, \quad (2.1026a\text{–}c)$$

$$\frac{\bar{v}_n}{c_*} = \frac{284(317)}{310} = 0.916(1.023), \quad \frac{\bar{v}_n}{c_0} = \frac{284(317)}{340} = 0.835(0.932), \quad (2.1027a\text{–}d)$$

and is: (i) not necessarily supersonic (2.1027a, b) relative to the critical sound speed as would be the case for a normal shock (2.804e), because (2.801a) is replaced for an oblique shock by (2.889b), that reduces the normal velocities; (ii) is subsonic (2.1027c, d) relative to the stagnation sound speed. Since any normal (oblique) shock [Section 2.7 (2.8)] is an irreversible thermodynamic process some stagnation flow variables must decrease, and the remaining may be conserved but not increased (subsection 2.8.29).

2.8.29 CONSERVED OR DECREASED STAGNATION FLOW VARIABLES

Across the oblique shock are conserved: (i–iv) the stagnation sound speed (2.991a) ≡ (2.875a, b), stagnation temperature (2.1002a) ≡ (2.875c, d) and internal energy (2.875e–h) [enthalpy (2.875i–l)] per unit mass; (v–viii) the critical sound speed (2.994c) ≡ (2.1028a, b), tangential velocity (2.993d) ≡ (2.1028c, d) and hence the tangential stagnation (2.998k) ≡ (2.1029a, b) and critical (2.998g) ≡ (2.1029c, d) Mach numbers:

$$c_* = 310 m s^{-1} = c_*^\pm \quad v_t = 359(253) m s^{-1} = v_t^\pm, \quad (2.1028a\text{–}d)$$

$$M_{0t} = 1.056(0.744) = M_{0t}^\pm, \quad M_{*t} = 1.158(0.816) = M_{*t}^\pm. \quad (2.1029a\text{–}d)$$

The downstream pressure (2.1019c) specifies (2.666e) ≡ (2.1030a) the down-stream stagnation pressure (2.1030b, c) using (2.1008c; 2.991a; 2.990a):

$$p_0^+ = p^+ \left| 1 - \frac{\gamma-1}{2} \left(\frac{v^+}{c_0}\right)^2 \right|^{\gamma/(1-\gamma)} = \left| 1 - 0.2 \times \left[\frac{415(319)}{340} \right]^2 \right|^{-3.5}$$

$$\times 0.277(0.424) atm = 0.955(0.835) atm < p_0 = 1 atm; \qquad (2.1030a\text{–}c)$$

the ratio of the upstream stagnation pressure (2.1030a) ≡ (2.1031a) is the same (2.1031b, c) for the mass density since the stagnation temperature is conserved:

$$p_0^- = 1 atm \quad \frac{p_0^+}{p_0^-} = 0.955(0.835) = \frac{\rho_0^+}{\rho_0^-} \qquad (2.1031a\text{–}c)$$

Thus the upstream stagnation mass density (2.1001a) ≡ (2.1032a) specifies the downstream stagnation mass density (2.1032b, c):

$$\rho_0^- = 1.225 \, kg \, m^{-3} \quad \rho_0^+ = 0.955(0.835) \times 1.225 \, kg \, m^{-3}$$

$$= 1.170(1.023) \, kg \, m^{-3} < \rho_0^-. \qquad (2.1032a\text{–}d)$$

The downstream stagnation pressure (mass density) is smaller (2.1033a, b) [(2.1034c, d)] than the upstream value because an oblique shock is an irreversible thermodynamic process:

$$p_0^+ - p_0^- = \left[0.955(0.835) - 1.000 \right] atm = -0.045(-0.165) atm \qquad (2.1033a, b)$$

$$\rho_0^+ - \rho_0^- = \left[1.170(1.023) - 1.225 \right] kg \, m^{-3} = -0.055(-0.202) kg \, m^{-3}. \qquad (2.1033c, d)$$

The flow variables for the oblique shock of a perfect gas with upstream Mach number (2.988a) and angles of incidence (2.989a, b) are listed in Table 2.22 showing which are conserved or changed. The sequence of calculations (subsections 2.8.23–2.8.29) is the simplest that specifies all flow variables for the oblique shock of a perfect gas. The normal (oblique) shocks [Section 2.7 (2.8)] can occur for a free (ducted) flow [Section 2.6 (2.9)].

2.9   THE "CHOKED" OR "SHOCKED" NOZZLE

The equations of three-dimensional fluid dynamics (Section 2.6) can be specialized for the quasi-one-dimensional flow in a duct of gradually varying cross-section or nozzle (subsection 2.9.1). For incompressible flow the volume flux is conserved, and the velocity increases (decreases) with decreasing (increasing) cross-sectional area (subsection 2.9.2), that is for a convergent (divergent) nozzle. This result extends to a compressible subsonic flow, but is reversed for a supersonic flow, whose velocity increases (decreases) in a divergent (convergent) nozzle, because the variation of mass density dominates in the conservation of the mass flux (subsection 2.9.3). Thus an adiabatic flow may be accelerated out of a reservoir into a convergent nozzle up to at most sonic speed at the exit (subsection 2.9.4). To accelerate adiabatically a flow to supersonic speed it is necessary to use a convergent–divergent or throated nozzle, with subsonic flow in the convergent section, sonic conditions at the throat, and supersonic flow in the divergent section (subsection 2.9.5); thus the maximum mass flux through a convergent–divergent nozzle corresponds to the "choked" condition of sonic flow at the throat (subsection 2.9.7) in contrast with the case of an incompressible flow (subsection 2.9.6).

In the nozzle exhaust to the atmosphere or another ambient fluid three cases (subsection 2.9.11) can occur depending on the atmospheric and nozzle exit pressures: (i) they are equal in the case of a perfect expansion, forming a jet with shear layer separating the nozzle exhaust from the atmosphere (subsections 2.9.8–2.9.10); (ii) in the underexpanded case of excess nozzle pressure relative to the atmosphere a supersonic exhaust may form oblique shocks, reflected by the shear layers into a diamond pattern (subsection 2.9.11); (iii) in the overexpanded case the exhaust pressure below the atmospheric pressure cannot be matched and a shock wave must form upstream in the nozzle (subsection 2.9.12). In the last case (iii) of an overexpanded "shocked" nozzle the location of the shock must be such (subsection 2.9.13) that the pressure at the exit matches the atmospheric pressure.

As a numerical example it is considered a convergent- divergent "choked" nozzle of a jet engine (subsection 2.9.18) in two conditions in the International Standard Atmosphere (subsections 2.9.14–2.9.17): (i) at take-off or landing at sea level the flow is underexpanded, that is the exhaust pressure exceeds the atmospheric pressure, and there is no internal shock (subsection 2.9.19–2.9.22); (ii) in cruise flight at higher altitude in the atmosphere (subsections 2.9.23–2.9.25) the nozzle is underexpanded and an internal shock forms. In the last case of a "choked" and "shocked" convergent–divergent nozzle there are three flow regimes: (i) adiabatic in the convergent subsonic section, through the sonic throat and in the divergent supersonic section upstream of the shock; (ii) non-adiabatic in the irreversible transition across the normal shock in the divergent supersonic section; (iii) adiabatic downstream of the shock ensuring that the exhaust pressure matches the atmospheric pressure.

A turbojet engine only needs fuel, because it uses atmospheric air (subsections 2.9.14–2.9.25), and thus cannot operate in the vacuum of space. A rocket engine carries two propellants, a fuel and an oxidizer, and can operate in the atmosphere or in space (2.9.26–2.9.39). Whereas a subsonic (supersonic) turbojet uses a convergent (convergent-divergent nozzle), a rocket engine (subsections 2.9.26–2.9.28) uses a throated nozzle with a large expansion ratio leading to a bell-type exit. The flow in the bulged (throated) nozzle of a turbojet (rocket) engine is: (i) underexpanded [subsections 2.9.23–2.9.26 (2.9.29–2.9.32)] at low pressure at high altitude (in the vacuum of space); (ii) overexpanded [subsections 2.9.19–2.9.22 (2.9.34–2.9.39)] with oblique shock waves at the exit in the high-pressure at sea level for take-off of an aircraft (lift-off of a rocket or satellite launcher); (iii) perfect expansion would occur at an intermediate altitude (subsection 2.9.32).

2.9.1 Unsteady Quasi-One-Dimensional Flow

A **quasi-one-dimensional nozzle** is understood as (Figure 2.40) a duct with straight axis x, along which the cross-sectional area $A(x)$ varies gradually, so that the flow velocity is parallel to the axis (2.1034a) and like all other flow variables is uniform over the cross-section and thus depends only on the axial coordinate x, and also on time for unsteady flow:

$$\vec{v} = \vec{e}_x\, v(x,t): \quad \rho = \frac{dm}{dV} \rightarrow \frac{dm}{dx} = \frac{dm}{dV}\frac{dV}{dx} = \rho\, A. \qquad (2.1034a\text{–}d)$$

The equations of fluid mechanics (subsections 2.6.1–2.6.6 and Notes 2.1–2.5) apply to the unsteady flow in a quasi-one-dimensional nozzle (Figure 2.40) in one-dimensional form (2.1034a) with the mass density per unit volume (2.1034b) replaced by the mass density per unit length (2.1034c) that equals the product of the cross-sectional area by the mass density per unit volume (2.1034d). The substitutions (2.1034a–d) can be made in: (i) continuity equation (2.1034d) of mass conservation (2.1035a); (ii) inviscid momentum equation (2.598b) balancing the inertia force against minus the pressure gradient (2.1035b) and neglecting viscosity; (iii) the equation of state (2.606c) for an adiabatic flow (2.606b) stating that the material derivatives of the pressure and mass density are proportional through the square of the sound speed (2.1035c), in the absence of any dissipative phenomena, like viscosity or heat conduction:

$$\frac{\partial \rho}{\partial t} + \frac{\partial}{\partial x}(\rho A v) = 0, \quad \frac{\partial v}{\partial t} + v\frac{\partial v}{\partial x} = -\frac{1}{\rho}\frac{\partial p}{\partial x}, \quad \frac{\partial p}{\partial t} + v\frac{\partial p}{\partial x} = c^2\left(\frac{\partial \rho}{\partial t} + v\frac{\partial \rho}{\partial x}\right). \quad (2.1035a\text{–}c)$$

Thus *the unsteady adiabatic flow in a quasi-one-dimensional nozzle (2.1034a–d) with straight x-axis and slowly varying cross-section A satisfies the continuity (2.1035a), momentum (2.1035b) and adiabatic (2.1035c) equations relating the velocity v, pressure p and mass density ρ, and involving the adiabatic sound speed c.* These equations are considered next in the steady incompressible (compressible) cases [subsection 2.9.2 (2.9.3)].

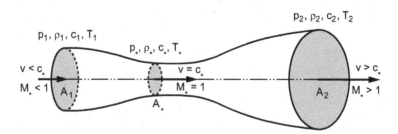

FIGURE 2.40 In order to accelerate an adiabatic flow from subsonic to supersonic it is necessary to use a convergent–divergent nozzle, with subsonic (supersonic) flow in the convergent (divergent) section, and sonic flow at the throat, that determines the mass flow rate.

2.9.2 Steady Flow in a Duct of Varying Cross-Section

The unsteady Equations (2.594a–e) describe linear and non-linear sound in a nozzle (Notes IV.7.16–IV.7.17). In the steady case (2.1036a) it is assumed that: (i) the cross-sectional area does not vary with time, that is the nozzle has rigid walls, but may vary with position along the axis, for example, for a convergent, divergent or throated nozzle (Figures 2.40–2.41); (ii) all flow variables are independent of time and thus depend only on axial position, so that (2.1035a–c) simplify to (2.1036b–d):

$$\frac{\partial}{\partial t} = 0: \quad \rho A v = const \equiv \dot{m}, \quad \rho v \, dv = -dp, \quad dp = c^2 \, d\rho. \quad (2.1036a\text{–}d)$$

Thus *the steady (2.1036a) quasi-one-dimensional (2.1034a–d) adiabatic flow in a nozzle satisfies the conservation of the **mass flux** (2.1036b), the balance of inertia and pressure forces (2.1036c) and absence of heat exchanges (2.1036d).* For low Mach number (2.1037a) the mass density (2.666d) is constant (2.1037b), and non-zero pressure perturbations (2.1037c) imply (2.1036d) an infinite sound speed (2.1037d):

$$M^2 \equiv \left(\frac{v}{c}\right)^2 \ll 1: \quad \rho(x) = \rho_0 = const, \quad dp \neq 0 \;\Rightarrow\; c = \infty, \quad (2.1037a\text{–}d)$$

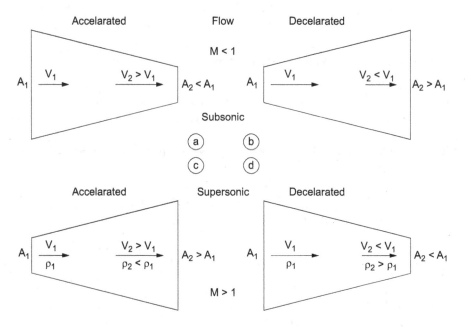

FIGURE 2.41 A subsonic flow $M < 1$, including an incompressible flow at low Mach number $M^2 \ll 1$, is accelerated (decelerated) in a convergent (divergent) nozzle. Due to the compressibility the reverse applies to a supersonic flow $M > 1$ that is accelerated (decelerated) in a divergent (convergent) nozzle.

that is, that perturbations reach instantaneously all points of the flow.

2.9.3   INCOMPRESSIBLE FLOW IN A CONVERGENT OR DIVERGENT NOZZLE

The two remaining equations (2.1036b, c) state that: *(i) the steady (2.1036a) incompressible (2.1038a) flow not subject to external forces conserves (2.1037b) ≡ (2.1038b) the stagnation pressure (2.1038c) and limits the maximum flow velocity to the **cavitation velocity** (2.1038d) when the pressure is zero:*

$$\rho = const: \quad 0 = d\left(\rho\frac{v^2}{2} + p\right) \implies p_0 = p + \frac{1}{2}\rho v^2 \implies v \le v_{max} = \sqrt{\frac{2\,p_0}{\rho}}; \quad (2.1038a\text{--}d)$$

*(ii) in the steady (2.1036a) incompressible (2.1039a) quasi-one-dimensional flow (2.1034a), the **volume flux** (2.1039b) is conserved (2.1039c) implying that velocity changes have opposite sign to changes in cross-sectional area (2.1039d):*

$$\rho = const: \quad A v = j \equiv \frac{\dot{m}}{\rho} = const \iff \frac{dv}{dA} = -\frac{A}{v} < 0, \quad (2.1039a\text{--}d)$$

that is the velocity increase (decrease) when the cross-section decreases (increases) for a convergent (divergent) nozzle [Figure 2.41a (2.41b)]. This conclusion remains valid for a compressible subsonic flow but is reversed in a supersonic flow because changes in mass density become dominant (subsection 2.9.4).

2.9.4   COMPRESSIBLE FLOW IN A THROATED NOZZLE (KÖRTING 1878; LAVAL 1888)

In the case of a compressible flow: (i) the **volume flux** (2.1039b), that is the volume of fluid that transverses a cross-section is no longer conserved because of density changes; (ii) the density changes are accounted for in the **mass flux** (2.1036b) that is the mass of fluid transversing the cross-section per unit time, is conserved, leading by logarithmic (2.1040a) differentiation to (2.1040b):

$$\log \rho + \log v + \log A = \log \dot{m} = const \iff \frac{dA}{A} = -\frac{dv}{v} - \frac{d\rho}{\rho}. \quad (2.1040a, b)$$

The last term on the r.h.s. of (2.1040b) is evaluated (2.1041b–d) from the adiabatic (2.1036d) and momentum (2.1036c) relations:

$$M \equiv \frac{v}{c}: \quad \frac{d\rho}{\rho} = \frac{dp}{\rho c^2} = -\frac{v}{c^2}dv = -M^2\frac{dv}{v}, \quad (2.1041a\text{--}d)$$

using the local Mach number (2.1041a) ≡ (2.1037a); substituting (2.1041d) in (2.1040b) follows (2.1042a):

$$\frac{dA}{A} = \frac{dv}{v}\left(M^2 - 1\right): \quad \frac{dA}{dv} \begin{cases} < 0 \text{ if } M < 1: subsonic, \\ = 0 \text{ if } M = 1: sonic, \\ > 0 \text{ if } M > 1: supersonic, \end{cases} \quad (2.1042\text{a–d})$$

showing that *for the quasi-one-dimensional (2.1034a–d), steady (2.1036a) adiabatic (2.1036d) flow: (i) in subsonic conditions (2.1042b), as in the incompressible (2.1039a) case, the flow is accelerated (decelerated) in a convergent (divergent) nozzle [Figure 2.41a (2.41b)]; (ii) in supersonic conditions (2.1042d) the reverse happens, that is the flow is accelerated (decelerated) in a divergent (convergent) nozzle [Figure 2.41c (2.41d)], because the reduction in mass density compensates the volume flux (2.1039b) in the mass flux (2.1036b); (iii) the sonic condition (2.1042c) can be reached only at the minimum area, that is at a "throat" matching convergent and divergent sections (Figure 2.40).* In (2.1042c) it was shown that the sonic condition corresponds to a stationary cross-sectional area $dA = 0$; it remains to prove that the stationary condition is a minimum (not a maximum) area (subsection 2.9.5), leading to the consideration of the maximum flow rate through a nozzle (subsection 2.9.6).

2.9.5 SONIC CONDITION IN A CONVERGENT – DIVERGENT NOZZLE

In order to check that the sonic condition corresponds indeed to a minimum of the cross-sectional area, the second-order derivative of the cross-sectional area with regard to the velocity is calculated from (2.1042a):

$$\frac{d^2A}{dv^2} = \frac{d}{dv}\left[\frac{A}{v}\left(M^2 - 1\right)\right] = \frac{M^2 - 1}{v}\left(\frac{dA}{dv} - \frac{A}{v}\right) + \frac{2A}{v}M\frac{dM}{dv}. \quad (2.1043)$$

The last term on the r.h.s. of (2.1043) follows (2.1044a–c):

$$\frac{dM}{dv} = \frac{d}{dv}\left(\frac{v}{c}\right) = \frac{1}{c} - \frac{v}{c^2}\frac{dc}{dv} = \frac{1}{c} - \frac{v}{2c^3}\frac{d\left(c^2\right)}{dv} = \frac{1}{c} + \frac{\gamma-1}{2}\frac{v^2}{c^3}, \quad (2.1044\text{a–d})$$

using the adiabatic sound speed (2.661c) $\equiv$ (2.1045a) in (2.1045b) for substitution in (2.1044d):

$$c^2 = c_0^2 - \frac{\gamma-1}{2}v^2: \quad \frac{d\left(c^2\right)}{dv} = -\left(\gamma-1\right)v \quad (2.1045\text{a, b})$$

Substitution of (2.1044d) in (2.1043) specifies *the second-order derivative of the cross-sectional area with regard to the velocity (2.1043):*

$$\frac{d^2A}{dv^2} = \left(\frac{dA}{A} - \frac{A}{v}\right)\left(M^2 - 1\right) + \frac{AM}{vc}\left[2 + \left(\gamma-1\right)M^2\right], \quad (2.1046)$$

showing that it is positive (2.1047c, d) at critical sonic conditions:

$$\frac{v_*}{c_*} = M_* = 1: \lim_{M \to 1} \frac{d^2A}{dv^2} = \frac{(\gamma+1)A}{c_*^2} > 0, \qquad (2.1047a\text{–}d)$$

confirming that the cross-sectional area is minimum at the throat. Replacing the variable area nozzle by a flow tube, (2.1042b–d) implies that in the quasi-one-dimensional (2.1034a–d) steady (2.1036a) adiabatic flow, the streamlines: (i) converge (diverge) for accelerated subsonic (supersonic) flow [Figure 2.42a (2.42b)]; (ii) diverge (converge) for decelerated subsonic (supersonic) flow [Figure 2.42c (2.42d)]; (iii) are closest at the sonic condition (Figure 2.40). Thus the acceleration of a flow from subsonic to supersonic involves (Figure 2.43) the transition from converging to diverging streamlines at the sonic condition. The maximum volume (mass) flow rate of the incompressible (compressible) flow through a nozzle is considered next [subsection 2.9.6 (2.9.7)].

2.9.6 MAXIMUM VOLUME FLUX FOR INCOMPRESSIBLE NOZZLE FLOW

In the incompressible case: (i) the mass density is constant (2.1039a) $\equiv$ (2.1048a); (ii) the maximum velocity (2.1038d) $\equiv$ (2.1048c) occurs at zero pressure (2.1048b),

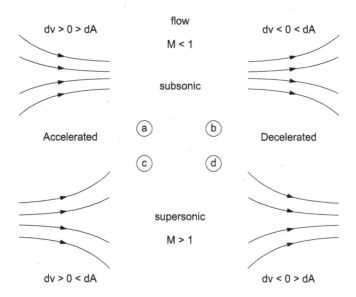

FIGURE 2.42   As for a nozzle (Figure 2.41) the streamlines of an adiabatic flow in a subsonic (supersonic) regime converge (diverge) for an accelerated flow and diverge (converge) for a decelerated flow. The nozzle (streamline) cases [Figure 2.41 (2.42)] explain why the adiabatic acceleration of a flow from subsonic to supersonic requires a convergent–divergent nozzle (Figure 2.40) with sonic flow at the "choked" throat.

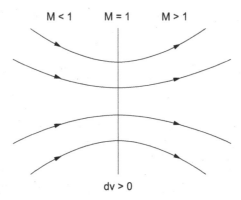

FIGURE 2.43 As for the throated nozzle (Figure 2.40) the adiabatic acceleration of a flow from subsonic to supersonic implies changing from converging to diverging streamlines with minimum flow cross-section at the sonic condition.

causing cavitation in the case of a liquid; (iii) if this condition occurs at the cross-sectional area A the volume (mass) flow rate is (2.1048d, e) [(2.1048f, g):

$$\rho = const, p = 0: \quad v_{max} = \sqrt{\frac{2\,p_0}{\rho}}, \quad j = v_{max}\, A = A\sqrt{\frac{2\,p_0}{\rho}},$$
$$\dot{m} = \rho\, j = A\sqrt{2\,p_0\,\rho}. \tag{2.1048a–g}$$

In the case of an ideal gas the incompressibility condition (2.1048a) $\equiv$ (2.1049a) limits the velocity to (2.1049b–e):

$$\rho = const: \quad v^2 \ll c_0^2 = \frac{\gamma\,p_0}{\rho} < \frac{2\,p_0}{\rho} \equiv \left(v_{max}\right)^2. \tag{2.1049a–e}$$

Thus *for the quasi-one-dimensional (2.1035a–d) steady (2.1036a) adiabatic (2.1036d) flow of an incompressible fluid (2.1048a) the maximum velocity (2.1048c) corresponds to zero pressure (2.1048b) and leads to a volume (mass) flow rate (2.1048d, e) [(2.1048f, g)] for a cross-sectional area A. For an ideal gas the incompressibility condition (2.1049a) limits the velocity (2.1049b–e) to values well below the maximum.* A distinct situation applies to the mass flow rate of a compressible fluid in a nozzle (subsection 2.9.7).

2.9.7 MAXIMUM MASS FLOW RATE FOR COMPRESSIBLE NOZZLE FLOW

In the compressible case the mass flow rate is constant and can be calculated at any section regardless of whether the nozzle is convergent (Figures 2.41a, d), divergent (Figures 2.41b, c), or convergent–divergent with one (Figure 2.40) or several throats;

the maximum velocity at the narrowest cross section is the sound speed, implying a mass flow rate (2.1050a):

$$\dot{m}_{max} = A_{min}\, c_*\, \rho_* = A_{min}\, c_0\, \rho_0 \left(\frac{2}{\gamma+1}\right)^{1/2+1/(\gamma-1)}, \qquad (2.1050a, b)$$

leading to (2.1050b) using the relations between critical and stagnation values of the sound speed (2.669c) and mass density (2.669e). It follows that *the maximum mass flow rate of the steady (2.1036a), adiabatic (2.1036d) quasi-one-dimensional (2.1034a–d) flow of a perfect gas in a nozzle corresponds to the critical or sonic condition at the narrowest cross-section (2.1050a, b), where the mass flow rate per unit area is given by (2.1050a)* $\equiv$ *(2.1051a):*

$$\dot{m} = \rho_*\, c_*\, A_* = \rho_0\, c_0\, A_0. \qquad (2.1051a, b)$$

*Introducing a **virtual stagnation area** A_0 for which (2.1051b) the flow rate would be the same with velocity equal to the stagnation sound speed, and the ratio to the throat area is given by (2.1052a–c), respectively (2.475a–c) for monoatomic/diatomic/polyatomic perfect gases (2.1052d–f):*

$$\frac{A_0}{A_*} \equiv \frac{\rho_*\, c_*}{\rho_0\, c_0} = \left(\frac{2}{\gamma+1}\right)^{1/(\gamma-1)+1/2} = \left(\frac{2}{\gamma+1}\right)^{(\gamma+1)/(2\gamma-2)} = \left(\frac{f}{f+1}\right)^{(1+f)/2}$$

$$= \left\{\left(\frac{3}{4}\right)^{3.5}, \left(\frac{5}{6}\right)^{3}, \left(\frac{6}{7}\right)^{2}\right\} = \{0.56250, 0.57870, 0.58302\}. \qquad (2.1052a\text{–}f)$$

In a supersonic convergent–divergent wind tunnel operated at different flow rates from the same stagnation values at a fluid reservoir the cross-sectional area of the throat must be adjusted with movable panels or a perforated wall used to keep sonic flow at the throat. The variation of flow variables with cross-sectional area can be specified in terms of stagnation (critical) values [subsection 2.9.8 (2.9.9)].

2.9.8   NOZZLE FLOW VARIABLES IN TERMS OF STAGNATION VALUES

At all cross-sections of a nozzle containing a quasi-one-dimensional adiabatic flow, the relation between cross-sectional area and the flow variables is specified by the mass flow rate (2.1053a) and stagnation values (2.1053b) for: (i) the velocity (2.1053c); (ii) mass density (2.1053d); (iii) the pressure (2.1053e); (iv) the temperature (2.1053f); (v) local sound speed (2.1053g); (vi) stagnation Mach number (2.1053h); (vii) local Mach number (2.1053i):

$$\frac{\dot{m}}{A} = \rho_0\, c_0\, \frac{A_0}{A} = \rho\, v = \rho_0\, v \left\{1 - \frac{\gamma-1}{2}\left(\frac{v}{c_0}\right)^2\right\}^{1/(\gamma-1)}, \qquad (2.1053a\text{–}c)$$

$$= c_0\,\rho\left|\frac{2}{\gamma-1}\left[1-\left(\frac{p}{p_0}\right)^{\gamma-1}\right]\right|^{1/2}, \tag{2.1053d}$$

$$= c_0\,\rho_0\left(\frac{p}{p_0}\right)^{1/\gamma}\left|\frac{2}{\gamma-1}\left[1-\left(\frac{p}{p_0}\right)^{1-1/\gamma}\right]\right|^{1/2}, \tag{2.1053e}$$

$$= c_0\,\rho_0\left(\frac{T}{T_0}\right)^{1/(\gamma-1)}\left|\frac{2}{\gamma-1}\left(1-\frac{T}{T_0}\right)\right|^{1/2}, \tag{2.1053f}$$

$$= c_0\,\rho_0\left(\frac{c}{c_0}\right)^{1/[2(\gamma-1)]}\left|\frac{2}{\gamma-1}\left(1-\sqrt{\frac{c}{c_0}}\right)\right|^{1/2}, \tag{2.1053g}$$

$$= \rho_0\,c_0\,M_0\left[1+\frac{\gamma-1}{2}(M_0)^2\right]^{1/(\gamma-1)}, \tag{2.1053h}$$

$$= \rho_0\,c_0\,M\left(1+\frac{\gamma-1}{2}M^2\right)^{-(\gamma+1)/[2(\gamma-1)]}, \tag{2.1053i}$$

where were used: (i) in (2.1053c) the density in terms of the velocity (2.666d); (ii) in (2.1053c) the inverse of (2.666d), that is the velocity in terms of the density:

$$\frac{v}{c_0}=\left|\frac{2}{\gamma-1}\left[1-\left(\frac{p}{p_0}\right)^{\gamma-1}\right]\right|^{1/2}; \tag{2.1054}$$

(iii/iv) in (2.1053e) [(2.1053f)] the adiabatic relation between mass density and pressure (2.478d) [temperature (2.478i)]; (v) in (2.1053g) the relation between sound speed and temperature (2.492e); (vi/vii) in (2.1053h) the stagnation Mach number (2.1055a) and in (2.1053i) its relation (2.677a) $\equiv$ (2.1055c, d) with the local Mach number (2.1055b):

$$M_0\equiv\frac{v}{c_0},\quad M\equiv\frac{v}{c}:\quad 1-\frac{\gamma-1}{2}(M_0)^2=1-\frac{(\gamma-1)M^2}{2+(\gamma-1)M^2}=\frac{2}{2+(\gamma-1)M^2}, \tag{2.1055a-d}$$

implying (2.1055e, f) from (2.677a; 2.1055d):

$$M_0\left[1+\frac{\gamma-1}{2}(M_0)^2\right]^{1/(\gamma-1)}=M\left(1+\frac{\gamma-1}{2}M^2\right)^{-1/2-1/(\gamma-1)}$$

$$=M\left(1+\frac{\gamma-1}{2}M\right)^{-(\gamma+1)/[2(\gamma-1)]}. \tag{2.1055e, f}$$

Thus *for the steady adiabatic unidimensional flow in a nozzle is specified the velocity (2.1053c), mass density (2.1053d), pressure (2.1053e), temperature (2.1053f), local sound speed (2.1053g), stagnation (2.1053h) and local (2.1053i) Mach number in terms of: (i) stagnation values* $(v_0, \rho_0, p_0, c_0, T_0)$; *(ii) the area A of the cross section; (iii) the constant mass flow rate (2.1053a). The latter cannot exceed the value (2.1050a,b) for a choked nozzle with sonic flow at the throat.* In the latter case of a "choked" convergent–divergent nozzle, the stagnation values in (2.1053a–h) can be replaced by the critical values at the sonic condition at the throat (subsection 2.9.9).

2.9.9   Nozzle Flow Variables in Terms of Critical Values

The adiabatic relations (2.668a–e) between local and critical flow variables can be combined with the constant mass flux (2.1050a, b) in a "choked" convergent–divergent nozzle leading to alternative relations in terms of stagnation (2.1053a–i) [critical (2.1056a–i)] values as stated next: *for the quasi-one-dimensional (2.1034a–d) steady (2.1036a) adiabatic (2.1036d) flow of a perfect gas, at all cross-sections of a "choked" convergent–divergent nozzle with mass flow rate (2.1051a, b)* ≡ *(2.1056a, b) all flow variables (2.1053b–h) can be related to the critical values of the sonic flow at the throat, for: (i) the velocity (2.1056c); (ii) the mass density (2.1056d); (iii) the pressure (2.1056e); (iv) the temperature (2.1056f); (v) the local sound speed (2.1056g); (vi/vii) the critical (2.1056h) and local (2.1056i) Mach numbers:*

$$\frac{\dot{m}}{A} = \rho_* c_* \frac{A_*}{A} = \rho v = \rho_* v \left(\frac{\gamma+1}{2} - \frac{\gamma-1}{2} \frac{v^2}{c_*^2} \right)^{1/(\gamma-1)}, \qquad (2.1056\text{a–c})$$

$$= c_* \rho \left| \frac{2}{\gamma-1} \left[\frac{\gamma+1}{2} - \left(\frac{\rho}{\rho_*} \right)^{\gamma-1} \right] \right|^{1/2}, \qquad (2.1056\text{d})$$

$$= c_* \rho_* \left(\frac{p}{p_*} \right)^{1/\gamma} \left| \frac{2}{\gamma-1} \left[1 - \frac{\gamma+1}{2} \left(\frac{p}{p_*} \right)^{1-1/\gamma} \right] \right|^{1/2}, \qquad (2.1056\text{e})$$

$$= c_* \rho_* \left(\frac{T}{T_*} \right)^{1/(\gamma-1)} \left| \frac{2}{\gamma-1} \left(1 - \frac{\gamma+1}{2} \frac{T}{T_*} \right) \right|^{1/2}, \qquad (2.1056\text{f})$$

$$= c_* \rho_* \left(\frac{c}{c_*} \right)^{1/[2(\gamma-1)]} \left| \frac{2}{\gamma-1} \left(1 - \frac{\gamma+1}{2} \sqrt{\frac{c}{c_*}} \right) \right|^{1/2}, \qquad (2.1056\text{g})$$

$$= \rho_* c_* M_* \left| \frac{\gamma+1}{2} - \frac{\gamma-1}{2} (M_*)^2 \right|^{1/(\gamma-1)}, \qquad (2.1056\text{h})$$

$$= \rho_* c_* M \left[\frac{\gamma+1}{2+(\gamma-1)M^2} \right]^{(\gamma+1)/[2(\gamma-1)]} . \tag{2.1056i}$$

The relations (2.1056c–g) follow from (2.1056b; 2.668a–e) and in the passage from (2.1056h) to (2.1056i) was used the relation (2.677c) $\equiv$ (2.1057c–e) between the critical (2.1057a) and local (2.1057b) Mach numbers:

$$M_* \equiv \frac{v}{c_*}, \quad M \equiv \frac{v}{c}: \quad \frac{\gamma+1}{2} - \frac{\gamma-1}{2}(M_*)^2 = \frac{\gamma+1}{2} - \frac{\gamma-1}{2}\frac{(\gamma+1)M^2}{2+(\gamma-1)M^2}$$

$$= \frac{\gamma+1}{2}\left[1 - \frac{(\gamma-1)M^2}{2+(\gamma-1)M^2} \right] \tag{2.1057a–e}$$

$$= \frac{\gamma+1}{2+(\gamma-1)M^2},$$

implying (2.1057f, b) from (2.677c; 2.1057e):

$$M_* \left| \frac{\gamma+1}{2} - \frac{\gamma-1}{2}(M_*)^2 \right|^{1/(\gamma-1)} = M \left[\frac{\gamma+1}{2+(\gamma-1)M^2} \right]^{1/2+1/(\gamma-1)}$$

$$= M \left[\frac{\gamma+1}{2+(\gamma-1)M^2} \right]^{(\gamma+1)/[2(\gamma-1)]} . \tag{2.1057f, g}$$

The consistency of (2.1053i) $\equiv$ (2.1056i) can be checked using (2.669d, f) in (2.1052a, b):

$$\rho_* c_* \left[\frac{\gamma+1}{2+(\gamma-1)M^2} \right]^{(\gamma+1)/[2(\gamma-1)]}$$

$$= \rho_0 c_0 \left(\frac{\gamma+1}{2} \right)^{-1/2-1/(\gamma-1)} \left[\frac{2+(\gamma-1)M^2}{\gamma+1} \right]^{-(\gamma+1)/[2(\gamma-1)]}$$

$$= \rho_0 c_0 \left(1 + \frac{\gamma-1}{2}M^2 \right)^{-(\gamma+1)/[2(\gamma-1)]} . \tag{2.1058a, b}$$

The flow variables for the adiabatic flow in a "choked" nozzle can be expressed in terms of stagnation (critical) values [subsection 2.9.8 (2.9.9)] or a combination of the two (subsection 2.9.10).

2.9.10 Mass Flow Rate and Variables in a "Choked" Nozzle

Considering the adiabatic flow in a "choked" nozzle with sonic conditions at the throat (Figure 2.40) instead of calculating all flow variables directly from stagnation (2.1053a–i) [critical (2.1056a–i)] values, it is possible to use combinations, such as: (i) in the convergent section (2.1059a) before the throat the stagnation values may be used (2.666b, d) to calculate the flow rate (2.1059b) ≡ (2.1036b) and the velocity (2.1053c) ≡ (2.1059c):

$$A_1 \geq A \geq A_* : \quad \frac{\dot{m}}{A} = \rho v = \rho_0\, v \left[1 - \frac{\gamma-1}{2}\left(\frac{v}{c_0}\right)^2\right]^{\frac{1}{\gamma-1}}, \qquad (2.1059\text{a–c})$$

with the other flow variables specified by the velocity using (2.666a–e; 2.1055d) ≡ (2.1060a–f):

$$1 - \frac{\gamma-1}{2}\frac{v^2}{(c_0)^2} = \frac{T}{T_0} = \frac{c^2}{c_0^2} = \left(\frac{\rho}{\rho_0}\right)^{\gamma-1} = \left(\frac{p}{p_0}\right)^{1-1/\gamma}$$

$$= 1 - \frac{\gamma-1}{2}(M_0)^2 = \left(1 + \frac{\gamma-1}{2}M^2\right)^{-1}; \qquad (2.1060\text{a–f})$$

(ii) in the divergent section (2.1061a) after the throat the critical values at the sonic condition may be used to calculate the mass flow rate (2.1061b) ≡ (2.1036b) and velocity (2.1061c) ≡ (2.1056c):

$$A_* \leq A \leq A_2 : \quad \frac{\dot{m}}{A} = \rho v = \rho_*\, v \left(\frac{\gamma+1}{2} - \frac{\gamma-1}{2}\frac{v^2}{c_*^2}\right)^{\frac{1}{\gamma-1}}, \qquad (2.1061\text{a–c})$$

with the remaining flow variables calculated (2.668a–e; 2.1057e) ≡ (2.1062a–f) from the velocity:

$$\frac{\gamma+1}{2} - \frac{\gamma-1}{2}\frac{v^2}{c_*^2} = \frac{T}{T_*} = \frac{c^2}{c_*^2} = \left(\frac{\rho}{\rho_*}\right)^{\gamma-1} = \left(\frac{p}{p_*}\right)^{1-1/\gamma}$$

$$= \frac{\gamma+1}{2} - \frac{\gamma-1}{2}M_*^2 = \frac{\gamma+1}{2+(\gamma-1)M^2}. \qquad (2.1062\text{a–f})$$

The cases of nozzles with parabolic or catenoidal profile are considered in example 10.8. The preceding relations (subsections 2.9.3–2.9.10) specify the exit in terms of the inlet flow of the nozzle. If the nozzle exhausts to the atmosphere or ambient medium, depending on the relative pressures, normal (oblique) shocks may be formed inside (outside) the nozzle.

2.9.11 PERFECTLY EXPANDED AND UNDER(OVER)EXPANDED FLOWS

The relation (2.1053e) specifies the exit pressure under the assumption of adiabatic flow throughout the nozzle:

$$\frac{\dot{m}}{A_2} = c_a\, \rho_a \left(\frac{p_2}{p_a}\right)^{1/\gamma} \left| \frac{2}{\gamma-1}\left[1-\left(\frac{p_2}{p_a}\right)^{1-1/\gamma}\right] \right|^{1/2}. \qquad (2.1063)$$

Depending on the ratio p_2/p_a to the ambient pressure three cases can arise:

$$p_2 \begin{cases} > p_a : underexpanded : expansion\ waves, \\ = p_a : perfectly\ expanded : shear\ layers, \\ < p_a : overexpanded : shock\ waves. \end{cases} \qquad (2.1064a\text{--}c)$$

The flow is **perfectly expanded** if the nozzle exhaust pressure matches (2.1064b) the atmospheric pressure: (i) the jet exits the nozzle lip tangentially as if continuing the nozzle (Figure 2.44a) and the whole nozzle is used to accelerate the jet and recover the wall pressure as thrust; (ii) the jet is separated from ambient air by two shear layers (Figure 2.45a) that are generally unsteady and turbulent, and broaden downstream mixing the jet with its surroundings.

The flow is **underexpanded** if the nozzle exit pressure exceeds (2.1064a) the atmospheric pressure: (i) the **bulged jet** spreads out of the nozzle lip (Figure 2.44b),

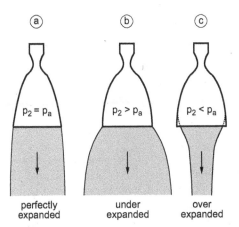

FIGURE 2.44   The flow out of a nozzle is perfectly expanded (a) when the exit pressure equals the atmospheric pressure: (i) the streamlines issue tangentially from the nozzle lip; (ii) the nozzle length extracts the maximum thrust from the internal pressure. In the case of an underexpanded (b) [overexpanded (c)] flow the exit pressure is larger (smaller) than the atmospheric pressure: (i) the streamlines bulge out (converge away) from the nozzle lip; (ii) the nozzle is too short (long) and does not extract the maximum thrust (is longer than needed to extract maximum thrust).

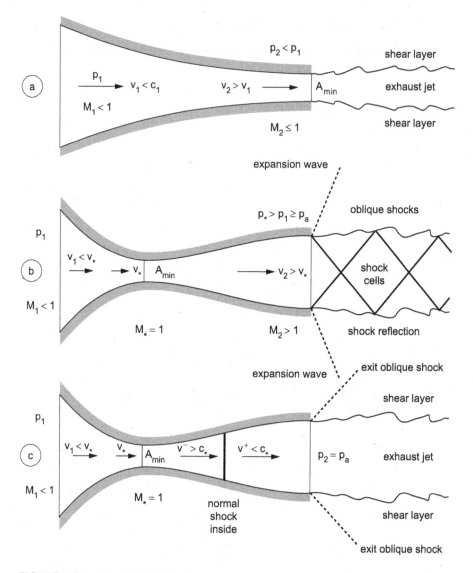

FIGURE 2.45 The adiabatic flow in a convergent nozzle (a) reaches the maximum velocity at the exit, and since it is subsonic there are no shock waves, just a shear layer separating the exhaust jet from the ambient medium. In the case (b) of a convergent–divergent nozzle with supersonic exit flow in the underexpanded case (Figure 2.44b) of exit pressure larger than the atmospheric pressure, oblique shocks may form from the nozzle lips, and their reflection from the shear layers leads to a diamond shock cell pattern. In the case (c) of a convergent–divergent nozzle for which the supersonic exit flow would be overexpanded (Figure 2.44c) and have an exit pressure lower than the atmospheric, the flow cannot be adiabatic in the whole nozzle, and a normal shock must form at a location between the throat and exit such that the exit pressure equals the atmospheric pressure, and thus there is no shock in the subsonic exhaust flow, although oblique shocks may form outside from the nozzle lips.

as the nozzle is too short to fully accelerate the flow and use the wall pressure for thrust; (ii) expansion waves may form around the nozzle lips or oblique shocks are generated (Figure 2.45b) that are reflected between the shear layers forming a **diamond shock cell pattern**.

The flow is **overexpanded** if the nozzle exhaust pressure is lower (2.1064c) than atmospheric pressure: (i) the jet may detach from the nozzle before the lip (Figure 2.44c) so that the **pinched jet** does not use the full length of the nozzle, that could be replaced by a shorter one; (ii) the nozzle exit pressure has to be raised to the atmospheric pressure by a shock (Figure 2.45c) at the correct location within the nozzle or at the nozzle lip that is thus both **choked** (sonic flow) at the throat and **shocked** (normal shock wave between the throat and the exit).

The atmospheric pressure decreases with altitude, and if the nozzle with fixed area ratio A_2/A_* has for **design point** a perfect expansion at altitude z_1 with equal exhaust and atmospheric pressures, it will be underexpanded (overexpanded) at higher (lower) altitude where the atmospheric pressure is lower (higher). A **variable area nozzle** varies the exit area so that it remains perfectly expanded over an altitude range.

A **variable area inlet** can also be used to regulate the mass flow for different velocities and mass densities as a function of altitude. The complexity and cost of variable area inlets and nozzles (Figure 2.46a) is most justified for supersonic aircraft with a wide speed and altitude range. A variable area inlet with **splitter plates** can be

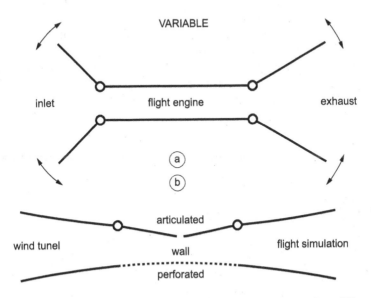

FIGURE 2.46 In order to adjust the flow in the jet engine of an aircraft at different flight speeds and altitudes (a) inlets (nozzles) with variable area may be used to adjust the mass flow rate (exhaust pressure). Similarly, in a throated supersonic wind tunnel the mass flow rate may be adjusted (b) by varying the throat area using movable panels or keeping the throat area with perforated walls to extract part of the flow.

used to place the oblique shock in supersonic flight for maximum compression prior to admission to a turbine engine.

In the case of a **wind tunnel** (Figure 2.46b) accelerating the air or another fluid from rest in a pressurized reservoir to a supersonic speed in the **test section** requires a convergent-divergent nozzle with sonic flow at the throat fixing the mass flow rate. The use of perforated walls or plates allows adjusting the mass flow rate to obtain different velocities downstream of the throat.

A rocket engine, with **fixed or extensible nozzle** has the widest operating range from sea level atmospheric pressure to the vacuum of space: in space the jet is always underexpanded because the external pressure is zero. If the design point for perfect expansion is at high altitude conditions a rocket engine with a fixed nozzle will be underexpanded at lift-off and low altitude and have an internal shock.

The location of the shock in an underexpanded nozzle can be determined by using the adiabatic relations before and after the shock together with one shock relation; two alternatives are the variation of pressure (mass density) across the shock [subsection 2.9.12 (2.9.13)] together with the pressure (mass density) at the throat and the exit stagnation Mach number. In both cases the normal shock is located by specifying the value of the upstream local Mach number M^-.

2.9.12   LOCATION OF THE NORMAL SHOCK IN AN OVEREXPANDED NOZZLE

In an overexpanded nozzle the adiabatic relations apply on each side of the shock so all that is needed is to locate the shock. Its position is uniquely specified by the upstream Mach number M^-; substitution in (2.1053i) or (2.1056i) then determines the cross-section A_3 where the shock is located. Across a normal shock the upstream v^- and downstream v^+ velocities satisfy (2.801a) $\equiv$ (2.1065a) implying that the upstream M_*^- and downstream M_*^+ critical Mach numbers are inverse (2.1065b) and the shock location is specified by the dimensionless parameter (2.1065c, d):

$$1 = \frac{v^- v^+}{c_*^2} = M_*^- M_*^+ \quad N = \left(M_*^-\right)^2 = \left(M_*^+\right)^{-2}. \qquad (2.1065\text{a--d})$$

The pressure is related to the stagnation pressure in adiabatic conditions by (2.668a, e) $\equiv$ (2.1066a, b):

$$\left(\frac{p}{p_*}\right)^{1-1/\gamma} = \frac{\gamma+1}{2} - \frac{\gamma-1}{2}\left(\frac{v}{c_*}\right)^2 = \frac{\gamma+1}{2} - \frac{\gamma-1}{2} M_*^2. \qquad (2.1066\text{a, b})$$

The flow is adiabatic from the throat to upstream of the shock leading to (2.1067a, b):

$$\left(\frac{p^-}{p_*}\right)^{1-1/\gamma} = \frac{\gamma+1}{2} - \frac{\gamma-1}{2}\left(M_*^-\right)^2 = \frac{\gamma+1}{2} - \frac{\gamma-1}{2} N; \qquad (2.1067\text{a, b})$$

likewise the flow is adiabatic from downstream of the shock to the nozzle exit leading to (2.1068a, b):

$$\left(\frac{p^+}{p_e}\right)^{1-1/\gamma} = \frac{\gamma+1-(\gamma-1)\left(\dfrac{v^+}{c_*}\right)^2}{\gamma+1-(\gamma-1)\left(\dfrac{v_e}{c_*}\right)^2} = \frac{1}{N}\cdot\frac{1-\gamma+(\gamma+1)N}{\gamma+1-(\gamma-1)\left(M_{*e}\right)^2}. \qquad (2.1068a, b)$$

The ratio of pressures across the shock is specified by the ratio of (2.1068b) to (2.1067b) leading to (2.1069):

$$\left(\frac{p^+}{p^-}\right)^{1-1/\gamma} = \left(\frac{p_e}{p_*}\right)^{1-1/\gamma}\times\frac{(\gamma+1)N+1-\gamma}{\gamma+1-(\gamma-1)N}\times\frac{2/N}{\gamma+1-(\gamma-1)\left(M_{*e}\right)^2}. \qquad (2.1069)$$

The ratio of pressures across the shock is given by (2.832b, a) $\equiv$ (2.1070a, b) in terms of the local upstream Mach number, or using (2.678c) by (2.1070b) in terms of the upstream critical Mach number, leading (2.1065c) to (2.1070c):

$$\frac{p^+}{p^-} = \frac{2\gamma\left(M^-\right)^2-\gamma+1}{\gamma+1} = \frac{(\gamma+1)\left(M_*^-\right)^2+1-\gamma}{\gamma+1-(\gamma-1)\left(M_*^-\right)^2} = \frac{(\gamma+1)N+1-\gamma}{\gamma+1-(\gamma-1)N}. \qquad (2.1070a\text{--}c)$$

The comparison of (2.1070c) and (2.1069) leads to:

$$f_1(N)\equiv\left[\frac{\gamma+1-(\gamma-1)N}{1-\gamma+(\gamma+1)N}\right]^{1/\gamma} \quad N=\left(\frac{p_e}{p_*}\right)^{1-1/\gamma}\frac{2}{\gamma+1-(\gamma-1)\left(M_{*e}\right)^2}, \qquad (2.1071a, b)$$

that *determines the root N > 1 of (2.1071b) the upstream critical Mach number (2.1065c) and hence the location of the shock in a "choked" and "shocked" convergent–divergent nozzle (Figure 2.45c) in terms of: (i) the critical pressure at the throat; (ii) the exit pressure; (iii) the exit critical Mach number.* The location of the shock in a "choked" and "shocked" nozzle can be determined from the exit critical Mach number using alternatively the pressure (mass density) at the throat and exit [subsection 2.9.12 (2.9.13)].

2.9.13 Shock Location via Pressure or Mass Density

The conservation of mass flux across a normal shock (2.762b) $\equiv$ (2.1072a) specifies the ratio of mass densities (2.1072b–e) in terms (2.1065a) of the upstream Mach number (2.1065c):

$$\frac{\rho^+}{\rho^-} = \frac{v^-}{v^+} = \frac{v^-}{\left(c_*\right)^2/v^-} = \left(\frac{v^-}{c_*}\right)^2 = \left(M_*^-\right) = N. \qquad (2.1072a\text{--}e)$$

The flow upstream (downstream) of the shock is adiabatic, and thus the mass density is related by (2.668d) ≡ (2.1073a) to the critical value at the throat (2.1073b) [the value at the exit of the nozzle (2.1074a, b)]

$$\left(\frac{\rho^-}{\rho_*}\right)^{\gamma-1} = \frac{\gamma+1}{2} - \frac{\gamma-1}{2}\left(\frac{v^-}{c_*}\right)^2 = \frac{\gamma+1}{2} - \frac{\gamma-1}{2}N, \qquad \text{(2.1073a, b)}$$

$$\left(\frac{\rho^+}{\rho_e}\right)^{\gamma-1} = \frac{\gamma+1-(\gamma-1)\left(\dfrac{v^+}{c_*}\right)^2}{\gamma+1-(\gamma-1)\left(\dfrac{v_e}{c_*}\right)^2} = \frac{1}{N}\frac{1-\gamma+(\gamma+1)N}{\gamma+1-(\gamma-1)(M_{*e})^2}. \qquad \text{(2.1074a, b)}$$

The ratio of mass densities after (2.1074b) and before (2.1073b) the shock is (2.647a–d):

$$\left(\frac{\rho^+}{\rho^-}\right)^{\gamma-1} = \left(\frac{\rho_e}{\rho_*}\right)^{\gamma-1}\frac{1-\gamma+(\gamma+1)N}{\gamma+1-(\gamma-1)N}\frac{2/N}{\gamma+1-(\gamma-1)(M_{*e})^2}; \qquad \text{(2.1075)}$$

using (2.1072e) leads to:

$$f_2(N) \equiv N^\gamma \frac{\gamma+1-(\gamma-1)N}{1-\gamma+(\gamma+1)N} = \left(\frac{\rho_e}{\rho_*}\right)^{\gamma-1}\frac{2}{\gamma+1-(\gamma-1)(M_{*e})^2}, \qquad \text{(2.1076a, b)}$$

that *specifies as the root N > 1 of (2.1076b) upstream Mach number or location of the shock (2.1065c) in a "choked" and "shocked" nozzle in terms of: (i–ii) the mass density at the throat and exit; (iii) the exit critical Mach number. The former (i–ii) may be replaced (2.668a–e) by the stagnation and exit sound speeds (2.1077b) [temperatures (2.1077c)] in (2.1077d, e):*

$$\left[f_1(N)\right]^\gamma = f_2(N), \qquad \text{(2.1077a)}$$

$$\left(\frac{p_e}{e_*}\right)^{\gamma-1} = \left(\frac{c_e}{c_*}\right)^2 = \frac{T_e}{T_*} = \frac{p_e}{p_*}\frac{\rho_*}{\rho_e} = \frac{\gamma+1}{2} - \frac{\gamma-1}{2}M_{*e}^2, \qquad \text{(2.1077b–d)}$$

from (2.1071a; 2.1076a) follows (2.1077a) and (2.1071b) is equivalent to (2.1076b) in agreement with (2.478d). Although (2.1076b) is somewhat simpler, (2.1071b) involves directly the exit pressure that must match the atmospheric pressure for an overexpanded nozzle (Figure 2.45c). The nozzle of a jet or rocket engine exhausts to the atmosphere (subsection 2.9.14), whose properties vary with altitude, for example, in the two lowest layers [subsection 2.9.15 (2.9.16)] namely the troposphere (stratosphere).

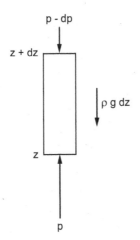

FIGURE 2.47   The hydrostatic equilibrium balances the weight of a column of fluid of density ρ under the acceleration g of gravity against the decrease of pressure with height.

2.9.14 HYDROSTATIC EQUILIBRIUM OF INCOMPRESSIBLE/COMPRESSIBLE FLUIDS

Consider column of fluid in hydrostatic equilibrium (Figure 2.47): (i/ii) the pressure below (above) at altitude z ($z + dz$) is higher p (lower $p - dp$) and directed upward (downward); (iii) the pressure difference is balanced (2.1078b) by the weight, equal to the product of the acceleration of gravity g by the mass, that equals the mass density multiplied by the altitude difference (2.1078a):

$$\rho = \frac{dm}{dz}: \quad p - (p - dp) = dp = -g \, dm = -\rho \, g \, dz \quad \Leftrightarrow \quad \frac{dp}{dz} = -\rho \, g. \quad (2.1078a\text{–}e)$$

Thus *in hydrostatic equilibrium the **pressure gradient** defined as the rate of change of pressure with altitude equal (2.1078e) minus the product of the mass density ρ by the acceleration of y g. The simplest case is an incompressible liquid (2.1079a) in a uniform gravity field (2.1079b) for which holds the **Archimedes law (250 B.C.)** stating alternatively that: (i) the pressure decreases linearly with altitude (2.1079c):*

$$\rho = const, \quad g = const; \quad p(z) = p(0) - \rho \, g \, z \quad \Leftrightarrow \quad p(0) - p(z) = \rho \, g \, z \quad (2.1079a\text{–}d)$$

(ii) the difference in pressure over a height z is obtained (2.1079d) multiplying by the constant mass density (2.1079a) and acceleration of gravity (2.1079b). For a perfect gas (2.473g) ≡ (2.1080b) in a uniform gravity field (2.1080a) the pressure (2.1078e) ≡ (2.1080c) depends only on the temperature

$$g = const, \quad p = \rho \overline{\overline{R}} T: \quad \frac{dp}{p} = -\frac{g}{\overline{\overline{R}} T} \, dz, \quad (2.1080a\text{–}c)$$

*and the **temperature profile** as a function of altitude specifies the dependence on altitude of pressure (2.1081a) [mass density (2.1081b)]:*

$$\exp\left\{-\int_0^z \frac{g}{RT(\xi)}d\xi\right\} = \frac{p(z)}{p(0)} = \frac{\rho(z)}{\rho(0)}\frac{T(z)}{T(0)}. \qquad (2.1081a, b)$$

The pressure and mass density profiles of a perfect gas under a uniform gravity field are obtained next for constant temperature and constant temperature gradient (subsection 2.9.15).

2.9.15 ISOTHERMAL/POLYTROPIC ATMOSPHERE OF A PERFECT GAS

*In the case of an **isothermal atmosphere** (2.1082b) under a uniform gravity field (2.1082a) both the pressure (2.1081a) [mass density (2.1081b)] decay exponentially (2.1082c) [(2.1082d)] on the **scale height** (2.1082e):*

$$g = const, \quad T(z) = const : \quad \frac{p(z)}{p(0)} = \frac{\rho(z)}{\rho(0)} = \exp\left(-\frac{z}{L}\right), \quad L \equiv \frac{\overline{\overline{R}}T}{g}. \qquad (2.1082a\text{–}e)$$

If instead the temperature decreases linearly with altitude (2.1083b) over a limited altitude range (2.1083a) the temperature gradient is constant (2.1083c):

$$0 \le z \le \ell: \quad T(z) = T_0\left(1 - \frac{z}{\ell}\right), \quad \frac{dT}{dz} = -\frac{T_0}{\ell}, \qquad (2.1083a\text{–}c)$$

and the pressure (2.1081a) is given by (2.657a–e):

$$\frac{p(z)}{p(0)} = -\frac{g}{RT_0}\exp\left\{-\int_0^z \frac{d\xi}{1-\frac{\xi}{\ell}}\right\} = \frac{g\ell}{RT_0}\exp\left\{\log\left(1-\frac{z}{\ell}\right)\right\} = \left(1-\frac{z}{\ell}\right)^{\frac{\ell}{L_0}}. \qquad (2.1084a\text{–}c)$$

Thus *the hydrostatic equilibrium (2.1078e) in a uniform gravity field (2.1080a) of a perfect gas (2.1080b) with a constant temperature gradient (2.1083a–c) leads to the pressure (2.1084c) ≡ (2.1085a) [mass density (2.1081b) ≡ (2.1085b):*

$$\frac{p(z)}{p(0)} = \left(1-\frac{z}{\ell}\right)^{\ell/L_0}, \quad \frac{\rho(z)}{\rho(0)} = \left(1-\frac{z}{\ell}\right)^{\ell/L_0-1}, \qquad (2.1085a, b)$$

*corresponding to a **polytropic atmosphere** (2.1086c) with **polytropic exponent** (2.1086a, b):*

$$n = \frac{\ell/L_0}{\ell/L_0-1} = \frac{\ell}{\ell-L_0} > 1: \quad \frac{p(z)}{p(0)} = \left[\frac{\rho(z)}{\rho(0)}\right]^n. \qquad (2.1086a\text{–}c)$$

The polytropic (isothermal) atmospheric model is applied next to the two lowest layers of the atmosphere (subsection 2.9.16), namely the troposphere (stratosphere) and their altitude limits, namely: (i) sea level at the base of the troposphere; (ii) the tropo(strato)pause at the coincident top (bottom) of the tropo(strato)sphere; (iii) at the top of the stratosphere.

2.9.16 SEA LEVEL, TROPOSPHERE, TROPOPAUSE AND STRATOSPHERE

The physical properties of the atmosphere of the earth vary with latitude (from equator to poles), season of the year and local weather conditions. For purposes such as calibration of instruments or comparison of the performance of aircraft it is necessary to have a reference such as the **International Standard Atmosphere (I.S.A.)** that corresponds to average conditions at intermediate latitudes. I.S.A. is defined (Figure 2.48) by the temperature profile (2.1087a, b):

$$T(z) = \begin{cases} T_0\left(1 - \dfrac{z}{\ell}\right) & \text{if } 0 \le z \le \overline{z_0}: \quad troposphere, \\ \overline{T_0} & \text{if } \overline{z_0} \le z \le \overline{\overline{z_0}}: \quad stratosphere, \end{cases} \qquad (2.1087\text{a--b})$$

in the two lowest layers (2.661a–d):

$$\{z_0, \overline{z_0}, \overline{\overline{z_0}}\} = \{0, 11, 25\}\,km = \{0.0, 1.1, 2.5\} \times 10^4\,m, \qquad (2.1088\text{a--c})$$

$$T_0 = 15\,C = 289.15\,K, \quad \overline{T_0} = -56.5\,C = 216.65\,K, \qquad (2.1088\text{d--g})$$

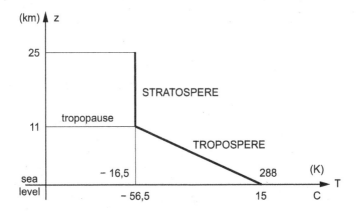

FIGURE 2.48 In order to assess aircraft performance in comparable conditions, the International Standard Atmosphere (I.S.A.) is used for a perfect gas, specified by the temperature profile as a function of altitude: (i) linear decay of the temperature from 15 C at sea level at a lapse rate of −6.5C/km in the troposphere up to the tropopause at 11 km altitude, corresponding to a polytropic atmosphere; (ii) the temperature −56.5 C at the tropopause is constant in the isothermal stratosphere up to the altitude of 25 km. The I.S.A. corresponds to the Earth's atmosphere at mid latitudes and moderate seasonal conditions, and may be considered for higher or lower temperatures or altitudes.

namely: (i) the **troposphere** (2.1087a) is a polytropic with the temperature varying linearly from (2.716a, b) ≡ (2.1088d, e) at **sea level** (2.1088a) to (2.1088f, g) at the **tropopause** (2.1088b); (ii) the **stratosphere** (2.1087b) is isothermal with the temperature (2.1089f, g) at the tropopause (2.1088b) unchanged up to the top (2.1088c) of the stratosphere. The I.S.A. may be considered: (i) with a temperature difference to represent different regions, for example, *ISA* + 15 *C* (*ISA* − 20 *C*) for tropical (polar) regions; (ii) at any altitude, for example, $z = 3$ *km* for an airport at a plateau or an helicopter hovering near a mountain; (iii) a combination of (i) and (ii), for example, *ISA* + 20 *C* at $z = 4$ *km* for hot-and-high conditions.

From the temperature at sea level (2.1088d, e) and the tropopause (2.1088f, g) and the height or thickness of the troposphere (2.1088b) follows the constant temperature gradient or **lapse rate** (2.663a–c):

$$0 \leq z \leq \bar{z}_0 : \quad \frac{dT}{dz} = \frac{\bar{T}_0 - T_0}{\bar{z}_0} = -\frac{71.5\,K}{11\,km} = -6.5\,K.km^{-1} = -6.5 \times 10^{-3}\,K\,m^{-1}, \quad (2.1089a\text{–}e)$$

and (2.1083c) the length scale (2.1089f–i)

$$\ell = -\frac{T_0}{dT/dz} = -\frac{288.15\,K}{6.5\,K.km^{-1}} = 44.33\,km = 4.433 \times 10^4\,m > \bar{z}_0, \quad (2.1089f\text{–}j)$$

that exceeds (2.1089j) the height of the troposphere (2.1088b), in agreement (2.1083a) with a positive temperature (2.1087a) at all altitudes. The acceleration of gravity (2.713a) ≡ (2.1090a) and perfect gas constant (2.718e) ≡ (2.1090b) determine the scale height (2.1082e) at sea level and at the troposphere (2.1090c–f):

$$g = 9.80665\,m\,s^{-2}, \quad R_0 = 287.11\,m^2\,s^{-2}\,K^{-1}, \quad (2.1090a, b)$$

$$\{L_0, \bar{L}_0\} = \frac{R_0}{g}\{T_0, \bar{T}_0\} = \frac{287.11\,m^2\,s^2\,K^{-1}}{9.80665\,m\,s^{-2}} \times \{288.15, 216.65\} \quad (2.1090c, d)$$

$$= 8.44(6.34) \times 10^3\,m = 8.44(6.34)\,km. \quad (2.1090e, f)$$

The exponents for the pressure (2.1085a) and mass density (2.1085b) and polytropic exponent (2.1086a, b) are, respectively (2.1091a, b)

$$\left\{ \frac{\ell}{L_0}, \frac{\ell}{L_0} - 1, \frac{\ell}{\ell - L_0} \right\} \equiv n \} = \left\{ \frac{44.33}{8.44}, \frac{35.89}{8.44}, \frac{44.33}{35.89} \right\} = \{5.25, 4.25, 1.237\}, \quad (2.1091a\text{–}e)$$
$$1 < n \equiv 1.235 < \gamma \geq 1.33,$$

where the polytropic exponent lies between unity and (2.475a–c) the adiabatic exponent (2.1091d, e). The preceding results specify the pressure and mass density at all altitudes in the troposphere and stratosphere from the values at sea level, for example, the corresponding values at the tropopause and top of the stratosphere (subsection 2.9.17).

2.9.17 PRESSURE, MASS DENSITY, TEMPERATURE AND SOUND SPEED

The atmospheric values at sea level (subsection 2.6.29) specify the values at all altitudes in the troposphere (stratosphere) using the polytropic (isothermal) relations. For example, Table 2.23 compares the values (index "0") at sea level (2.1088a) with: (i) the values (one overbar) at the tropopause (2.1088b), that is the top (bottom) of the troposphere (stratosphere); (ii) the values (two overbars) at the top (2.1088c) of the stratosphere. The temperature profile (Figure 2.48) determines the square of the sound speed. The sound speed c_0 is (2.721a–g) at sea level (2.1088a) and its square varies linearly with altitude like the temperature to a constant value (2.1092a–e) at the tropopause (2.1088b):

$$\bar{c}_0 = \sqrt{\gamma\, R \bar{T}_0} = \sqrt{1.4 \times 287.11\, m^2\, s^{-1}\, K^{-1} \times 216.65\, K} = 295\, m\, s^{-1} \qquad (2.1092\text{a–c})$$

$$= 295 \times 3.6\, km\, h^{-1} = 1062\, km\, h^{-1}. \qquad (2.1092\text{d, e})$$

The polytropic factor:

$$1 - \frac{\bar{z}_0}{\ell} = \frac{\bar{T}_0}{T_0} = \frac{216.65}{288.15} = 0.752, \qquad (2.1093)$$

TABLE 2.23
Physical Properties of the International Standard Atmosphere (ISA)

Variable			Condition		
Name	Symbol	Unit	Sea Level (a)	Tropopause (b)	Top of Stratosphere (c)
Altitude	z	$\times 10^3\, m$	0	11	25
Temperature	T	C	15	−56.5	−56.5
		K	288.15	216.65	216.65
Sound speed	c	$m\, s^{-1}$	340	295	295
		$km\, h^{-1}$	1224	1062	1062
Pressure	p	atm	1	0.224	0.0246
		bar	1.014	0.227	0.0250
		kPa	101.4	22.7	2.50
Mass density	ρ	$kg\, m^{-3}$	1.225	0.365	0.0402

Note: Temperature, sound speed, pressure and mass density in the International Standard Atmosphere (ISA) at three altitudes: (a) sea level; (b) tropopause separating the troposphere from the stratosphere; (c) top of the stratosphere.

specifies with (2.1091a–c) the pressure (2.672a–g) [mass density (2.1094a–c)] at the tropopause from their sea level values (2.716e–m) [(2.721a–c)]:

$$\bar{p}_0 = p_0 \left(1 - \frac{\bar{z}_0}{\ell}\right)^{\ell/L_0} = 0.752^{5.25} \, atm = 0.224 \, atm$$

$$= 0.224 \times 1.014 \times 10^5 \, kg \, m^{-1} \, s^{-2} = 2.27 \times 10^4 \, kg \, m^{-1} \, s^{-2} = 22.7 \, kPa,$$

(2.1094a–e)

$$\bar{\rho}_0 = \rho_0 \left(1 - \frac{\bar{z}_0}{\ell}\right)^{\ell/L_0 - 1} = 0.752^{4.25} \times 1.225 \, kg \, m^{-3} = 0.365 \, kg \, m^{-3}. \quad (2.1095a–c)$$

Across the stratosphere, from the bottom at the tropopause (2.1088b) to the top at altitude (2.1088c) both the pressure and mass density decay (2.655a–e) by a factor (2.1096a, b):

$$\frac{\bar{\bar{p}}_0}{\bar{p}_0} = \frac{\bar{\bar{\rho}}_0}{\bar{\rho}_0} = \exp\left(-\frac{\bar{\bar{z}}_0 - \bar{z}_0}{L_0}\right) = \exp\left(-\frac{14}{6.34}\right) = 0.110; \quad (2.1096a–d)$$

the values of the pressure (2.672a–g) [mass density (2.673a–h)] at the tropopause lead to the corresponding values (2.675a–c) [(2.1098a, b)] at the top of the stratosphere:

$$\bar{\bar{p}}_0 = 0.110 \times 0.224 \, atm = 0.0246 \, atm = 2.46 \times 10^{-2} \times 1.014 \times 10^5 \, kg \, m^{-1} \, s^{-2}$$

$$= 2.50 \times 10^3 \, kg \, m^{-1} \, s^{-2} = 2.50 \, kPa,$$

(2.1097a–d)

$$\bar{\bar{\rho}}_0 = 0.110 \times 0.365 \, kg \, m^{-3} = 0.0402 \, kg \, m^{-3}. \quad (2.1098a, b)$$

The I.S.A. values for the troposphere and stratosphere are used next to consider the nozzles of turbine (rocket) engines [subsections 2.9.19–2.9.25 (2.9.26–2.9.39)], for the three cases of perfect/over/under expansion (subsection 2.9.18) of a throated or convergent–divergent nozzle.

2.9.18   Six Flow Regimes in a Throated Nozzle

Consider (Figure 2.49c) a convergent–divergent nozzle, and relative to the stagnation sound speed c_*, that is conserved across shocks: (i) the inlet velocity $v_i < c_*$ is subsonic; (ii) at the throat the flow is sonic $v_* = c_*$ if the nozzle is choked; (iii) in the divergent section there is a normal shock with discontinuous upstream v^- and downstream v^+ velocities $v^- > c_* > v^+$ in the one overexpanded case; (iv) in the latter (iii) case the subsonic flow after the shock in the diverging section of the nozzle further decreases the velocity and the exit section has the lowest velocity $v_e < v^+$. The full sequence (i) to (iv) corresponds to an overexpanded flow (2.1064c) in the case V with a normal shock inside the nozzle (Figure 2.50); for perfect expansion (2.1064b) [underexpansion (2.1064a)] there is no shock in case II (III),

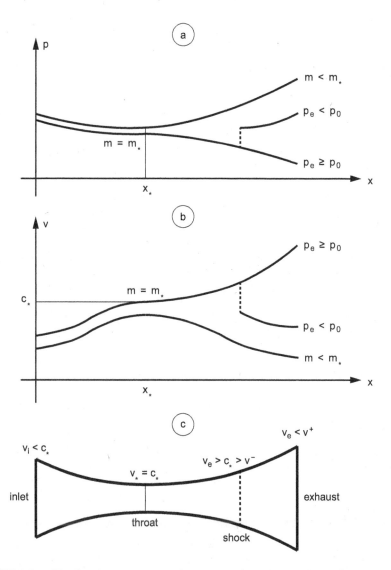

FIGURE 2.49 The flow in a convergent–divergent nozzle (c) has maximum flow rate m_* for the "choked" condition of sonic flow at the throat (Figure 2.40), and a "shock" may form between the throat and exit in the overexpanded case (Figure 2.45c). The pressure (a) [velocity (b)] in the case of (i) "unchoked" nozzle $m < m_*$, with less than the maximum flow rate, corresponding to subsonic flow up to the exit, decreases (increases) in the convergent section and increases (decreases) in the divergent section; (ii) "choked" nozzle $m = m_*$ with sonic flow at the throat, perfect expansion $p_e = p_0$ or underexpansion $p_e > p_0$ at the exit, decreases (increases) monotonically from inlet to exit; (iii) "choked" $m = m_*$ and "shocked" nozzle with overexpansion at the exit $p_e < p_0$, there is a normal shock (Figure 2.49c) between the throat and exit. In the case (iii) the monotonic decrease in pressure (increase in velocity) from the inlet, through the throat until upstream of the shock is reversed by an upward (downward) jump across the shock front, followed by a further increase in pressure (decrease in velocity) in the subsonic flow downstream of the shock in the divergent part of the nozzle.

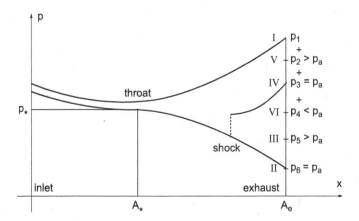

FIGURE 2.50 There are five cases of flow in a throated nozzle (Figures 2.40 and 2.44) illustrated by the pressure variation (Figure 2.50) along its length: (I) "unchoked", that is with less than the maximum mass flow rate, so that the flow is subsonic from entry to exit, and that the pressure decreases (increases) in the convergent (divergent) section before (after) the throat with an exit value p_1; (II) "choked" that is with maximum mass flow rate and perfect expansion (Figure 2.44a) at the exit pressure equal to the atmospheric pressure $p_6 = p_a$ leading to tangential shear layers issuing from the nozzle lips (Figure 2.45a); (III) if the exit pressure is higher than the atmospheric pressure $p_5 > p_a$ underexpanded case leads to a bulged jet (Figure 2.44b), a diamond shock cell pattern forms between the shear layers (Figure 2.45b) and expansion waves issue from the nozzle lips; (IV) if the adiabatic exit pressure is lower than the atmospheric pressure, then a shock must form, and it occurs at the nozzle exit for a particular pressure downstream of the shock $p_3^+ = p_a$; (V) if the pressure is higher $p_2^+ > p_a$ the shock moves inside the nozzle (Figure 2.45c), that is both "choked" and "shocked"; (VI) if the pressure is lower $p_4^+ < p_a$ an oblique shock is formed and the overexpanded exhaust is pinched (Figure 2.44c).

and the nozzle is not choked in Case I of subsonic flow. There are two more shock cases V (VI), namely normal (oblique) shock at the exit, for a total of six cases considered next.

Consider a convergent–divergent nozzle and assume that the inlet and exit pressures are equal so there is no flow. A flow down the nozzle is produced increasing the inlet pressure or decreasing the exit pressure. The option of fixed inlet pressure and variable exit pressure is sufficient to illustrate the six cases of flow (Figure 2.50):

$$case \begin{cases} I : m < m_* - subsonic\ flow : p_1 \\ II : m = m_*, p_e = p_a - perfectly\ expanded \\ III : m = m_*, p_2 > p_a - underexpanded \\ IV : m = m_*, p_a < p_3 - normal\ shock\ at\ nozzle\ exit \\ V : m = m_*, p_a < p_3 < p_2 - normal\ shock\ inside\ nozzle \\ VI : m = m_*, p_a < p_4 < p_3 - oblique\ shock\ at\ nozzle\ exit, \end{cases} \qquad (2.1099a\text{–}f)$$

In Case I of mass flow rate less (2.1099a) than the "choked" value (2.1050a, b), the flow pressure decreases and velocity increases in the convergent section, remaining subsonic at the throat, where the velocity is maximum and the pressure is minimum, and in the following divergent section the velocity decreases and pressure increases. In all other cases II to VI the nozzle is "choked", with subsonic (supersonic) flow before (after) the throat and continuously decreasing pressure leading to: (i) perfect expansion if the exit pressure equals (2.1099b) the atmospheric pressure (case II) leading to a jet with shear layers issuing tangentially from the nozzle lips (Figures 2.44a and 2.45a); (ii) if the atmospheric pressure is less (2.1099c) than the exit pressure (case III) the jet expands out of the nozzle (Figure 2.44b) and oblique shocks may be reflected between the shear layers leading to the diamond shock cell pattern (Figure 2.45b); (iii) an exit pressure lower than the atmospheric pressure is not possible implying the existence of shock waves to match the pressures, and leading to three cases IV to VI. The intermediate case IV is a normal shock at the nozzle exit for a reference pressure p_3 in (2.1099d). For a higher (2.1099e) exit pressure (case V) the shock moves inside the nozzle (Figure 2.45c) that is both "choked" and "shocked". For a lower (2.1099f) exit pressure (case VI) an oblique shock is formed, and the exhaust is pinched (Figure 2.44c). Cases of perfect expansion and over(under)expansion are considered for turbojet (rocket) engines [subsections 2.9.19–2.9.25 (2.9.26–2.9.39)] operating at different altitudes in the atmosphere (and also in the vacuum of space).

2.9.19 INLET AND EXHAUST OF A JET ENGINE AT SEA LEVEL

The jet engine considered (subsections 2.6.29–2.6.35), as an example of the Brayton cycle (subsections 2.6.23–2.6.28) is re-considered with: (i) the same compression and combustion phase; (ii) the expansion through the turbine leading to a convergent–divergent nozzle to obtain a supersonic exhaust. It is assumed that at the exit of the turbine that is coincident with the nozzle inlet the pressure is the geometric mean (2.1100a) of the turbine inlet (2.728d) and atmospheric (2.728a) pressures leading to the value (2.1100b–f)

$$p_i = \sqrt{p_+ \, p_-} = \sqrt{29.7} \, atm = 5.45 \, atm = 5.45 \times 1.014 \times 10^5 \, kg \, m^{-1} \, s^{-2}$$
$$= 5.53 \times 10^5 \, kg \, m^{-1} \, s^{-2} = 553 \, kPa. \tag{2.1100a–f}$$

Assuming adiabatic conditions for I.S.A. at sea level with stagnation pressure (2.716e–m) and mass density (2.720a–c) leads to the inlet mass density (2.679a–e):

$$\rho_i = \rho_0 \left(\frac{p_i}{p_0} \right)^{1/\gamma} = 5.45^{1/1.4} \times 1.225 \, kg \, m^{-3} = 4.113 \, kg \, m^{-3}. \tag{2.1101a–c}$$

The mass flow rate (2.722h–k) ≡ (2.1102a) for a nozzle inlet area (2.1102b) smaller than that of the intake (2.722a–c) leads to an inlet velocity (2.1102c–e):

$$\dot{m} = 76.7 \, kg \, s^{-1}, \quad A_i = 0.180 \, m^2 :$$
$$v_i = \frac{\dot{m}}{\rho_i \, A_i} = \frac{76.7 \, kg \, s^{-1}}{4.113 \, kg \, m^{-3} \times 0.180 \, m^2} = 104 \, m \, s^{-1}. \tag{2.1102a–e}$$

The pressure (2.1100f) and mass density (2.1101c) lead to the inlet sound speed (2.1103b–d) for a diatomic perfect gas (2.475c) ≡ (2.1103a):

$$\gamma = 1.4 \quad c_i = \sqrt{\gamma \frac{p_i}{\rho_i}} = \sqrt{1.4 \times \frac{5.53 \times 10^5 \, kg \, m^{-1} \, s^{-2}}{4.113 \, kg \, m^{-3}}} = 435 \, m \, s^{-2}, \quad (2.1103a\text{–}d)$$

and to the (2.661d) stagnation sound speed (2.1104a–c):

$$c_0 = \left| (c_i)^2 + \frac{\gamma-1}{2}(v_i)^2 \right|^{1/2} = \sqrt{435^2 + 0.2 \times 105^2} \, m \, s^{-1} = 438 \, m \, s^{-1}. \quad (2.1104a\text{–}c)$$

Using (2.664c) [(2.479d; 2.718g) the corresponding critical sound speed (stagnation temperature) is (2.1105a–c) [(2.1105d–f)]:

$$c_* = \sqrt{\frac{2}{\gamma+1}} \, c_0 = \frac{43}{\sqrt{1.2}} \, m \, s^{-1} = 400 \, m \, s^{-1}, \quad (2.1105a\text{–}c)$$

$$T_0 = \frac{(c_0)^2}{\gamma R_0} = \frac{43^2 \, m^2 \, s^{-2}}{1.4 \times 287 \, m^2 \, s^{-2} \, K^{-1}} = 477 \, K. \quad (2.1105d\text{–}f)$$

The stagnation sound speed (2.1104a–c) and temperature (2.1105d–f) are conserved by the adiabatic flow in the nozzle and also across a normal shock in the overexpanded case; the critical sound speed is also conserved (2.1105a–c) and coincides with the flow velocity at the throat. The inlet temperature is (2.1106a–c):

$$T_i = T_0 \left(\frac{c_i}{c_0} \right)^2 = \left(\frac{435}{43} \right)^2 \times 477 \, K = 470 \, K, \quad (2.1106a\text{–}c)$$

and the local (2.1103c), critical (2.1105c) and stagnation (2.1104c) inlet (2.1102e) Mach numbers are (2.1107a–c):

$$\{M_i, M_{i*}, M_{i0}\} = \frac{v_i}{\{c_i, c_*, c_0\}} = \frac{104}{\{435, 400, 438\}} = \{0.239, 0.260, 0.237\}. \quad (2.1107a\text{–}c)$$

The inlet conditions (subsection 2.9.19) specify all flow variables at the throat (exit), for example, for a perfect expansion [subsection 2.9.20 (2.9.21)].

2.9.20   CONDITIONS AT THE THROAT OF A CONVERGENT – DIVERGENT NOZZLE

The mass density at the throat is given by (2.481d) ≡ (2.1108a) leading (2.1101c; 2.1103a, d; 2.1105c) to (2.1108b, c):

$$\rho_* = \rho_i \left(\frac{c_*}{c_i}\right)^{2/(\gamma-1)} = \left(\frac{400}{435}\right)^5 \times 4.113 \, kg \, m^{-3} = 2.704 \, kg \, m^{-3}. \quad (2.1108\text{a–c})$$

The mass flow rate (2.1102a) and critical sound speed (2.1105c) at the throat speci-fies the area (2.1109a–e):

$$A_* = \frac{\dot{m}}{\rho_* c_*} = \frac{76.7 \, kg \, s^{-1}}{2.704 \, kg \, m^{-3} \times 400 \, m \, s^{-1}} = 0.0709 \, m^2, \quad \frac{A_*}{A_i} = \frac{0.0709}{0.180} = 0.394, \quad (2.1109\text{a–e})$$

that is less than half (2.1108d, e) of the inlet area (2.1102b). The temperature at throat (2.1105c, f; 2.1104c) is (2.1110a–c:)

$$T_* = T_0 \left(\frac{c_*}{c_0}\right)^2 = \left(\frac{400}{438}\right)^2 \times 477 \, K = 396 \, K, \quad (2.1110\text{a–c})$$

$$M_{0*} \equiv \frac{v_*}{c_0} = \frac{c_*}{c_0} = \sqrt{\frac{2}{\gamma+1}} = \frac{1}{\sqrt{1.2}} = 0.913, \quad (2.1110\text{d–h})$$

and the stagnation Mach number (2.1104c; 2.1105c) or (2.1103a) is (2.1110d–h), whereas the local and critical Mach numbers are unity. The pressure at the throat is given (2.1100f; 2.1101c; 2.1106c; 2.1108c; 2.1110c) by (2.1111a–c):

$$p_* = p_i \frac{\rho_* \, T_*}{\rho_i \, T_i} = \frac{2.704}{4.113} \times \frac{398}{470} \times 5.45 \, atm = 3.03 \, atm. \quad (2.1111\text{a–c})$$

The flow conditions at the throat (subsection 2.9.20) specify those at the exit, includ-ing thrust and power, for example, for a perfect expansion (subsection 2.9.21) con-sidered in static conditions or taking into account the low flight velocity at approach to land or climb after take–off.

2.9.21 APPROACH TO LAND AND CLIMB AFTER TAKE-OFF

A perfect expansion corresponds to an exit pressure equal to the atmospheric pres-sure (2.1112a, b) specifying (2.668e) ≡ (2.1112c) the exit velocity (2.1112d, e):

$$p_e = p_0 = 1 \, atm : \quad v_e = c_* \left| \frac{\gamma+1}{\gamma-1} - \frac{2}{\gamma-1}\left(\frac{p_e}{p_*}\right)^{1-1/\gamma} \right|^{1/2}$$

$$= \left|6 - 5 \times (3.03)^{-0.4/1.4}\right|^{1/2} \times 400 \, m \, s^{-1} = 614 \, m \, s^{-1}. \quad (2.1112\text{a–e})$$

The corresponding exit mass density $(2.478d) \equiv (2.1113a)$ is $(2.1113b, c)$ using $(2.1103a; 2.1108c; 2.1111c; 2.1112a)$:

$$\rho_e = \rho_* \left(\frac{p_e}{p_*} \right)^{1/\gamma} = 3.03^{-1/1.4} \times 2.704 \, kg \, m^{-3} = 1.225 \, kg \, m^{-3}, \quad (2.1113a\text{–}c)$$

and equals the atmospheric mean density $(2.720a\text{–}c)$; this implies $(2.1102a; 2.1112e; 2.1113c)$ a nozzle exit area $(2.1114a\text{–}c)$:

$$A_e = \frac{\dot{m}}{\rho_e v_e} = \frac{76.7 \, kg \, s^{-1}}{1.225 \, kg \, m^{-3} \times 614 \, m \, s^{-1}} = 0.1020 \, m^2, \quad (2.1114a\text{–}c)$$

$$\frac{A_e}{\{A_i, A_*\}} = \frac{0.1020}{\{0.180, 0.0709\}} = \{0.567, 1.439\}, \quad (2.1114d, e)$$

smaller $(2.1114d)$ [larger $(2.1114e)$] than the inlet $(2.1102b)$ [throat $(2.1109c)$] area. The exit local sound speed $(2.661d) \equiv (2.1115a)$ given $(2.1104c; 2.1103a; 2.1112e)$ by $(2.1115b, c)$:

$$c_e = \left| c_0^2 - \frac{\gamma - 1}{2} (v_e)^2 \right|^{1/2} = \sqrt{438^2 - 0.2 \times 614^2} \, m \, s^{-1} = 341 \, m \, s^{-1}, \quad (2.1115a\text{–}c)$$

in agreement with the atmospheric sound speed $(2.721a\text{–}g)$ within a small truncation error. The exit temperature is given $(2.1110c; 2.1105c; 2.1115c)$ by $(2.1116b, c)$:

$$T_e = T_* \left(\frac{c_e}{c_*} \right)^2 = \left(\frac{341}{400} \right)^2 \times 398 \, K = 289 \, K, \quad (2.1116a\text{–}c)$$

again close to the atmospheric value $(2.716a,b)$.

The local $(2.1115c)$, critical $(2.1105c)$ and stagnation $(2.1104c)$ exit Mach numbers are $(2.1112e)$ given by $(2.1117a\text{–}c)$:

$$\{M_e, M_{*e}, M_{0e}\} \equiv \frac{v_e}{\{c_e, c_*, c_0\}} = \frac{614}{\{341, 400, 438\}} = \{1.801, 1.535, 1.402\}. \quad (2.1117a\text{–}c)$$

The mass flow rate $(2.1102a)$ and exit velocity $(2.1112e)$ specify the thrust $(2.1118a\text{–}d)$ [power $(2.1118e\text{–}h)$]:

$$F_0 = \dot{m} v_e = 76.7 \, kg \, s^{-1} \times 614 \, m \, s^{-1} = 4.71 \times 10^4 \, kg \, m \, s^{-2} = 47.1 \, kN, \quad (2.1118a\text{–}d)$$

$$\dot{W}_0 = F v_e = \dot{m} (v_e)^2 = 4.71 \times 10^4 \, kg \, m \, s^{-2} \times 614 \, m \, s^{-1} \quad (2.1118e\text{–}g)$$

$$= 2.89 \times 10^7 \, kg \, m^2 \, s^{-3} = 28.9 \, MW. \qquad (2.1118h, \, i)$$

Taking into account the velocity of the aircraft (2.722d) ≡ (2.1119a), the relative velocity (2.1119b–d) reduces the thrust (power) in a moving frame to (2.1119e–h) [(2.1119i–l)]:

$$v_0 = 80 \, m \, s^{-1}: \quad \Delta v_e = v_e - v_0 = (614 - 80) m \, s^{-1} = 534 \, m \, s^{-1}, \qquad (2.1119a–d)$$

$$F = \dot{m} \Delta v_e = 76.7 \, kg \, s^{-1} \times 534 \, m \, s^{-1} = 4.10 \times 10^4 \, kg \, m \, s^{-2} = 41.0 \, kN, \quad (2.1119e–h)$$

$$\dot{W} = F \Delta v_e = 4.10 \times 10^4 \, kg \, m^{-2} \times 534 \, m \, s^{-1} \qquad (2.1119i, \, j)$$

$$= 2.19 \times 10^7 \, kg \, m^2 \, s^{-3} = 21.9 \, MW. \qquad (2.1119k, \, l)$$

The low velocity (2.1119a) corresponding (subsection 2.6.30) to approach to land or climb after take-off at sea level (subsections 2.9.19–2.9.21) is replaced by a higher cruise speed at the tropopause (subsections 2.9.22–2.9.25), with some adaptations of the convergent–divergent nozzle.

2.9.22 Low(High)-Speed Flight at Sea Level (The Tropopause)

The cruise flight at the tropopause (2.1088b) ≡ (2.1120a), where the stagnation sound speed is (2.1092c) ≡ (2.1120b), at a stagnation Mach number (2.1120c) corresponds to the velocity (2.445a, b):

$$\bar{z}_0 = 1.1 \times 10^4 \, m, \quad \bar{c}_0 = 295 \, m \, s^{-1}, \quad \bar{M}_0 = 0.85: \qquad (2.1120a–c)$$

$$\bar{v}_0 = \bar{M}_0 \, \bar{c}_0 = 0.85 \times 295 \, m \, s^{-1} = 251 \, m \, s^{-1}; \qquad (2.1120d–f)$$

the same mass flow rate (2.1102a) for the mass density at the tropopause (2.1095c) ≡ (2.1121a) corresponds to an inlet area (2.1121b–d):

$$\bar{\rho}_0 = 0.365 \, kg \, m^{-3}: \quad \bar{A}_0 = \frac{\dot{m}}{\bar{\rho}_0 \, \bar{v}_0} = \frac{76.7 \, kg \, s^{-1}}{0.365 \, kg \, m^{-3} \times 251 \, m \, s^{-1}} = 0.837 \, m^2, \quad (2.1121a–d)$$

that is not too different (2.1122b, c) from the sea level case (2.722c) ≡ (2.1122a) because the larger velocity (2.1122d, e) compensates for the lower density (2.1122f, g):

$$A_0 = 0.785 \, m^2: \quad \frac{\bar{A}_0}{A_0} = \frac{0.837}{0.785} = 1.066, \qquad (2.1122a, \, c)$$

$$\frac{\bar{v}_0}{v_0} = \frac{251}{80} = 3.14 \quad, \frac{\bar{\rho}_0}{\rho_0} = \frac{0.365}{1.225} = 0.298. \qquad (2.1122d–g)$$

The pressure at the tropopause (2.1094c) $\equiv$ (2.1123a, b) for the same compression ratio (2.728a, d) $\equiv$ (2.1123c–e) leads under the earlier assumption (2.1100a) $\equiv$ (2.1123f) to the pressure (2.1123g–l) at the inlet of the convergent–divergent nozzle:

$$\bar{p}_0 = 0.224\,atm = \bar{p}^- : \quad \bar{p}^+ = 29.7 \times \bar{p}^- = 29.7 \times 0.224\,atm = 6.65\,atm,$$

$$\bar{p}_i = \sqrt{\bar{p}^+\,\bar{p}^-} = \sqrt{6.65 \times 0.224}\;atm = 1.220\,atm$$

$$= 1.220 \times 1.014 \times 10^5\,kg\,m^{-1}\,s^{-2} = 1.238 \times 10^5\,kg\,m^{-1}\,s^{-2}$$

$$= 124\,kPa. \qquad (2.1123a\text{–}l)$$

The pressure (2.1123i) specifies the inlet (throat) conditions [subsection 2.9.23 (2.9.24)] and those at the exit, for example, for perfect expansion (subsection 2.9.25).

2.9.23   NOZZLE INLET CONDITIONS FOR CRUISE FLIGHT AT THE TROPOPAUSE

Assuming adiabatic conditions the inlet mass density (2.449a–c) follows (2.478d) from (2.1121a; 2.1123a, h):

$$\bar{\rho}_i = \bar{\rho}_0 \left(\frac{\bar{p}_i}{\bar{p}_0} \right)^{1/\gamma} = \left(\frac{1.220}{0.224} \right)^{1/1.4} \times 0.365\,kg\,m^{-3} = 1.225\,kg\,m^{-3}. \quad (2.1124a\text{–}c)$$

The mass flow rate (2.1102a), nozzle inlet area equal to triple the engine inlet area (2.1102b) $\equiv$ (2.1125a, b) and mass density (2.1124c) lead (2.1036b) $\equiv$ (2.1124c) to the inlet velocity (2.1125d, e):

$$\bar{A}_i = 0.540\,m^2 = 3\,A_i \quad \bar{v}_i = \frac{\dot{m}}{\bar{\rho}_i\,\bar{A}_i} = \frac{76.7\,kg\,s^{-1}}{1.225\,kg\,m^{-3} \times 0.540\,m^2} = 116\,m\,s^{-1}, \quad (2.1125a\text{–}e)$$

that is larger than in the sea level case (2.1102e). The local sound speed (2.479d) $\equiv$ (2.1126a) is (2.1103a; 2.1123j; 2.1124c) given by (2.1126b, c):

$$\bar{c}_i = \sqrt{\gamma \frac{\bar{p}_i}{\bar{\rho}_i}} = \sqrt{1.4 \times \frac{1.238 \times 10^5\,kg\,m^{-1}\,s^{-2}}{1.225\,kg\,m^{-3}}} = 376\,m\,s^{-1}, \quad (2.1126a\text{–}c)$$

smaller than in the sea level case (2.1103b–d), and corresponds to the stagnation (2.661d) $\equiv$ (2.1127a) [critical (2.664c) $\equiv$ (2.1127d)] sound speeds (2.1127b, c) [(2.1127e, f)] using (2.1103a; 2.1126c; 2.1125e) [(2.1120b)]:

$$\bar{c}_0 = \left[(\bar{c}_i)^2 + \frac{\gamma-1}{2}(\bar{v}_i)^2 \right]^{1/2} = \sqrt{376^2 + 0.2 \times 116^2}\;m\,s^{-1} = 380\,m\,s^{-1}, \quad (2.1127a\text{–}c)$$

$$\bar{c}_* = \sqrt{\frac{2}{\gamma+1}}\,\bar{c}_0 = \frac{380}{\sqrt{1.2}}\,m\,s^{-1} = 346\,m\,s^{-1}. \tag{2.1127d–f}$$

Thus follows the stagnation (2.454a–c) [(local (2.455a–c)] temperature from (2.479d) ≡ (2.1128a) [≡ (2.1128d)] using (2.1120b; 2.1103a; 2.718g) [(2.1126b; 2.1126c)]:

$$\bar{T}_0 = \frac{\left(\bar{c}_0\right)^2}{\gamma\,R_0} = \frac{380^2\,m^2\,s^{-2}}{1.4\times287\,m^2\,s^{-2}\,K^{-1}} = 359\,K, \tag{2.1128a–c}$$

$$\bar{T}_i = \bar{T}_0\left(\frac{\bar{c}_i}{\bar{c}_0}\right)^2 = \left(\frac{376}{380}\right)^2 \times 359\,K = 351\,K. \tag{2.1128d–f}$$

The local (2.1126c), critical (2.1127f) and stagnation (2.1127c) inlet (2.1125e) Mach numbers (2.456a–c):

$$\left\{\bar{M}_i,\bar{M}_{*i},\bar{M}_{0i}\right\} \equiv \frac{\bar{v}_i}{\left\{\bar{c}_i,\bar{c}_*,\bar{c}_0\right\}} = \frac{116}{\left\{376,346,380\right\}} = \left\{0.309,0.335,0.305\right\}, \tag{2.1129a–c}$$

are higher than at sea level (2.1107a–c). The high nozzle inlet velocity (subsection 2.9.23) implies a smaller variation towards the throat (subsection 2.9.24).

2.9.24 CHOKED THROATED NOZZLE AT THE CRUISE CONDITION

The mass density at the throat (2.1130b, c) is given (2.1124c) by the adiabatic condition (2.669d) ≡ (2.1130a) for a diatomic perfect gas (2.475b) ≡ (2.1103a):

$$\bar{\rho}_* = \bar{\rho}_i\left(\frac{2}{\gamma+1}\right)^{1/(\gamma-1)} = 1.2^{-2.5}\times1.225\,kg\,m^{-3} = 0.777\,kg\,m^{-3}, \tag{2.1130a–c}$$

and specifies from (2.1036b) the mass flow rate (2.1102a) and critical sound speed (2.1127f) the throat area (2.1131a–c)

$$\bar{A}_* = \frac{\dot{m}}{\bar{\rho}_*\,v_*} = \frac{\dot{m}}{\bar{\rho}_*\,\bar{c}_*} = \frac{76.7\,kg\,s^{-1}}{0.777\,kg\,m^{-3}\times346\,m\,s^{-1}} = 0.285\,m^2, \quad \frac{\bar{A}_*}{\bar{A}_i} = \frac{0.284}{0.540} = 0.528, \tag{2.1131a–f}$$

that is smaller than (2.1131d, f) the inlet area (2.1125a, b). The temperature (2.479d) ≡ (2.1132a) at the throat is (2.1128c; 2.1127c, f) given by (2.1132b, c):

$$\bar{T}_* = \bar{T}_0\left(\frac{\bar{c}_*}{\bar{c}_0}\right)^2 = \left(\frac{346}{380}\right)^2 \times 359\,K = 298\,K, \tag{2.1132a–c}$$

and is lower than (2.1110a–c). The stagnation Mach number at the throat is always (2.1110d–h) and the critical and local Mach numbers at the throat are always unity. The pressure at the throat is (2.473g) $\equiv$ (2.1133a) given by (2.1133b, c) using (2.1123h; 2.1124c; 2.1130c; 2.1128f; 2.1132c):

$$\bar{p}_* = \bar{p}_i \, \frac{\bar{\rho}_i}{\bar{\rho}_*} \, \frac{\bar{T}_i}{\bar{T}_*} = \frac{1.225}{0.777} \times \frac{351}{290} \times 1.220 \, atm = 2.266 \, atm. \qquad (2.1133a\text{–}c)$$

The Table 2.24 compares the convergent–divergent nozzle for low (high) speed flight [subsection 2.9.19 (2.9.22)] at sea level (the tropopause) including inlet [subsection 2.9.20 (2.9.23)], throat [subsection 2.9.21 (2.9.24)] and exhaust [subsection 2.9.22 (2.9.25)] flow variables, the latter for a perfect expansion.

2.9.25 THRUST AND POWER FOR A PERFECT EXPANSION

For a perfect expansion the exit pressure (2.1134a, b) equals the value at the tropopause (2.1094e) and the exit velocity is given by (2.1134d, e) using (2.1103a; 2.1127f; 2.1133c):

$$\bar{p}_e = \bar{p}_0 = 0.224 \, atm:$$

$$\bar{v}_e = \bar{c}_* \left| \frac{\gamma+1}{\gamma-1} - \frac{2}{\gamma-1} \left(\frac{\bar{p}_e}{\bar{p}_*} \right)^{1-1/\gamma} \right|^{1/2} \qquad (2.1134a\text{–}e)$$

$$= \left| 6 - 5 \times \left(\frac{0.224}{2.266} \right)^{0.4/1.4} \right|^{1/2} \times 346 \, m \, s^{-1} = 640 \, m \, s^{-1};$$

the corresponding (2.478d) $\equiv$ (2.1135a) exit mass density is (2.1135b, c) using (2.1130c; 2.1134a; 2.1133c):

$$\bar{\rho}_e = \bar{\rho}_* \left(\frac{\bar{p}_e}{\bar{p}_*} \right)^{1/\gamma} = \left(\frac{0.224}{2.266} \right)^{1/1.4} \times 0.777 \, kg \, m^{-3} = 0.149 \, kg \, m^{-3}, \qquad (2.1135a\text{–}c)$$

implying (2.1036b) $\equiv$ (2.1136a) a nozzle exit area (2.1136b, c) using (2.1102a; 2.1135c; 2.1134e):

$$\bar{A}_e = \frac{\dot{m}}{\bar{\rho}_e \bar{v}_e} = \frac{76.7 \, kg \, s^{-1}}{0.149 \, kg \, m^{-3} \times 640 \, m \, s^{-1}} = 0.804 \, m^2, \qquad (2.1136a\text{–}c)$$

$$\frac{\bar{A}_e}{\bar{A}_i} = \frac{0.804}{0.540} = 1.489, \qquad \frac{\bar{A}_e}{\bar{A}_*} = \frac{0.804}{0.285} = 2.821, \qquad (2.1136d\text{–}g)$$

TABLE 2.24
Flow in a Convergent–Divergent Nozzle at Different Flight Conditions

Condition	Variable	Symbol	Unit	Low-speed*	High-speed**
Atmosphere	Altitude	z	m	0	1.1×10^4
	Pressure	p_0	atm	1.000	0.224
	Density	ρ_0	$kg\ s^{-1}$	1.225	0.365
	Sound speed	c_0	$m\ s^{-1}$	340	295
	Temperature	T_0	K	288	216.5
Inlet	Pressure	p_i	atm	5.45	1.220
	Density	ρ_i	$kg\ m^{-3}$	4.113	1.225
	Temperature	T_i	K	470	351
	Area	A_i	m^2	0.180	0.540
	Flow rate	$\dot{m}$	$kg\ s^{-1}$	76.7	76.7
	Velocity	v_0	$m\ s^{-1}$	104	116
	Stagnation temperature	T_0	K	477	359
	Stagnation sound speed	c_0	$m\ s^{-1}$	438	380
	Critical sound speed	$c*$	$m\ s^{-1}$	400	346
	Local sound speed	c	$m\ s^{-1}$	435	376
	Local Mach number	M	-	0.239	0.309
	Critical Mach number	$M*$	-	0.263	0.335
	Stagnation Mach number	M_0	-	0.237	0.305
Throat	Area	$A*$	m^2	0.0709	0.285
	Velocity	$v*$	$m\ s^{-1}$	400	346
	Pressure	$p*$	atm	3.03	2.266
	Density	$\rho*$	$kg\ m^{-1}$	2.704	0.777
	Temperature	$T*$	K	398	298
Exit	Area	A_e	m^2	0.1026	0.801
	Pressure	p_e	atm	1.000	0.224
	Velocity	v_e	$m\ s^{-1}$	614	640
	Density	ρ_e	$kg\ m^{-3}$	1.225	0.149
	Temperature	T_e	K	289	155
	Local sound speed	c_e	$m\ s^{-1}$	341	250
	Local Mach number	M_e	-	1.801	2.560
	Critical Mach number	M_e*	-	1.535	1.850
	Stagnation Mach number	M_{e0}	-	1.402	1.684
	Flight velocity	v_0	$m\ s^{-1}$	80	251
	Relative velocity	$v_e - v_0$	$m\ s^{-1}$	534	389
	Static thrust	F_0	kN	47.1	49.1
	Net thrust	F	kN	41.0	29.8
	Static power	$\dot{W}_0$	MW	28.9	31.4
	Net power	$\dot{W}$	MW	21.9	11.6

* Approach to land or climb after take-off at sea level

** Cruise at the tropopause

Note: Comparison of the flow in the convergent-divergent nozzle of a turbojet engine in two very different flight conditions: (a) low speed at sea level for approach to land or climb after take-off; (b) high-speed cruise at the tropopause. For the two flight conditions (a) and (b) are compared: (i–v) atmospheric conditions: altitude, pressure, density, sound speed, temperature;

TABLE 2.24 (Continued)
Flow in a Convergent–Divergent Nozzle at Different Flight Speed

(vi–xi) inlet flow conditions: pressure, density, temperature, area, flow rate, velocity; (xii–xviii) inlet flow parameters: stagnation temperature, and stagnation, critical and local sound speeds and Mach numbers; (xix–xxiii) conditions at the throat of the nozzle: area, velocity, pressure, density and temperature (xxiv–xxviii) exhaust conditions: area, pressure, velocity, density and temperature; (xxix–xxxii) local sound speed and local, critical and stagnation Mach numbers; (xxxiii–xxxviii) flight velocity (exhaust velocity relative to flight velocity) and corresponding static (net) thrust and power.

that is larger (2.1136d, e) [(2.1136f, g)] than the inlet (2.1125a, b) [throat (2.1131a–c)] area. The local exit sound speed (2.661d) ≡ (2.1137a) is (2.1137b, c) using (2.1103a; 2.1127c; 2.1134e):

$$\bar{c}_e = \left| (\bar{c}_0)^2 - \frac{\gamma-1}{2}(\bar{v}_e)^2 \right|^{1/2} = \sqrt{380^2 - 0.2 \times 640^2}\ m\,s^{-1} = 250\,m\,s^{-1}, \quad (2.1137a\text{–}c)$$

$$\bar{T}_e = \bar{T}_0 \left(\frac{\bar{c}_e}{\bar{c}_0}\right)^2 = \left(\frac{250}{381}\right)^2 \times 359\,K = 155\,K, \quad (2.1137d\text{–}f)$$

implying (2.479d) ≡ (2.1137d) the exit temperature (2.1137e, f) using (2.1127c; 2.1137c; 2.1128c):

The local (2.1137c), critical (2.1127f) and stagnation (2.1127c) exit (2.1134e) Mach numbers are (2.466b):

$$\{\bar{M}_e,\bar{M}_{*e},\bar{M}_{0e}\} \equiv \frac{\bar{v}_e}{\{\bar{c}_e,\bar{c}_*,\bar{c}_0\}} = \frac{640}{\{250,346,380\}} = \{2.560,1.850,1.684\}, \quad (2.1138a\text{–}c)$$

The static thrust (2.466c) [power (2.467a–e)] follows from (2.1102a; 2.1134e):

$$\bar{F}_0 = \dot{m}\bar{v}_e = 76.7\,kg\,m^{-1} \times 640\,m\,s^{-1} = 4.91\times10^4\,kg\,m^{-2} = 49.1\,kN, \quad (2.1139a\text{–}d)$$

$$\dot{W}_0 = \bar{F}_0\bar{v}_e = 4.91\times10^4\,kg\,m\,s^{-2}\times640\,m\,s^{-1} \quad (2.1139e\text{–}h)$$
$$= 3.14\times10^7\,kg\,m^2\,s^{-3} = 31.4\,MW,$$

that are hardly relevant, because the high flight velocity (2.1120f) reduces the relative (2.472a–e) velocity (2.1134e):

$$\Delta\bar{v}_e = \bar{v}_e - \bar{v}_0 = (640-251)\,m\,s^{-1} = 389\,m\,s^{-1}, \quad (2.1140a\text{–}c)$$

leading to a much lower net thrust (2.467j–l) [power (2.468a, 2.445a, b)] in a reference frame moving at cruise speed:

$$\bar{F} = \dot{m}\Delta\bar{v}_e = 76.7\,kg\,s^{-1} \times 389\,m\,s^{-1} = 2.98 \times 10^4\,kg\,m\,s^{-2} = 29.8\,kN, \quad (2.1141a\text{–}d)$$

$$\dot{\bar{W}} = \bar{F}\,\Delta\bar{v}_e = 2.98 \times 10^4\,kg\,m\,s^{-2} \times 389\,m\,s^{-1} \quad\quad (2.1141e\text{–}h)$$
$$= 1.16 \times 10^7\,kg\,m^2\,s^{-3} = 11.6\,MW.$$

The convergent–divergent nozzle of a turbojet engine has been considered for [subsection 2.9.19–2.9.21 (2.9.22–2.9.25)] perfect expansion in low (high) speed flight at sea level (the tropopause). Next is considered a rocket engine (subsections 2.9.26–2.9.27) for: (i) conditions at the combustion chamber and nozzle throat and exit (subsection 2.9.28); (ii) an underexpanded/perfectly expanded/overexpanded convergent–divergent nozzle [subsection 2.9.29–2.9.32/2.9.33/2.9.34–2.9.36)], respectively, in the vacuum of space/in the stratosphere/at sea level; (iii) in the last overexpanded case existence of a normal (oblique) shock inside (at the exit of) the nozzle [subsection(s) 2.9.36 (2.9.37–2.9.39)].

2.9.26 Specific Impulse and Thrust of a Rocket Engine

A turbojet engine produces thrust (subsections 2.6.24–2.6.34 and 2.9.18–2.9.25) by burning a fuel with oxygen from the atmosphere, limiting its operation to lower altitudes. A rocket engine carries two **propellants**, namely a fuel and an oxidizer, and can operate anywhere from the atmosphere to the vacuum of space or even underwater. The most important property of a pair of propellants is the **specific impulse** defined alternatively and equivalently as: (i) the thrust in kilograms-force produced by a kilogram of propellant during one second; (ii) the time in seconds that one kilogram-weight of propellant produces one kilogram-force of thrust. Thus *in perfectly expanded conditions, that exit pressure p_e equal (2.1142a) to the ambient pressure p_a, the thrust of a rocket engine equals: (i) the product (2.1142b) of the specific impulse I_{sp} by the weight flow rate of propellant, that is the mass flow rate $\dot{m}$ multiplied by the acceleration of gravity g:*

$$p_e = p_a: \quad I_{sp}\,\dot{m}\,g = F = \dot{m}\,v_e; \quad\quad (2.1142a\text{–}c)$$

(ii) the product (2.1142c) of the exhaust velocity v_e by the mass flow rate. From (2.1142b) ≡ (2.1142c) it follows that the exit velocity is the product (2.1143a) of the specific impulse by the acceleration of gravity:

$$v_e = I_{sp}\,g, \quad \dot{m} = \rho_e\,v_e\,A_e = \rho_e\,I_{sp}\,A_e\,g, \quad\quad (2.1143a\text{–}c)$$

and the mass flow rate (2.11423b) is obtained multiplying by the exit mass density ρ_e and nozzle exit area A_e leading to (2.1143c). In conditions other than ideal expansion (2.1141a), that is when the exit pressure differs from the ambient pressure (2.1144a), *the pressure difference multiplied by the nozzle area must be added to (2.1143b, c) to specify the total thrust (2.1144b, c):*

$$p_e \neq p_a: \quad F - \left(p_e - p_a\right) A_e = \dot{m} \, v_e = I_{sp} \, \dot{m} \, g. \qquad (2.1144\text{a--c})$$

Next is considered the operation of a rocket engine (subsections 2.9.27–2.9.39) using a minimum input data.

2.9.27 Minimum Data to Describe the Operation of a Rocket

Rockets may use: (i) solid propellants that are easier to store but generally have lower specific impulse; (ii) liquid propellants that generally have a higher specific impulse but must be loaded shortly before launch, apart from exceptional "storable" liquid propellants. Among the liquid propellants one of the highest specific impulses is for oxygen burning hydrogen to form water (2.694a), which is not a pollutant, apart from residues of other substances; water vapour is a greenhouse gas at same altitudes in the Earth's atmosphere. In order to occupy the least possible volume both oxygen (hydrogen) must be in a liquid state LOX (LH), requiring high pressure and low temperature of about 54 K (22 K). An alternative to avoid very low propellant temperatures is to replace hydrogen as fuel by a hydrocarbon (Figure 2.51).

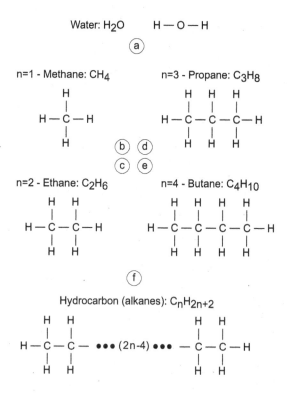

FIGURE 2.51 Valence bonds linking atoms in molecules for: (a) water; (f) general alkane hydrocarbons, including (b) methane $=1$, (c) ethane $n = 2$, (d) propane $n = 3$ and (e) butane $n = 4$.

The volatility of LOX and LH requires filling rocket propellant tanks just before launch. Since the molecular mass of oxygen (2.1145a) is 16 and that of hydrogen (2.1145b) is 1, the molecular mass of water (2.1145c) resulting from their combustion is 18, corresponding for a perfect gas (2.473h; 2.470f) to the gas constant (2.1145d–g):

$$m_O = 16, \quad m_H = 1, \quad m_{OH_2} = 18:$$
$$R_w = \frac{\bar{R}}{m_{OH_2}} = \frac{8.314\,J\,K^{-1}\,mole^{-1}}{18\,g\,mole^{-1}} = \frac{8314}{18}\,J\,kg^{-1}\,K^{-1} = 462\,m^2\,s^{-2}\,K^{-1}, \quad (2.1145a\text{–}g)$$

that is larger than for air (2.718a–e), that has a larger molecular mass (2.717a–g). An example of a rocket using LOX/LH is the Vulcain 2 engine used in the first stage of the Ariane 5 ECA/ES rocket for which the following data is publicly available: (i) nozzle area ratio (2.1146a); (ii) pressure in the combustion chamber (2.1146b, c):

$$\frac{A_e}{A_*} = 58.2, \quad p_0 = 117.3\,bar = 11.73\,MPa; \quad (2.1146a\text{–}c)$$

$$p_a = 0: \quad I_{sp} = 429\,s, \quad F = 1359\,kN, \quad (2.1146d\text{–}f)$$

(iii/iv) specific impulse (2.1146e) [thrust (2.1146f)] in vacuum (2.1146d). The preceding data will be used to determine the critical Mach number at the exit (subsection 2.9.28) as a preliminary step to specify the flow conditions at: (i/ii) the [subsection(s) 2.9.32 (2.9.29–2.9.30)] nozzle throat (exit); (iii) combustion chamber (subsection 2.9.31).

2.9.28 CRITICAL MACH NUMBER AT THE NOZZLE EXIT

The acceleration of gravity (2.713a) ≡ (2.1147a) and specific impulse (2.1146e) lead (2.1143a) ≡ (2.1147b) to the jet exhaust speed (2.1147c, d):

$$g = 9.81\,m\,s^{-1}: \quad v_e = I_{sp}\,g = 429\,s \times 9.81\,m\,s^{-2} = 4208\,m\,s^{-1}. \quad (2.1147a\text{–}d)$$

The area ratio of the nozzle between the throat and exit is related by the conservation of mass flux (2.1056b, c) ≡ (2.1148a, 2.475a–c) to the exit velocity and velocity at the throat, the latter coinciding with the critical sound speed:

$$\frac{A_*}{A_e} = \frac{\rho_e\,v_e}{\rho_*\,v_*} = \frac{v_e}{c_*}\left[\frac{\gamma+1}{2} - \frac{\gamma-1}{2}\left(\frac{v_e}{c_*}\right)^2\right]^{1/(\gamma-1)}. \quad (2.1148a, b)$$

Thus the exit critical Mach number (2.1149a) satisfies (2.1149c–f) where were used the area ratio (2.1146a) and the adiabatic exponent (2.475b) ≡ (2.1148b), since water is a linear polyatomic molecule (Figure 2.51):

$$M_{*e} \equiv \frac{v_e}{c_*}, \quad \gamma = 1.4: \quad j(M_{*e}) \equiv M_{*e}\left[1.2 - 0.2(M_{*e})^2\right]^{2.5} \qquad (2.1149a-c)$$

$$= \frac{A_*}{A_e} = \frac{1}{58.2} = 0.01718. \qquad (2.1149d-f)$$

For an adiabatic flow, that is heat and shock free flow, the term in square brackets in (2.1149c) must be positive, and thus the critical exit Mach number cannot exceed the value (2.1149h–j) corresponding to an infinite area ratio (2.1149g):

$$\frac{A_*}{A_e} \geq 0 \quad \Rightarrow \quad M_{*e} \leq \sqrt{\frac{1.2}{0.2}} = \sqrt{6} = 2.449; \qquad (2.1149g-j)$$

the result (2.1149j) agrees with (2.683f–j) in the second case of a linear polyatomic molecule. The exit critical Mach number is a root of the transcendental equation (2.1149f), and since the area ratio (2.1146a) is large it must be close to the maximum possible value (2.1149j). Choosing a bissonic critical exit Mach number (2.1150a) gives (2.1150b) for the function:

$$M_{*e} = 2.000 \rightarrow 2.300 \rightarrow 2.301 \rightarrow 2.302, \qquad (2.1150a)$$

$$j(M_{*e}) = 0.2024 \rightarrow 0.01748 \rightarrow 0.01720 \rightarrow 0.01693, \qquad (2.1150b)$$

and further iterations lead to an exact value with 3 digits (2.1151a) in (2.1150a) for which (2.1150b) is closest to (2.1149f):

$$M_{*e} = 2.301: \quad c_* = \frac{v_e}{M_*} = \frac{4208}{2.301}\,m\,s^{-1} = 1829\,m\,s^{-1}, \qquad (2.1151a-c)$$

the corresponding critical (2.1147d; 2.1151a) sound speed is (2.1151b, c) and is conserved in an adiabatic flow and across shock waves. Together with the stagnation pressure at the combustion chamber (2.1146b) this is sufficient to determine all flow variables at the nozzle exit (subsection 2.9.29).

2.9.29   NOZZLE EXIT FLOW IN VACUO

Besides the critical sound speed (2.1151b, c) the stagnation sound speed (2.664c) is also conserved in the adiabatic flow and across shocks (2.482a–e):

$$c_0 = c_*\sqrt{\frac{\gamma + 1}{2}} = \sqrt{1.2} \times 1829\,m\,s^{-1} = 2003\,m\,s^{-1}. \qquad (2.1152a-c)$$

Assuming that the fluid velocity is negligible in the combustion chamber (2.1146b) $\equiv$ (2.1153a,b) is the stagnation pressure:

$$p_0 = 117.3\,bar = 1.173 \times 10^7\,kg\,m^{-1}\,s^{-2}. \qquad (2.1153a,\ b)$$

The nozzle exit pressure follows from (2.666e) ≡ (2.1154a–d) using (2.1147d; 2.1149b; 2.1152c; 2.1153b):

$$p_e = p_0 \left[1 - \frac{\gamma-1}{2} \left(\frac{v_e}{c_0} \right)^2 \right]^{\frac{\gamma}{\gamma-1}} = 117.3 \times \left[1 - 0.2 \times \left(\frac{4208}{2003} \right)^2 \right]^{3.5} bar \qquad (2.1154a\text{–}d)$$

$$= 6.48 \times 10^{-2} \, bar = 6.48 \times 10^3 \, N \, m^{-2}.$$

The local exit sound speed (2.662b) ≡ (2.1155a) is given (2.1149b; 2.1147d; 2.1152c) by (2.1155b, c):

$$c_e = \left| c_0^2 - \frac{\gamma-1}{2} (v_e)^2 \right|^{1/2} = \sqrt{2003^2 - 0.2 \times 4208^2} \, m \, s^{-1} = 686 \, m \, s^{-1}. \qquad (2.1155a\text{–}c)$$

The exit temperature (2.1156b, c) follows (2.479d) ≡ (2.1156a) from (2.1145g; 2.1149b; 2.1155c):

$$T_e = \frac{(c_e)^2}{\gamma \, R_w} = \frac{\left(686 \, m \, s^{-1} \right)^2}{1.4 \times 462 \, m^2 \, s^{-2} \, K^{-1}} = 728 \, K, \qquad (2.1156a\text{–}c)$$

and the exit mass density (2.473g) ≡ (2.1157a) is given (2.1145g; 2.1154e; 2.1156c) by (2.1157b, c):

$$\rho_e = \frac{p_e}{R_w \, T_e} = \frac{6.48 \times 10^3 \, kg \, m^{-1} \, s^{-2}}{462 \, m^2 \, s^{-2} \, K^{-1} \times 728 \, K} = 1.93 \times 10^{-2} \, kg \, m^{-3}. \qquad (2.1157a\text{–}c)$$

The preceding data together with the thrust specifies the nozzle exit area (subsection 2.9.30).

2.9.30 TOTAL THRUST AND NOZZLE EXIT AREA

In vacuo the ambient pressure is zero (2.1158a), and the thrust (2.1144b) simplifies to (2.1158b) leading (2.1143b) to (2.1158c):

$$p_a = 0: \quad F = \dot{m} \, v_e + p_e \, A_e = \left(p_e + \rho_e \, v_e^2 \right) A_e. \qquad (2.1158a\text{–}c)$$

From (2.1147d; 2.1157c) follows twice the dynamic pressure at the nozzle exit (2.488a–i):

$$\rho_e \, v_e^2 = 1.93 \times 10^{-2} \, kg \, m^{-3} \times \left(4208 \, m \, s^{-1} \right)^2 \quad = \quad 3.42 \times 10^5 \, kg \, m^{-1} \, s^{-2}$$
$$\qquad (2.1159a\text{–}c)$$
$$= \quad 3.42 \times 10^5 \, N \, m^{-2},$$

that appears in (2.1158c), leading from (2.1146f; 2.1154e) to the nozzle exit area (2.489a–f):

$$A_e = \frac{F}{p_e + \rho_e (v_e)^2} = \frac{1.359 \times 10^6 \, N}{(6.48 \times 10^3 + 3.42 \times 10^5) \, N \, m^{-2}} = 3.90 \, m^2. \qquad (2.1160a\text{–}c)$$

The mass flow rate (2.1143b) is given (2.1147d; 2.1157c; 2.1160c) by (2.490a–g):

$$\dot{m} = \rho_e \, v_e \, A_e = 1.93 \times 10^{-2} \, kg \, m^{-2} \times 4208 \, m \, s^{-1} \times 3.90 \, m^2 = 317 \, kg \, s^{-1}. \qquad (2.1161a\text{–}c)$$

The local (2.1155c), stagnation (2.1152c) and critical (2.1151c) Mach numbers at the exit (2.1147d) are given by (2.1162a–c):

$$\{M_e, M_{0e}, M_{*e}\} \equiv \frac{v_e}{\{c_e, c_0, c_*\}} = \frac{4208}{\{686, 2003, 1829\}} = \{6.134, 2.101, 2.301\}, \qquad (2.1162a\text{–}c)$$

with the last value confirming (2.1151a). The underexpanded flow in the nozzle gives rise to an expansion wave at the exit, and the Prandtl – Meyer function (2.967a–c) specifies (2.1149b; 2.1162b) the angle by which the shock front is turned (2.1163a–k):

$$a = \frac{\gamma - 1}{\gamma + 1} = \frac{0.4}{2.4} = \frac{1}{6}, \quad M = 6.134: \qquad (2.1163a\text{–}c)$$

$$\varphi_1(M) = \arctan\left(\sqrt{M^2 - 1}\right) = \arctan\left(\sqrt{6.134^2 - 1}\right) = 80.6°, \qquad (2.1163d\text{–}f)$$

$$\varphi_2(M) = \arctan\left[\sqrt{a(M^2 - 1)}\right] = 68.0°, \qquad (2.1163g, h)$$

$$\varphi(M) = \varphi_1(M) - \frac{\varphi_2(M)}{\sqrt{a}} = 80.6° - \frac{68.0°}{\sqrt{6}} = 52.8°. \qquad (2.1163j\text{–}k)$$

The physical conditions in the combustion chamber are considered next (subsection 2.9.31).

2.9.31   PHYSICAL CONDITIONS IN THE COMBUSTION CHAMBER

The stagnation pressure (2.1153b) and sound speed (2.1152c) specify (2.492d) $\equiv$ (2.1164a) using (2.1149b) the stagnation mass density (2.1164b, c):

$$\rho_0 = \frac{\gamma \, p_0}{c_0^2} = 1.4 \times \frac{1.173 \times 10^7 \, kg \, m^{-1} \, s^{-2}}{(2003 \, m \, s^{-1})^2} = 4.09 \, kg \, m^{-3}, \qquad (2.1164a\text{–}c)$$

neglecting the flow velocity in the combustion chamber. The stagnation temperature (2.1165b, c) follows (2.492e) $\equiv$ (2.1164a) from (2.1145g; 2.1149b; 2.1152c)

$$T_0 = \frac{(c_0)^2}{\gamma R_w} = \frac{(2003 \, m \, s^{-1})^2}{1.4 \times 462 \, m^2 \, s^{-2} \, K^{-1}} = 6203 \, K, \qquad (2.1165\text{a--c})$$

and the corresponding internal energy (2.471a; 2.264d) [enthalpy (2.474a; 2.264e)] per unit mass are given by (2.1166a–e) [(2.1167a–f)] using (2.1165c; 2.1149b; 2.1145g):

$$u_0 = c_V \, T_0 = \frac{R_w}{\gamma-1} T_0 = \frac{(c_0)^2}{\gamma(\gamma-1)} = \frac{462 \, m^2 \, s^{-2} \, K^{-1} \times 6203 \, K}{0.4} = 7.16 \times 10^6 \, m^2 \, s^{-2}, \qquad (2.1166\text{a--e})$$

$$h_0 = c_p \, T_0 = \frac{\gamma}{\gamma-1} R_w \, T_0 = \gamma \, u_0 = \frac{c_0^2}{\gamma-1} = \frac{(2003 \, m \, s^{-1})^2}{0.4} = 1.003 \times 10^7 \, m^2 \, s^{-2}. \qquad (2.1167\text{a--f})$$

The very high temperatures in the combustion chamber are reduced by thermal conduction and radiation that are not taken into account in the adiabatic calculation, so (2.1165c) is a gross overestimate or an upper bound for the possible values. The flow conditions at the nozzle throat are obtained next (subsection 2.9.32).

2.9.32 FLOW CONDITIONS AT THE THROAT OF THE NOZZLE

The area ratio (2.1146a) and nozzle exit area (2.1160c) specify the smallest cross-sectional area of the nozzle at the throat (2.1168a)

$$A_* = \frac{A_e}{58.2} = \frac{3.90}{58.2} m^2 = 0.0670 \, m^2, \qquad (2.1168\text{a--c})$$

The temperature (2.1169b, c) follows (2.479d) $\equiv$ (2.1169a) and (2.664c) $\equiv$ (2.1169b) from (2.1165c; 2.1149b) leading to (2.1169c, d):

$$T_* = T_0 \left(\frac{c_*}{c_0}\right)^2 = T_0 \sqrt{\frac{2}{\gamma+1}} = \frac{6203 \, K}{\sqrt{1.2}} = 5663 \, K. \qquad (2.1169\text{a--d})$$

The mass density (2.669e) $\equiv$ (2.1170a) is (2.1164c; 2.1149b) given by (2.1170b, c):

$$\rho_* = \rho_0 \left(\frac{2}{\gamma+1}\right)^{1/(\gamma-1)} = (1.2)^{-2.5} \times 4.09 \, kg \, m^{-3} = 2.59 \, kg \, m^{-3}, \qquad (2.1170\text{a--c})$$

and the pressure (2.669f) ≡ (2.1171a) is given (2.1146b; 2.1149b) by (2.1051a) ≡ (2.1172d):

$$p_* = p_0 \left(\frac{2}{\gamma+1} \right)^{\gamma/(\gamma-1)} = (1.2)^{-3.5} \times 117.3\,bar = 62.0\,bar. \qquad (2.1171a\text{–}c)$$

Using (2.1151c; 2.1170c; 2.1168c) it can be checked that the mass flow rate (2.1051a) ≡ (2.1172d):

$$\dot{m} = \rho_* c_* A_* = 2.59\,kg\,m^{-3} \times 1829\,m\,s^{-1} \times 0.0670\,m^2 = 317\,kg\,s^{-1} \qquad (2.1172a\text{–}c)$$

agrees (2.1172b, c) with (2.1161c). The flow conditions at the nozzle throat/combustion chamber/nozzle exit (respectively, subsections 2.9.32/2.9.31/2.9.30) are compared in Table 2.25 and were obtained for space vacuum (subsections 2.9.26–2.9.29), that is zero ambient pressure, so that the flow is always underexpanded. The adiabatic exit pressure (2.1154c) is less than the atmospheric pressure at sea level, implying that the flow is overexpanded, and a shock wave forms inside the nozzle (subsections

---

TABLE 2.25

Flow Conditions at Three Locations of Rocket Nozzle

Variable	Symbol	Unit	Combustion Chamber	Throat	Exit
			Location		
Area	A	m^2	–	0.0670	3.90
Pressure	p	bar	117.3	62.0	0.0648
Density	ρ	$kg\,m^{-3}$	4.09	2.59	0.0193
Temperature	T	K	6203	5663	728
Velocity	v	$m\,s^{-1}$	0.0	1829	4208
Local sound speed	c	$m\,s^{-1}$	2003	1829	686
Critical sound speed	$c*$	$m\,s^{-1}$	1829	1829	1829
Stagnation sound speed	c_0	$m\,s^{-1}$	2003	2003	2003
Local Mach number	M	–	0.0	1.00	6.134
Critical Mach number	$M*$	–	0.0	1.00	2.301
Stagnation Mach number	M_0	–	0.0	0.913	2.101
Internal energy per unit mass	u	$\times10^6\,m^2\,s^{-2}$	7.16	6.54	0.841
Enthalpy per unit mass	h	$\times10^6\,m^2\,s^{-2}$	10.03	9.16	1.177

Gas constant (LH-LOX): $R_w = 462\,m^2\,s^{-2}\,K^{-1}$
Mass flow rate: $m = 317\,kg\,s^{-1}$
Specific impulse: $I_{sp} = 429\,s$
Thrust: $F = 1359\,kN$
Adiabatic turn angle at exit: $\varphi(M_e) = 52.8°$

Note: Flow conditions for a rocket nozzle in vacuum at three locations: (a) combustion chamber; (b) throat; (c) exit. For each location (a, b, c) are indicated: (i–iv) area, pressure, density and temperature; (v–viii) velocity and local, critical and stagnation sound speed; (ix–xi) local, critical and stagnation Mach numbers; (xii–xiii) internal energy and enthalpy.

2.9.34–2.9.37). There is an intermediate altitude for which the ambient pressure equals the adiabatic exit pressure corresponding to a perfectly expanded nozzle flow (subsection 2.9.33).

2.9.33 ALTITUDE FOR PERFECTLY EXPANDED NOZZLE FLOW

The adiabatic nozzle exit pressure (2.1154c) $\equiv$ (2.1173b) lies between the atmospheric pressure at the tropopause (2.1094c) = (2.1173a) and at the top of the stratosphere (2.1097b) $\equiv$ (2.1173c):

$$\bar{p}_0 = 0.227\,bar > p_e = 0.0648 > 0.0250 = \bar{p}_{0/}, \qquad (2.1173a–c)$$

implying that there is an altitude in the stratosphere corresponding to perfectly expanded nozzle flow. Since the stratosphere is isothermal, the altitude is determined by the pressure (2.1082c) equal to the adiabatic nozzle exit pressure (2.1174a) $\equiv$ (2.1174b):

$$p_e = \bar{p}_0 \exp\left(-\frac{z_e - \bar{z}_0}{\bar{L}_0}\right) \iff z_e = \bar{z}_0 + \bar{L}_0 \log\left(\frac{\bar{p}_0}{p_e}\right). \qquad (2.1174a, b)$$

Using (2.1088b; 2.1090f; 2.1173a; 2.1154c) the altitude in the stratosphere for perfectly expanded nozzle flow is (2.507a–c):

$$z_e = 1.1\times10^4\,m + 6.34\times10^3\,m \times \log\left(\frac{0.227}{0.0648}\right) = 1.90\times10^4\,m = 19.0\,km. \quad (2.1175a–c)$$

At this altitude (2.1175c) the thrust (2.1142c) equals (2.1176a–c) the mass flow rate (2.1161c) multiplied by the nozzle exit velocity (2.1147c)

$$F_e = \dot{m}\,v_e = 317\,kg\,s^{-1} \times 4208\,m\,s^{-1} = 1.334\times10^6\,N: \qquad (2.1176a–c)$$

$$\frac{F_e}{F} = \frac{1.334}{1.359} = 0.982. \qquad (2.1176d, e)$$

The thrust (2.1176c) at the intermediate altitude (2.1175c) is close to and slightly less (2.1176d, 2.509a–d) than the thrust in vacuo (2.1146f) because in that case (2.1177a) the ratio of thrust (2.1144b) to (2.1142c) is (2.1177b) given (2.1154e; 2.1146f; 2.1160c) by (2.1177c) in agreement with (2.1176e) to within rounding-off errors:

$$p_a = 0: \quad 1 - \frac{p_e\,A_e}{F} = 1 - \frac{6.48\times10^3\,N\,m^{-2}\times3.90\,m^2}{1.359\times10^6} = 0.981. \qquad (2.1177a–e)$$

This assumes the same specific impulse (2.1146e) and exit velocity (2.1147d) in the vacuum of space and in the stratosphere at the altitude (2.1175c). The perfect expansion in the latter case would lead to a larger exit velocity for the same pressure in the combustion chamber, easily outweighting the small differences in (2.1177e) = (2.1176e). The nozzle is assumed to be fixed and the flow conditions inside [subsections 2.9.31 (2.9.32)] are assumed to be the same for the nozzle at sea level, with overexpanded flow implying the existence of a shock wave (subsections 2.9.36–2.9.39). The nozzle exit conditions in the overexpanded case are different from the underexpanded case in that the atmospheric pressure must be matched (subsections 2.9.34–2.9.35).

2.9.34   EXIT CONDITIONS FOR OVEREXPANDED NOZZLE AT ATMOSPHERIC PRESSURE

For an overexpanded flow the pressure at the exit is the atmospheric pressure equal to 1 atm (2.716a–m) for lift-off at sea level, and thus the exit mass density ρ_a and local sound speed c_a satisfy (2.492d) ≡ (2.1178a) leading (2.1149b) to (2.1179b–d):

$$\rho_a (c_a)^2 = \gamma\, p_a = 1.4\, atm = 1.4 \times 1.014 \times 10^5\, kg\, m^{-1}\, s^{-2} \qquad (2.1178a\text{–}d)$$
$$= 1.42 \times 10^5\, kg\, m^{-1}\, s^{-2}.$$

The mass density ρ_a and exit velocity v_a also satisfy the conservation of the mass flux (2.1161c) for the nozzle exit area (2.1160c) leading to (2.512a–e):

$$\rho_a v_a = \frac{\dot{m}}{A_e} = \frac{317\, kg\, s^{-1}}{3.90\, m^2} = 81.3\, kg\, m^{-2}\, s^{-1}. \qquad (2.1179a\text{–}c)$$

The exit velocity and local sound speed are also related (2.662b) ≡ (2.1180b) by the stagnation sound speed (2.1152c) leading to (2.1180a–d)

$$(c_a)^2 + 0.4 (v_a)^2 = (c_a)^2 + \frac{\gamma-1}{2}(v_a)^2 = (c_0)^2 \qquad (2.1180a,\ b)$$

$$= (2003\, m\, s^{-1})^2 = 4.01 \times 10^6\, m^2\, s^{-2}. \qquad (2.1180c,\ d)$$

The ratio of (2.1178d) to (2.1179c) is (2.1181a, 2.475a–c):

$$\frac{(c_a)^2}{v_a} = \frac{1.42 \times 10^5\, kg\, m^{-1}\, s^{-2}}{81.3\, kg\, m^{-2}\, s^{-1}} = 1747\, m\, s^{-1}. \qquad (2.1181a,\ b)$$

Substitution of (2.1181b) in (2.1180b) leads to a quadratic equation for the exit velocity (2.515a–e)

$$0.4 (v_a)^2 + 1747\, v_a - 4.01 \times 10^6 = 0, \qquad (2.1182)$$

whose positive root (2.1183a, 2.475a–c) specifies the exit velocity:

$$v_a = \frac{-1747 + \sqrt{1747^2 + 4 \times 0.4 \times 4.01 \times 10^6}}{2 \times 0.4} = 1663 \, m \, s^{-1}. \qquad \text{(2.1183a, b)}$$

The exit velocity (2.1183b) specifies other flow variables and the thrust for lift-off at sea level (subsection 2.9.35).

2.9.35 THRUST AT LIFT-OFF FOR OVEREXPANSION AT SEA LEVEL

The exit velocity (2.1183b) leads (2.1181b) to the local sound speed (2.517a–e)

$$c_a = \sqrt{1747 \, v_a} = \sqrt{1747 \times 1663} \, m \, s^{-1} = 1704 \, m \, s^{-1}, \qquad \text{(2.1184a–c)}$$

and hence to the local (2.1183a), critical (2.1151c) and stagnation (2.1152c) Mach numbers (2.1185a–):

$$\{M_a, M_{*a}, M_{0a}\} \equiv \frac{v_a}{\{c_a, c_*, c_0\}} = \frac{1663}{\{1704, 1829, 2003\}} = \{0.974, 0.909, 0.830\} < 1, \qquad \text{(2.1185a–c)}$$

that are all subsonic (2.1185d). The exit mass density (2.1179c) $\equiv$ (2.1186a) is given (2.1179c; 2.1183b) by (2.1186b, c):

$$\rho_a = \frac{\dot{m}}{v_a \, A_e} = \frac{81.3 \, kg \, m^{-2} \, s^{-1}}{1663 \, m \, s^{-1}} = 0.0489 \, kg \, m^{-3}, \qquad \text{(2.1186a–c)}$$

and the exit temperature (2.492e) $\equiv$ (2.1187a) by (2.1187b, c) using (2.1145g; 2.1149b; 2.1184c):

$$T_a = \frac{(c_a)^2}{\gamma \, R_w} = \frac{(1704 \, m \, s^{-1})^2}{1.4 \times 462 \, m^2 \, s^{-2} \, K^{-1}} = 4489 \, K. \qquad \text{(2.1187a–c)}$$

Since the exit pressure matches the atmospheric pressure (2.1142a) the thrust (2.1142c) $\equiv$ (2.1188a) equals the mass flow rate (2.1161c) multiplied by the exit velocity for overexpanded flow at sea level (2.1183a) leading to a value (2.1188b–d):

$$F_a = \dot{m} \, v_a = 317 \, kg \, s^{-1} \times 1663 \, m \, s^{-1} = 5.27 \times 10^5 \, kg \, m^{-2} = 527 \, kN, \qquad \text{(2.1188a–d)}$$

of the lift-off thrust at sea level much lower (2.1189a, b) than the thrust in vacuo (2.1146f) because: (i) the mass flow rate (2.1161c) is the same; (ii) the exhaust velocity is lower at lift-off (2.1183c) [higher in vacuo (2.1147d)] in the ratio (2.1189d, e):

$$0.387 = \frac{527}{1359} = \frac{F_a}{F} = \frac{v_a}{v_e} = \frac{1663}{4208} = 0.395; \qquad \text{(2.1189a–e)}$$

the ratios coincide (2.1189a–e) to within accumulated rounding-off errors. The large thrust ratio (2.1189a–e) is due to the presence (absence) of a shock in the nozzle excluding (allowing) a supersonic exhaust (2.1185a–) [(2.1162a–c)] for an underexpanded (overexpanded) nozzle flow. The lower thrust at lift-off and altitudes below ideal expansion (2.1175a–c) affects the initial acceleration in the denser layers of the atmosphere that cause higher aerodynamic drag, that may be overcome with the aid of strap-on boosters. The shock wave in the nozzle responsible for the thrust loss is considered next, showing (subsection 2.9.36) that a normal shock either inside the nozzle (case V) or at the nozzle exit (case IV) is not possible, leaving as the only remaining possibility (case VI) the oblique shock at the nozzle exit (subsections 2.9.37–2.9.39).

2.9.36 INEXISTENCE OF A NORMAL SHOCK INSIDE OR AT THE NOZZLE EXIT

The adiabatic exit pressure (2.1154c) is less than the atmospheric pressure, implying that the pressure must be raised by a shock. Of the three possible cases IV–VI the normal shocks inside (2.1099e) [at the exit (2.1099d)] of the nozzle are not possible [case V (IV)] because it cannot match the atmospheric pressure (subsection 2.9.36) leaving as the only possibility the case VI, an oblique shock at the nozzle exit (subsections 2.9.38–2.9.39). In the case V of a normal shock inside the divergent section of the nozzle (2.1099e), the upstream Mach number (2.1065c) = (2.1190a) must satisfy (2.1071a, b) ≡ (2.1190d, e) with the exit pressure equal to the atmospheric pressure (2.1190b, c), critical pressure (2.1171c) and critical exit Mach number (2.1185b) leading (2.1149b) to the value (2.1190e):

$$N \equiv \left(\frac{v^-}{c_*} \right)^2, \quad p_e = p_a = 1\,atm:$$

$$f_1(N) \equiv \left[\frac{2.4 - 0.4\,N}{2.4\,N - 0.4} \right]^{\frac{1}{1.4}} N = 2 \times \frac{62.0^{\frac{1}{3.5}}}{2.4 - 0.4 \times 0.909^2} = 0.306.$$

(2.1190a–e)

In order for the shock to occur within the divergent section of the nozzle the critical exit Mach number must lie between the exit value (2.1162c) and the value unity at the throat (2.1191a) implying (2.1191b) that (2.1190e) must have a root in the range (2.1191c)

$$1 < M_*^- < 2.30 \quad \Rightarrow \quad 1 < N \equiv \left(M_*^- \right) < 5.29 \qquad (2.1191a–c)$$

$$\Rightarrow \quad 1 = f_1(1) > f_1(N), \quad f_1(5.29) > 0.359. \qquad (2.1191d, e)$$

Since the value (2.1190e) lies outside the range (2.1191d, e) the case V of a normal shock inside the nozzle is not possible. In Case IV of a normal shock at the nozzle exit the upstream pressure would be (2.1154c) ≡ (2.1192a) and the upstream local Mach number (2.1162c) ≡ (2.1192b) implying (2.832c) ≡ (2.1192c) the downstream pressure (2.1192d, e):

$p^- = p_e = 6.48 \times 10^{-2}\, bar, \quad M^- = 6.134:$

$$p^+ = p^- \left\{ 1 + \frac{2\gamma}{\gamma+1} \left[\left(M^- \right)^2 - 1 \right] \right\}$$

$$= 6.48 \times 10^{-2} \times \left[1 + \frac{2.8}{2.4} \times \left(6.134^2 - 1 \right) \right] bar = 2.83\, bar,$$

(2.1192a–e)

that does not match the atmospheric pressure (2.1190b, c). Thus the only possible remaining case is an oblique shock at the nozzle exit (subsections 2.9.37–2.9.39).

2.9.37 STRONG OBLIQUE SHOCK AT THE NOZZLE EXIT

An oblique shock at the nozzle exit creates a two-dimensional flow, and thus the quasi-one-dimensional approximation is no more than an order of magnitude estimate near the nozzle exit plane. The upstream Mach number is large (2.1192b) and the transformation to a strong oblique shock with angle of incidence (2.941a) ≡ (2.1193a, b) leads from (2.1192c) to (2.1193c):

$$\left(M^- \right)^2 \to \left(M^- \sin\theta \right)^2 \gg 1 \qquad \frac{p^+}{p^-} = \frac{2\gamma}{\gamma+1} \left(M^- \sin\theta \right)^2 ; \qquad (2.1193a–c)$$

thus the angle of incidence of the oblique shock is approximated by (2.1193c) ≡ (2.1194a) with upstream pressure (2.1192b) and downstream atmospheric pressure (2.716k–m) ≡ (2.1195b–d):

$$\sin^2\theta = \frac{\gamma+1}{2\gamma} \frac{p^+/p^-}{\left(M^- \right)^2} : \quad p^+ = p_a = 1\,atm = 1.014\, bar. \qquad (2.1194a–d)$$

Substitution of (2.1149b; 2.1194d; 2.1192a) in (2.1194a) ≡ (2.1195a) leads to (2.1195b) specifying the approximate angle of incidence (2.1196c):

$$\sin^2\theta = \frac{2.4}{2.8} \times \frac{1.014}{6.134^2 \times 0.0648} = 0.356 \quad \Rightarrow \theta = 36.7°. \qquad (2.1195a–c)$$

The exact value of the angle of incidence, without using the strong shock approximation (2.1193a) follows from (2.1192c) ≡ (2.1196a) and is given by (2.1196b–d):

$$\sin^2\theta = \left(M^- \right)^{-2} \left[1 + \frac{\gamma+1}{2\gamma} \left(\frac{p^+}{p^-} - 1 \right) \right]$$

$$= \frac{1}{6.134^2} \left[1 + \frac{2.4}{2.8} \left(\frac{1.014}{0.0648} - 1 \right) \right] = 0.360 = 36.87°,$$

(2.1196a–d)

and (2.1196d) is quite close to (2.1195c). The angle of deflection (2.918b) ≡ (2.1197a) is given (2.1196c; 2.1149b; 2.1162c) by (2.1197b–d):

$$\cot\varphi = \left[\frac{1}{2} \frac{\gamma+1}{\sin^2\theta - \left(M^-\right)^{-2}} - 1 \right] \tan\theta$$

$$= \left[\frac{1.2}{0.360 - (6.134)^{-2}} - 1 \right] \times 0.751 = 1.952,$$

$$\varphi = 27.13°, \quad \theta - \varphi = 36.87° - 27.13° = 9.74°, \tag{2.1197e–f}$$

implying that the angle of incidence downstream is (2.1197e, f). The discussion of the rocket nozzle (subsections 2.9.26–2.9.39) concludes with the flow conditions upstream (downstream) of the oblique shock [subsection 2.9.38 (2.9.39)].

2.9.38   FLOW CONDITIONS UPSTREAM OF THE OBLIQUE SHOCK

The atmospheric flow conditions (subsections 2.9.34–2.9.35) apply upstream of the shock, with the velocity (2.1147d) = (2.1198a) at an angle of incidence (2.1196d) having normal and tangential components (2.1198b–d):

$$v^- \equiv v_e = 4208\, m\, s^{-1}:$$
$$\left\{v_n^-, v_t^-\right\} = v^- \left\{\sin, \cos\theta\right\} = 4208 \times \left\{\sin, \cos\left(36.87°\right)\right\} = \left\{2524, 3366\right\} m\, s^{-1}. \tag{2.1198a–d}$$

The local (2.1155c), critical (2.1151c) and stagnation (2.1152c) upstream normal (tangential) Mach numbers are (2.1119a–c) [(2.1200a–c)]:

$$\left\{M_n^-, M_{*n}^-, M_{0n}^-\right\} \equiv \frac{v_n^-}{\left\{c_e, c_*, c_0\right\}} = \frac{2524}{\left\{686, 1829, 2003\right\}} = \left\{3.679, 1.380, 1.260\right\}, \tag{2.1199a–c}$$

$$\left\{M_t^-, M_{*t}^-, M_{0t}^-\right\} \equiv \frac{v_t^-}{\left\{c_e, c_*, c_0\right\}} = \frac{3366}{\left\{686, 1829, 2003\right\}} = \left\{4.907, 1.840, 1.680\right\}. \tag{2.1200a–c}$$

The tangential velocity (2.1198d) is continuous, and the downstream normal velocity satisfies (2.889b) ≡ (2.1201a) leading (2.1049b; 2.1151c) to a large value (2.1201b, c):

$$v_n^+ = \frac{c_*^2 - \left[(\gamma-1)/(\gamma+1)\right]\left(v_t\right)^2}{v_n^-} = \frac{1829^2 - 3365^2/6}{2524}\, m\, s^{-1} = 577\, m\, s^{-1}. \tag{2.1201a–c}$$

The downstream tangential velocity (2.1202a–c) follows from (2.1201c; 2.1197f):

$$v_t^+ = v_n^+ \cot\left(\theta - \varphi\right) = 577\, m\, s^{-1} \times \cot\left(9.74°\right) = 3361\, m\, s^{-1}. \tag{2.1202a–d}$$

is continuous (2.1198d) and leads to the total value (2.536a–c):

$$v^+ \equiv \sqrt{\left(v_t^+\right)^2 + \left(v_n^+\right)^2} = \sqrt{3361^2 + 577^2} \, m\,s^{-1} = 3410 \, m\,s^{-1}. \qquad (2.1203\text{a--c})$$

This specifies the remaining flow variables downstream of the shock (subsection 2.9.39).

2.9.39   FLOW VARIABLES DOWNSTREAM OF THE OBLIQUE SHOCK

The downstream local sound speed (2.662b) ≡ (2.1204a) is given (2.1152c; 2.1149b; 2.1203c) by (2.1204b, c):

$$c^+ = \left| c_0^2 - \frac{\gamma-1}{2}\left(v^+\right)^2 \right|^{1/2} = \sqrt{2003^2 - 0.2 \times 3410^2} \, m\,s^{-1} = 1299 \, m\,s^{-1}, \qquad (2.1204\text{a--c})$$

corresponding (2.473g) ≡ (2.1205a) to (2.1152c; 2.1204c; 2.1165c) the downstream temperature (2.1205b, c):

$$T^+ = T_0 \left(\frac{c^+}{c_0}\right)^2 = \left(\frac{1299}{2003}\right)^2 \times 6203 \, K = 2609 \, K. \qquad (2.1205\text{a--c})$$

The downstream mass density (2.883a) ≡ (2.1206a) is given (2.1147d; 2.1203c; 2.1157c) by (2.1206b, c):

$$\rho^+ = \rho_e \frac{v_e}{v^+} = \frac{4208}{3410} \times 1.93 \times 10^{-2} \, kg\,m^{-3} = 2.38 \times 10^{-2} \, kg\,m^{-3}, \qquad (2.1206\text{a--c})$$

and the downstream pressure (2.473g) ≡ (2.1207a) follows (2.1207b, c) from (2.1154c; 2.1206c; 2.1157c; 2.1205c; 2.1156c):

$$p^+ = p_e \frac{\rho^+ \, T^+}{\rho_e \, T_e} = \frac{2.38}{1.93} \times \frac{2603}{728} \times 6.48 \times 10^{-2} \, bar = 2.86 \times 10^{-1} \, bar. \qquad (2.1207\text{a--c})$$

The local (2.1204c), critical (2.1151c) and stagnation (2.1152c) upstream total (2.1203c), normal (2.1201c) and tangential (2.1202d) downstream Mach numbers are, respectively (d)/(2.540a–c)/(2.540d–f):

$$\left\{M^+, M_*^+, M_0^+\right\} \equiv \frac{v^+}{\left\{c^+, c_*, c_0\right\}} = \frac{3410}{\left\{1299, 1829, 2003\right\}} = \left\{2.625, 1.864, 1.702\right\}, \qquad (2.1208\text{a--c})$$

$$\left\{M_n^+, M_{*n}^+, M_{0n}^+\right\} \equiv \frac{v_n^+}{\left\{c^+, c_*, c_0\right\}} = \frac{577}{\left\{1299, 1829, 2003\right\}} = \left\{0.444, 0.315, 0.288\right\}, \qquad (2.1209\text{a--c})$$

$$\left\{M_t^+, M_{*t}^+, M_{0t}^+\right\} \equiv \frac{v_t^+}{\left\{c^+, c_*, c_0\right\}} = \frac{3366}{\left\{1299, 1829, 2003\right\}} = \left\{2.591, 1.840, 1.680\right\}. \qquad (2.1210\text{a--c})$$

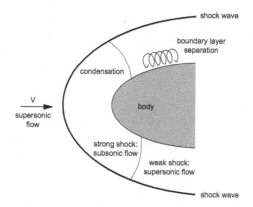

FIGURE 2.52 In front of a blunt body in a supersonic flow forms a bow shock wave, nearly normal in front of the nose and oblique elsewhere. The strong shock corresponds to a downstream subsonic flow, and the large preassure rise may cause condensation, rendering the shock wave visible. The weak shock further downstream could have supersonic downstream flow. The shock waves involve an increase in pressure, and the resulting counter-flow pressure gradient may cause boundary layer separation.

TABLE 2.26
Comparison of Flow Upstream and Downstream of an Oblique Shock

Variable	Symbol	Unit	Upstream	Downstream
Pressure	p	bar	0.0648	0.286
Density	ρ	$kg\ m^{-3}$	0.0195	0.0236
Temperature	T	K	728	2609
Local sound speed	c	$m\ s^{-1}$	686	1299
Critical sound speed	$c*$	$m\ s^{-1}$	1829	1829
Stagnation sound speed	c_0	$m\ s^{-1}$	2003	2003
Total velocity	v	$m\ s^{-1}$	4208	3410
Normal velocity	v_n	$m\ s^{-1}$	2524	577
Tangential velocity	v_t	$m\ s^{-1}$	3366	3361
Local Mach number	M	-	6.133	2.625
Critical Mach number	$M*$	-	2.300	1.864
Stagnation Mach number	M_0	-	2.100	1.702
Normal local Mach number	M_n	-	3.679	0.444
Normal critical Mach number	$M*_n$	-	1.380	0.315
Normal stagnation Mach number	M_{0n}	-	1.260	0.288
Tangential local Mach number	M_t	-	4.907	2.587
Tangential critical Mach number	$M*_t$	-	1.840	1.838
Tangential stagnation Mach number	M_{0t}	-	1.680	1.678
Angle of incidence	θ	-	36.89°	27.13°

Angle of deflection: $\varphi = \theta^- - \theta^+ = 9.74°$

Note: Comparison of flow upstream and downstream of an oblique shock indicating: (i–iii) pressure, density and temperature; (iv–vi) local, critical and stagnation sound speed; (viii–x) total, normal and tangential velocity; (xi–xiii/xiv–xvi/xvii–xix) total/normal/tangential local, critical and stagnation Mach numbers.

The normal Mach numbers downstream of the oblique shock are always subsonic (2.540a–c), and since the total Mach numbers are high supersonic (d), the tangential Mach numbers (2.540d–f) are also supersonic. The upstream (downstream) conditions in the oblique shock are compared in Table 2.26.

The consideration of adiabatic flow and shock waves (Sections 2.5–2.9) includes irreversibility only across shocks. More generally irreversibility is associated with dissipation, for example, (Figure 2.52) for a body in a supersonic flow: a shock forms ahead of the body, and: (a) the temperature rise may cause chemical effects, like condensation and ionization; (b) the pressure rise can cause the separation of the boundary layer and transition from laminar to turbulent flow; (c) thermal effects may include heat conduction and radiation. The equations of non-dissipative fluid mechanics (subsections 2.6.1–2.6.5) are extended next (Notes 2.1–2.10) to include dissipation by viscosity, thermal conduction, mass diffusion and chemical reactions.

NOTE 2.1 THE FIRST (SECOND) PRINCIPLES OF THERMODYNAMICS AND REVERSIBLE (IRREVERSIBLE) PROCESSES

The first (second) principle of thermodynamics specifies bilateral (unilateral) conditions of equilibrium (evolutionary inequalities), which apply to reversible (irreversible) processes, that is non-dissipative (dissipative). They have numerous applications in all areas of physics where energy conservation (dissipation) occurs. By contrast the third principle of thermodynamics is concerned with the limit of zero absolute temperature, and shows it is difficult to approach. An example of a reversible (irreversible) thermodynamic process in a fluid is the adiabatic flow (a shock wave), which is a continuous (discontinuous) solution of the equations of fluid mechanics.

The adiabatic flow is limited in the maximum velocity it can reach and maximum angle it can turn. The simplest shock waves are normal (oblique), and both have flat wave fronts. The oblique flat shock arises in a corner. A pointed nose gives rise to a conical shock issuing from the apex. A blunt body in a supersonic flow causes a **detached shock** (Figure 2.52) separated from the body, leading to a sequence of interconnected problems: (i) location of the shock wave; (ii) determination of the local strength of the shock, which may lead to a transition between a strong (weak) shock at the front (rear), with subsonic (supersonic) downstream flow; (iii) the mixed subsonic-supersonic flow behind the shock must satisfy the boundary conditions at the surface of the body. The overexpanded nozzle is a simple example of shock location.

The shock waves in ducts or around bodies can cause: (i) boundary layer separation due to the adverse pressure gradient, opposite to the flow direction; (ii) the flow separation may lead to turbulence, which affects heat and mass transfer; (iii) if the ambient humidity is high the shock wave may cause condensation, making its location visible; (iv) the very strong shocks may lead to high temperature, for which heat radiation can have a smoothing effect; (v) very high temperatures may lead to chemical reactions and ionization. The very strong shocks can be caused by: (i) explosions;

(ii) hypersonic vehicles; (iii) re-entry into the atmosphere from Earth orbit or return from interplanetary missions; (iv) entry into the atmospheres of other planets like Mars.

NOTE 2.2   THICKNESS, DISSIPATION AND INTERNAL STRUCTURE OF SHOCK WAVES

The simplest treatment of shock waves is as surfaces of discontinuity. The conservation of mass, momentum and energy is sufficient to relate flow variables on the upstream and downstream sides of the shock. These relations imply entropy growth, showing that a shock wave is an irreversible process, involving energy dissipation. The amount of energy which must be dissipated is determined by the external conditions across the shock. The dissipation of energy within the shock is due to viscosity, thermal conduction and radiation, and relaxation; the latter is due to changes in the vibrational state of the fluid molecules, dissociation and chemical reactions. The width of the shock wave is determined by the energy to be dissipated and the internal gradients of flow variables. If dissipation is weak, the entropy change corresponds to large gradients, and the shock wave is thin, justifying the approximation as a surface of discontinuity. If thickness of shock waves is comparable to the mean three paths of molecules, their internal structure is determined by microscopic effects. Conversely at high temperatures, with strong radiation, relaxation, ionization or chemical reactions, the gradients of flow variables are small, and the entropy increase corresponds to a wider, smoother transition front. Thus there are three levels of consideration of shock waves: (i) as surfaces of discontinuity in the non-dissipative equation of fluid mechanics to relate flow variables on both sides; (ii) as a continuous flow subject to dissipation by viscosity, thermal conduction and radiation and relaxation to determine the thickness; (iii) as statistics of many particles (molecules, atoms, and their combinations, vibrations, reactions and dissipations) as concerns the internal structure.

NOTE 2.3   EQUATIONS OF NON-DISSIPATIVE AND DISSIPATIVE FLUID MECHANICS

The equations of fluid mechanics are: (i) the equation of continuity (subsection 2.6.1) or mass conservation is the same in the absence of mass diffusion; (ii) in the equation of momentum or force balance the viscous stresses are added (subsection 2.4.11) relative to the inviscid case (subsection 2.6.2); (iii) the equation of state is unchanged, for example, for a perfect gas (subsection 2.5.4); (iv) the conservation of entropy along streamlines for an adiabatic flow is replaced by the entropy production by dissipative processes (Note 2.4), namely viscosity and thermal conduction; (v) the equation of energy relative to a non-dissipative fluid (subsection 2.6.4) adds a viscous contribution to the energy flux plus heat release by dissipation (Note 2.5). Thus the equations of non-dissipative fluid mechanics (subsection 2.6.5) are extended to include viscous and thermal dissipation, using either the entropy (energy) equation [Note(s) 2.4 (2.5–2.6)], and may be specialized for an ideal gas (Notes 2.7–2.8) and written in conservative form (Note 2.9).

The effects of mass diffusion are considered for a two-phase flow with different chemical species allowing coupling with heat exchanges in the mass, heat and energy fluxes (Note 2.10). These appear in the fundamental equations for a viscous, thermally conducting two-phase flow in the alternative entropy (energy) forms [Note 2.11 (2.12)]. The cross-coupling of the temperature and mass fraction gradients in the heat and mass fluxes (Note 2.13) must satisfy the conditions of entropy production (Note 2.14). The general explicit form of the equations of Newtonian viscous perfect gas in the presence of coupled heat and mass fluxes specified by the combined Fourier – Fick law (Note 2.15) are linearized for a steady one-dimensional flow (Note 2.16). This shows the effects of triple diffusion, that is viscous, thermal and mass diffusion (Note 2.17) on the temperature (Note 2.18), mass fraction (Note 2.19) and velocity, density and pressure (Note 2.20).

NOTE 2.4   EQUATIONS OF VISCOUS SINGLE-PHASE FLOW

*The fundamental equations of fluid mechanics are: (i) the **equation of continuity** (2.592c) $\equiv$ (2.1211a–c) stating the conservation of mass in the absence of mass sources or sinks and with no mass diffusion, involving the mass density ρ and velocity v_i:*

$$0 = \frac{\partial \rho}{\partial t} + \frac{\partial}{\partial x_i}(\rho v_i) = \frac{\partial \rho}{\partial t} + v_i \frac{\partial \rho}{\partial x_i} + \rho \frac{\partial v_i}{\partial x_i} = \frac{d\rho}{dt} + \rho \frac{\partial v_i}{\partial x_i}; \qquad (2.1211a\text{–}c)$$

*(ii) the **viscous equation of momentum** (2.384a) $\equiv$ (2.1212a,b) equation balancing the inertia force against minus the pressure gradient and the divergence of the viscous stresses in the absence of external forces:*

$$-\frac{\partial p}{\partial x_i} + \frac{\partial \tau_{ij}}{\partial x_j} = \rho \frac{dv_i}{dt} = \rho \left(\frac{\partial v_i}{\partial t} + v_j \frac{\partial v_i}{\partial x_j} \right), \qquad (2.1212a, b)$$

and in the case of a Newtonian fluid the viscous stresses are given by (2.377b) $\equiv$ (2.1212c):

$$\tau_{ij} = \zeta \left(\frac{\partial v_i}{\partial x_j} + \frac{\partial v_j}{\partial x_i} - \frac{2}{3}\frac{\partial v_k}{\partial x_k} \right) \delta_{ij} + \eta \frac{\partial v_k}{\partial x_k} \delta_{ij}, \qquad (2.1212c)$$

involving the shear ζ and bulk η static viscosities:

$$\rho \frac{dv_i}{dt} + \frac{\partial p}{\partial x_i} = \frac{\partial}{\partial x_j}\left[\zeta \left(\frac{\partial v_i}{\partial x_j} + \frac{\partial v_j}{\partial x_i} \right) \right] + \frac{\partial}{\partial x_i}\left[\left(\eta - \frac{2}{3}\zeta \right)\left(\frac{\partial v_j}{\partial x_j} \right) \right], \qquad (2.1213a)$$

and if they are constant (2.1213b, c) the momentum equation simplifies to (2.1213d):

$$\zeta, \eta = const: \quad \rho \frac{dv_i}{dt} = -\frac{\partial p}{\partial x_i} + \zeta \frac{\partial^2 v_i}{\partial x_j \partial x_j} + \left(\eta + \frac{\zeta}{3} \right)\frac{\partial^2 v_j}{\partial x_i \partial x_j}; \qquad (2.1213b\text{–}d)$$

*(iii) the **equation of state**, for example (2.1214a) relating pressure to mass density and temperature:*

$$p = p(\rho,T): \quad p = \rho \bar{R} T,$$

(2.1214a, b)

*a simple case being an ideal gas (2.473g) $\equiv$ (2.1214b) with gas constant $\bar{R}$; (iv) the **equation of entropy** (2.1215) stating that heat production:*

$$\rho T \frac{ds}{dt} = -\frac{\partial G_i^q}{\partial x_i} + \tau_{ij} \frac{\partial v_i}{\partial x_j}$$

(2.1215)

is due to: (iv-1) thermal conduction (1.323b) with the heat flux (1.320a) in the first term on the r.h.s. of (2.1215) involving in the case (1.324b) of the Fourier law the thermal conductivity k; (iv-2) viscous dissipation (2.380c) leading to the second term on the r.h.s. of (2.546a–e), that in the case of a Newtonian fluid (2.542a–d) with constant shear and bulk diffusivities (2.1213b, c) $\equiv$ (2.1216a, b) takes the explicit form (2.383c) $\equiv$ (2.1216c):

$$\zeta, \eta = const: \quad \rho T \frac{ds}{dt} = k \frac{\partial^2 T}{\partial x_i \partial x_i} + \eta \left(\frac{\partial v_i}{\partial x_i} \right)^2$$

$$+ \frac{\zeta}{2} \left[\frac{\partial v_i}{\partial x_j} + \frac{\partial v_j}{\partial x_i} - \frac{2}{3} \left(\frac{\partial v_k}{\partial x_k} \right) \delta_{ij} \right]^2 .$$

(2.1216a–c)

The equation of entropy (2.1215) may be replaced by the equation of energy (Note 2.5) including thermal conduction and shear and bulk viscosities.

NOTE 2.5   EQUATION OF ENERGY FOR A DISSIPATIVE FLUID

The general form of the energy equation is (2.1217a):

$$\frac{\partial E}{\partial t} + \frac{\partial G_i}{\partial x_i} = \dot{D}: \quad E = \rho \left(u + \frac{v^2}{2} \right),$$

(2.1217a, b)

where: (i) the energy density per unit volume (2.607b) $\equiv$ (2.1217b) in the stagnation internal energy per unit mass u, and applies both in the non-dissipative and dissipative cases; (ii) the energy flux (2.1218a) consists of a dissipative energy flux plus the convective energy flux (2.614g) $\equiv$ (2.1218b) involving the enthalpy per unit mass h:

$$G_i = G_i^d + G_i^c: \quad G_i^c = \rho \left(h + \frac{v^2}{2} \right) v_i;$$

(2.1218a, b)

(iii) there are no energy sources in the non-dissipative case, when the energy equation (2.1217a) simplifies to (2.607a). Thus to obtain the energy equation with dissipation it is necessary to specify: (a) the dissipative heat flux G_i^d in (2.1218a); (b) the energy source $\dot{D}$ in (2.1217a).

Both (a) and (b) can be determined by: (i) taking the local time derivative of the energy density (2.607b) ≡ (2.1217b) as in the non-dissipative case (2.608) ≡ (2.550a–c):

$$\frac{\partial E}{\partial t} = \left(u + \frac{v^2}{2} \right) \frac{\partial \rho}{\partial t} + \rho\, v_i \frac{\partial v_i}{\partial t} + \rho \frac{\partial u}{\partial t}, \tag{2.1219a}$$

where the internal energy per unit mass is given by (2.602b) leading to (2.1219b):

$$\frac{\partial E}{\partial t} = \left(u + \frac{p}{\rho} + \frac{v^2}{2} \right) \frac{\partial \rho}{\partial t} + \rho\, v_i \frac{\partial v_i}{\partial t} + \rho T \frac{\partial s}{\partial t}; \tag{2.1219b}$$

(ii) the substitution of the equations of continuity (2.1211a), momentum (2.1212a) and entropy (2.1215) leads to (2.1220a–e):

$$\frac{\partial E}{\partial t} + \left(u + \frac{p}{\rho} + \frac{v^2}{2} \right) \frac{\partial}{\partial x_i} (\rho\, v_i)$$

$$= \rho\, v_i \left(\frac{dv_i}{dt} - v_j \frac{\partial v_i}{\partial x_j} \right) + \rho T \left(\frac{ds}{dt} - v_i \frac{\partial s}{\partial x_i} \right)$$

$$= -v_i \frac{\partial p}{\partial x_i} + v_i \frac{\partial \tau_{ij}}{\partial x_j} - \rho\, v_j \frac{\partial}{\partial x_j} \left(\frac{v^2}{2} \right) - \frac{\partial G_i^q}{\partial x_i} + \tau_{ij} \frac{\partial v_i}{\partial x_j} - \rho T v_i \frac{\partial s}{\partial x_i}$$

$$= -\frac{\partial G_i^q}{\partial x_i} + \frac{\partial}{\partial x_j} (v_i \tau_{ij}) - \rho\, v_i \left[T \frac{\partial s}{\partial x_i} + \frac{1}{\rho} \frac{\partial p}{\partial x_i} + \frac{\partial}{\partial x_i} \left(\frac{v^2}{2} \right) \right]; \tag{2.1220a–c}$$

(iii) bringing the dissipative terms to the l.h.s. and introducing the enthalpy (2.615b, d) the energy balance (2.1220c) is rearranged (2.1221a–c):

$$\frac{\partial E}{\partial t} + \frac{\partial G_i^q}{\partial x_i} - \frac{\partial}{\partial x_j} (v_i \tau_{ij})$$

$$= -\left(u + \frac{p}{\rho} + \frac{v^2}{2} \right) \frac{\partial}{\partial x_i} (\rho\, v_i) - \rho\, v_i \left[T \frac{\partial s}{\partial x_i} + \frac{1}{\rho} \frac{\partial p}{\partial x_i} + \frac{\partial}{\partial x_i} \left(\frac{v^2}{2} \right) \right]$$

$$= -\left(h + \frac{v^2}{2} \right) \frac{\partial}{\partial x_i} (\rho\, v_i) - \rho\, v_i \frac{\partial}{\partial x_i} \left(h + \frac{v^2}{2} \right)$$

$$= -\frac{\partial}{\partial x_i} \left[\left(h + \frac{v^2}{2} \right) \rho\, v_i \right]; \tag{2.1221a–c}$$

(iv) comparing (2.1221c) with (2.1217a) it follows that: (i) there are no energy sources (2.1222a) in the energy equation (2.1222b, c):

$$\dot{D} = 0: \quad 0 = \frac{\partial E}{\partial t} + \frac{\partial G_i}{\partial x_i} = \frac{\partial E}{\partial t} + \frac{\partial G_i^c}{\partial x_i} + \frac{\partial G_i^d}{\partial x_i}; \qquad (2.1222a\text{--}c)$$

(v) the convective energy flux coincides with (2.1218b); (v) the dissipative heat flux (2.1223a, 2.475a–c) is due to thermal conduction and viscosity:

$$G_i^d = G_i^q - G_i^v = G_i^q - v_j\,\tau_{ij}. \qquad (2.1223a, b)$$

This leads to the equation of energy in general form (Note 2.6) including some particular cases.

NOTE 2.6   EQUATION OF ENERGY WITH VISCOSITY AND HEAT CONDUCTION

*The equation of energy per unit volume is (2.1222b, c) without heat sources (2.1217a; 2.1222a) involves: (i) the **energy density** (2.1217b) that is the stagnation internal energy (2.607b–d); (ii) the **convective energy flux** (2.1218b) that is the flux of the stagnation enthalpy (2.614a–h); (iii) the dissipative heat flux (2.1223a, b) ≡ (2.1224b) consisting of the conduction heat flux minus the flux of the viscous stresses, and using Fourier law (1.324b) ≡ (2.1224a) and the Newtonian viscous stresses (2.377b) ≡ (2.1712c) is specified by (2.1224b):*

$$G_i^q = -k\frac{\partial T}{\partial x_i}: \quad G_i^d = -k\frac{\partial T}{\partial x_i} - \zeta\, v_j\left(\frac{\partial v_i}{\partial x_j} + \frac{\partial v_j}{\partial x_i}\right) - v_i\left(\eta - \frac{2}{3}\zeta\right)\frac{\partial v_j}{\partial x_j}. \qquad (2.1224a, b)$$

Substitution of (2.1217b; 2.1218b; 2.1224b) in (2.1222c) specifies *the equation of energy for a viscous thermally conducting fluid (2.1225)*:

$$\frac{\partial}{\partial t}\left[\rho\left(u + \frac{v^2}{2}\right)\right] + \frac{\partial}{\partial x_i}\left[\rho\left(h + \frac{v^2}{2}\right)v_i\right]$$
$$= \frac{\partial}{\partial x_i}\left(k\frac{\partial T}{\partial x_i}\right) - \frac{\partial}{\partial x_i}\left[v_j\zeta\left(\frac{\partial v_i}{\partial x_j} + \frac{\partial v_j}{\partial x_i}\right)\right] - \frac{\partial}{\partial x_i}\left[\left(\eta - \frac{2}{3}\zeta\right)v_i\frac{\partial v_j}{\partial x_j}\right], \qquad (2.1225)$$

involving the shear ζ and bulk η viscosities (thermal conductivity k), that are not assumed to be constant, in the Newtonian viscous stresses (2.1212c) [Fourier law of heat conduction (2.1224a)]. The fundamental equations of a viscous thermally conducting flow are: (i) one vector equation, namely momentum (2.1212a, b), for a Newtonian viscous fluid (2.1212c), implying (2.1213a), and simplifying to (2.1213d) for constant shear and bulk viscosities (2.1213b, c); (ii–iv) three scalar equations,

namely (ii) continuity (2.1211a–c), (iii) state (2.1214a) for a perfect gas (2.1214b)
and (iv) entropy (2.1215) [or energy (2.1217a, b; 2.1218a, b; 2.1222a; 2.1224b)],
for Newtonian viscous stresses (2.1212c) and Fourier thermal conduction (2.1224a).
The variables are: (i) one vector, namely the velocity v_i; (ii–iv) three scalars, namely
the pressure p, density ρ and temperature T; (v) the internal energy u, enthalpy h and
entropy s per unit mass can be expressed in terms of any two of (p,ρ,T). This is done
next (Note 2.7) for an ideal gas.

NOTE 2.7　EQUATIONS OF FLUID MECHANICS FOR A PERFECT GAS

The fundamental equations of a viscous thermally conducting perfect gas are: (i)
the equation of state (2.1214b); (ii) the equation of continuity (2.1211a–c,), (iii)
the equation of momentum for a Newtonian viscous fluid (2.1213d) with constant
shear and bulk viscosities (2.1213b, c); (iv) the equation of energy (2.1225) ≡
(2.1226c) with the internal energy (enthalpy) per unit mass (2.623a) ≡ (2.1226a)
[(2.623b) ≡ (2.1226b)] involving the specific heat at constant volume c_V (pres-
sure c_p):

$$\{u,h\} = T\{c_V, c_p\}:$$

$$\frac{\partial}{\partial t}\left[\rho\left(c_V T + \frac{v^2}{2}\right)\right] + \frac{\partial}{\partial x_i}\left[\rho\left(c_p T + \frac{v^2}{2}\right)v_i\right] \qquad (2.1226\text{a–c})$$

$$= k\frac{\partial^2 T}{\partial x_i \partial x_i} - \varsigma\frac{\partial}{\partial x_i}\left[v_j\left(\frac{\partial v_i}{\partial x_j} + \frac{\partial v_j}{\partial x_i}\right)\right] - \left(\eta - \frac{2\varsigma}{3}\right)\frac{\partial}{\partial x_i}\left(v_i\frac{\partial v_j}{\partial x_j}\right),$$

with thermal conductivity (2.1224a) and shear and bulk viscosities (2.1212e) all con-
stant. With the same assumptions the equation of energy (2.1226c) may be replaced
(2.1216c) by the equation of entropy (2.1227a–d):

$$\dot{Q} \equiv k\frac{\partial^2 T}{\partial x_i \partial x_i} + \eta\left(\frac{\partial v_i}{\partial x_i}\right)^2 + \frac{\varsigma}{2}\left[\frac{\partial v_i}{\partial x_j} + \frac{\partial v_j}{\partial x_i} - \frac{2}{3}\left(\frac{\partial v_k}{\partial x_k}\right)\delta_{ij}\right]^2$$

$$= \frac{1}{\gamma - 1}\left(\frac{dp}{dt} - \gamma RT\frac{d\rho}{dt}\right) = \rho c_p \frac{dT}{dt} - \frac{dp}{dt} \qquad (2.1227\text{a–d})$$

$$= \rho c_V \frac{dT}{dt} - RT\frac{d\rho}{dt},$$

where in the r.h.s. of (2.1227a–d) was used: (i) the entropy per unit mass for a perfect
gas (2.476 a–c) ≡ (2.1228 a–c):

$$\frac{ds}{dt} = \frac{c_V}{p}\frac{dp}{dt} - \frac{c_p}{\rho}\frac{d\rho}{dt} = \frac{c_p}{T}\frac{dT}{dt} - \frac{R}{p}\frac{dp}{dt} = \frac{c_V}{T}\frac{dT}{dt} - \frac{R}{\rho}\frac{d\rho}{dt}; \qquad (2.1228\text{a–c})$$

(ii) multiplied by ρT with use of the equation of state (2.1214b) and gas constant (2.1229a) and adiabatic exponent (2.1229b) leading to (2.1229c–h):

$$R = c_p - c_V, \quad \gamma = \frac{c_p}{c_V}:$$

$$\dot{Q} \equiv \rho T \frac{ds}{dt} = c_V \frac{\rho T}{p} \frac{dp}{dt} - c_p T \frac{d\rho}{dt} = \frac{1}{\gamma - 1} \left(\frac{dp}{dt} - \gamma RT \frac{d\rho}{dt} \right)$$

$$= c_p \rho \frac{dT}{dt} - \frac{\rho RT}{p} \frac{dp}{dt} = \rho c_p \frac{dT}{dt} - \frac{dp}{dt}$$

$$= \rho c_V \frac{dT}{dt} - RT \frac{d\rho}{dt} = R \left(\frac{\rho}{\gamma - 1} \frac{dT}{dt} - T \frac{d\rho}{dt} \right).$$

(2.1229a–h)

The coefficients on the r.h.s. of (2.1227a–c) = (2.1229a–h) relate to the adiabatic and isothermal sound speeds (Note 2.8).

NOTE 2.8   ADIABATIC AND ISOTHERMAL SOUND SPEEDS

If there are no heat sources (2.1230a) the entropy conservation (2.1227b) leads to the adiabatic relation (2.1230b) involving the adiabatic sound speed (2.1230c–e):

$$\dot{Q} = 0: \quad \frac{dp}{dt} = (c_s)^2 \frac{d\rho}{dt}, \quad (c_s)^2 = \left(\frac{\partial p}{\partial \rho} \right)_s = \gamma RT = \gamma \frac{p}{\rho}. \quad (2.1230a\text{–}e)$$

If the pressure (mass density) is conserved along streamlines (2.1231a) [(2.1231c)] the heat production on the r.h.s. of (2.1227c) [(2.1227d)] involves the specific heat at constant pressure (2.1231b) [volume (2.1231d)]:

$$\frac{dp}{dt} = 0: \quad \dot{Q} = \rho c_p \frac{dT}{dt}; \quad \frac{d\rho}{dt} = 0: \quad \dot{Q} = \rho c_V \frac{dT}{dt}. \quad (2.1231a\text{–}d)$$

In the cases of pressure (2.1232a) and mass density (2.1232b) not conserved along streamlines the l.h.s. of (2.1227c) [(2.1227d)] involves (2.1232c, d):

$$\frac{dp}{dt} \neq 0 \neq \frac{d\rho}{dt}: \quad \rho c_p \frac{dT}{dt} - \frac{dp}{dt} = \dot{Q} = \rho c_V \frac{dT}{dt} - (c_t)^2 \frac{d\rho}{dt}, \quad (2.1232a\text{–}d)$$

the isothermal sound speed (2.1233a–c):

$$(c_t)^2 = \left(\frac{\partial p}{\partial \rho} \right)_T = RT = \frac{p}{\rho}, \quad (\gamma - 1)\dot{Q} = \frac{dp}{dt} - (c_s)^2 \frac{d\rho}{dt}, \quad (2.1233a\text{–}d)$$

instead of the adiabatic sound speed (2.1230c, d) in (2.1227b) ≡ (2.1233d). The equations of dissipative single-phase fluid mechanics can be put into conservative form (Note 2.9).

NOTE 2.9 CONSERVATIVE FORM FOR A VISCOUS, THERMALLY CONDUCTING FLOW

The conservative form of the equations of non-dissipative fluid mechanics (2.622) can be extended to include thermal conduction and shear and bulk viscosities as follows: *(i,ii) the equations of continuity (2.1211a–c) and energy (2.1225) are in conservative form; (iii) the equation of momentum (2.1212a,b) [for a Newtonian fluid (2.1212c)] can be put into a conservative form using the same transformation (2.620a–d) as in the inviscid case, leading to [(2.1234) (2.1235)]:*

$$0 = \frac{\partial}{\partial t}(\rho v_i) + \frac{\partial}{\partial x_j}(\rho v_i v_j - p\delta_{ij} - \tau_{ij}), \qquad (2.1234)$$

$$0 = \frac{\partial}{\partial t}(\rho v_i) + \frac{\partial}{\partial x_j}\left[\rho v_i v_j - p\delta_{ij} - \zeta\left(\frac{\partial v_i}{\partial x_j} + \frac{\partial v_j}{\partial x_i}\right) - \left(\eta - \frac{2}{3}\zeta\right)\delta_{ij}\frac{\partial v_k}{\partial x_k}\right]. \qquad (2.1235)$$

Thus *the conservative form of the equations of dissipative fluid mechanics including thermal conduction and shear and bulk viscosities is (2.1236):*

$$0 = \frac{\partial}{\partial t}\begin{bmatrix} \rho \\ \rho v_x \\ \rho v_y \\ \rho v_z \\ \rho u_0 \end{bmatrix} + \frac{\partial}{\partial x}\begin{bmatrix} \rho v_x \\ p + \rho v_x^2 - \tau_{xx} \\ \rho v_x v_y - \tau_{xy} \\ \rho v_x v_z - \tau_{xz} \\ \rho h_0 + G_x^q - v_i \tau_{ix} \end{bmatrix}$$

$$+ \frac{\partial}{\partial y}\begin{bmatrix} \rho v_y \\ \rho v_y v_x - \tau_{yx} \\ p + \rho v_y^2 - \tau_{yy} \\ \rho v_y v_z - \tau_{yz} \\ \rho h_0 + G_y^q - v_i \tau_{iy} \end{bmatrix} + \frac{\partial}{\partial z}\begin{bmatrix} \rho v_z \\ \rho v_z v_x - \tau_{zx} \\ \rho v_z v_y - \tau_{zy} \\ p + \rho v_z^2 - \tau_{zz} \\ \rho h_0 + G_z^q - v_i \tau_{iz} \end{bmatrix}. \qquad (2.1236)$$

where appear: (i) the stagnation internal energy (2.1237a) [enthalpy (2.1237b)]:

$$\{u_0, h_0\} = \{u, h\} + \frac{v^2}{2}; \qquad (2.1237a, b)$$

(ii) the viscous stresses (2.1238a–d) in the Newtonian case (2.542a–d):

$$\tau_{xx} = 2\zeta \frac{\partial v_x}{\partial x} + \left(\eta - \frac{2}{3}\zeta\right)\left(\frac{\partial v_x}{\partial x} + \frac{\partial v_y}{\partial y} + \frac{\partial v_z}{\partial z}\right)$$

$$= \left(\eta + \frac{4}{3}\zeta\right)\frac{\partial v_x}{\partial x} + \eta\left(\frac{\partial v_y}{\partial y} + \frac{\partial v_z}{\partial z}\right), \tag{2.1238a, b}$$

$$\tau_{xy} = \zeta\left(\frac{\partial v_x}{\partial y} + \frac{\partial v_y}{\partial x}\right), \quad \text{cyclic}(x,y,z), \tag{2.1238c, d}$$

where (2.1238c,d) apply cyclically in (x,y,z); (iii) the viscous energy flux is given by (2.1239a, b):

$$-G_i^v = v_j \tau_{ij} = \{ v_x \tau_{xx} + v_y \tau_{xy} + v_z \tau_{xz},$$

$$v_x \tau_{xy} + v_y \tau_{yy} + v_z \tau_{yz}, \tag{2.1238e, f}$$

$$v_x \tau_{xz} + v_y \tau_{yz} + v_z \tau_{zz} \}.$$

In (2.1236) there is no restriction on the viscous stresses, that need not be Newtonian (2.1212c), nor on the heat flux, that need not be specified by the Fourier law (2.1224a). The fundamental equations of fluid mechanics for a viscous flow (Notes 2.4–2.9) in the presence of thermal conduction (subsections 2.4.1–2.4.2) are extended (Notes 2.10–2.16) to coupling with mass diffusion (subsection 2.4.18) in a two-phase flow.

NOTE 2.10   HEAT, MASS AND ENERGY FLUXES

For a two-phase medium the internal energy (2.613a) [enthalpy (2.615d)] per unit mass gains an extra term (2.91h) involving the mass fraction (2.91a–c) and the relative affinity (2.91h) in (2.1240a) [(2.1240b)]:

$$du = T\,ds + \frac{p}{\rho^2}\,d\rho + \bar{v}\,d\xi, \quad dh = T\,ds + \frac{dp}{\rho} + \bar{v}\,d\xi. \tag{2.1240a, b}$$

The dissipative processes, besides (iii) viscosity, are associated with: (i) heat conduction (2.1241a) corresponding to the first term on the r.h.s. of (2.546a–e) where $\bar{G}^q$ is the heat flux (2.1224a): (ii) mass diffusion (2.414d) $\equiv$ (2.1241b) where $\bar{I}$ is the mass flux:

$$\rho T \frac{ds}{dt} = -\nabla . \bar{G}^q, \quad \rho \frac{d\xi}{dt} = -\nabla . \bar{I}, \tag{2.1241a, b}$$

The corresponding dissipative part in the internal energy (2.1240a) $\equiv$ (2.1242a) is (2.1242b):

$$du = \frac{p}{\rho^2} d\rho + d\tilde{u}, \quad d\tilde{u} = T \, ds + \bar{v} \, d\xi, \tag{2.1242a, b}$$

and may be expected to satisfy a conservation equation (2.1243a):

$$\rho \frac{d\tilde{u}}{dt} = -\nabla.\vec{G}^{qm}; \quad \vec{G}^{qm} = \vec{G}^{q} + \vec{G}^{m}, \tag{2.1243a, b}$$

where the diffusive energy flux (2.1243b) is due to thermal conduction and mass diffusion.

Substituting (2.1243b) in (2.1243a) leads to (2.1244a) and using (2.1242a) leads to (2.1244b) and substituting (2.1241a, 2.475a–c) yields (2.1244c)

$$\nabla.\vec{G}^{qm} = -\rho \frac{d\tilde{u}}{dt} = -\rho T \frac{ds}{dt} - \rho \bar{v} \frac{d\xi}{dt} = \nabla.\vec{G}^{q} + \bar{v}\nabla.\vec{I}. \tag{2.1244a–c}$$

Thus *the diffusive energy flux* $\vec{G}^{qm}$ *associated with the heat flux* $\vec{G}^{q}$ *and mass flux* $\vec{I}$ *has divergence (2.1244c)* $\equiv$ *(2.1245a,b):*

$$\nabla.\vec{G}^{qm} = \nabla.\vec{G}^{q} + \bar{v}\nabla.\vec{I} = \nabla.\left(\vec{G}^{q} + \bar{v}\,\vec{I}\right) - \vec{I}.\nabla\bar{v}. \tag{2.1245a, b}$$

In the case of constant relative affinity (2.1246a) the dissipative energy flux due to the heat flux and mass flux is given by (2.1246b):

$$\bar{v} = const : \quad \vec{G}^{qm} = \vec{G}^{q} + \bar{v}\,\vec{I}. \tag{2.1246a, b}$$

The relative affinity $\bar{v}$ *appears in (2.1245a,b) and (2.1246a,b) to match the different dimensions of the heat* $\vec{G}^{q}$ *and mass* $\vec{I}$ *fluxes.* The preceding simplified derivation (Note 2.10) of the energy flux (2.1245a,b) due to thermal conduction and mass diffusion is justified more thoroughly next (Note 2.12) from the fundamental equations of a viscous, thermally conducting two-phase flow (Note 2.11).

NOTE 2.11   VISCOUS, THERMALLY CONDUCTING TWO-PHASE FLOW

There are five fundamental equations for the two-phase flow of a viscous thermally conducting fluid: (i–ii) the equations of continuity (2.1211a–e) and state (2.1214a) are unaffected by dissipation; (iii) the momentum equation (2.1212a, b) is affected by the viscous stresses; (iv) the entropy equation (2.1215) is affected by viscosity and thermal conduction; (v) the equation for the mass fraction (2.1241b) is affected by mass diffusion. The equation of entropy (iv) may be replaced by the equation of energy, that is derived next including all three dissipative effects, namely viscous stresses, thermal conduction and mass diffusion. The energy density (2.1217b) is unaffected by dissipation, and thus (2.1219a) holds. Substituting (2.1240a) in the last term on the r.h.s. of (2.1219a) leads to (2.1247) that adds the last diffusion term to (2.1219b):

$$\frac{\partial E}{\partial t} = \left(u + \frac{p}{\rho} + \frac{v^2}{2}\right)\frac{\partial \rho}{\partial t} + \rho v_i \frac{\partial v_i}{\partial t} + \rho T \frac{\partial s}{\partial t} + \rho \bar{v}\frac{\partial \xi}{\partial t}. \tag{2.1247}$$

In the four terms on the r.h.s. of (2.1247) are substituted, respectively, the equations of continuity (2.1211a), viscous momentum (2.1212b), entropy (2.1215) and mass diffusion (2.1241b) leading to:

$$\frac{\partial E}{\partial t} + \left(u + \frac{p}{\rho} + \frac{v^2}{2}\right)\frac{\partial}{\partial x_i}(\rho v_i)$$

$$= v_i\left(-\frac{\partial p}{\partial x_i} - \rho v_j \frac{\partial v_i}{\partial x_j} + \frac{\partial \tau_{ij}}{\partial x_j}\right) - \frac{\partial G_i^q}{\partial x_i} + \tau_{ij}\frac{\partial v_i}{\partial x_i}$$

$$-\rho T v_i \frac{\partial s}{\partial x_i} - \bar{v}\frac{\partial I_i}{\partial x_i} - \rho \bar{v} v_i \frac{\partial \xi}{\partial x_i}, \tag{2.1248}$$

that generalizes (2.1220b) by adding mass diffusion to viscous stresses and thermal conduction. The dissipative terms in (2.1248) are collected on the l.h.s. of (2.1249a):

$$-\frac{\partial G_i^q}{\partial x_i} + \frac{\partial}{\partial x_j}(v_i \tau_{ij}) - \bar{v}\frac{\partial I_i}{\partial x_i} = \frac{\partial E}{\partial t} + \left(h + \frac{v^2}{2}\right)\frac{\partial}{\partial x_i}(\rho v_i)$$

$$+ \rho v_i\left[T\frac{\partial s}{\partial x_i} + \frac{1}{\rho}\frac{\partial p}{\partial x_i} + \bar{v}\frac{\partial \xi}{\partial x_i} + \frac{\partial}{\partial x_i}\left(\frac{v^2}{2}\right)\right]$$

$$= \frac{\partial E}{\partial t} + \left(h + \frac{v^2}{2}\right)\frac{\partial}{\partial x_i}(\rho v_i) + \rho v_i \frac{\partial}{\partial x_i}\left(h + \frac{v^2}{2}\right)$$

$$= \frac{\partial E}{\partial t} + \frac{\partial}{\partial x_i}\left[\rho\left(h + \frac{v^2}{2}\right)v_i\right]. \tag{2.1249a–c}$$

and the enthalpy (2.1240b) is used in (2.1249b, c).

NOTE 2.12   ENERGY DENSITY AND CONVECTIVE AND DIFFUSIVE FLUXES

Thus *the energy equation for a viscous, thermally conducting two-phase flow is (2.1249c) ≡ (2.1222b, c) where: (i) the energy density (2.1217b) ≡ (2.607b–d) is the stagnation internal energy per unit volume; (ii) the convective energy flux (2.1218b) ≡ (2.614a–i) is the stagnation enthalpy per unit volume multiplied by the velocity; (iii) the dissipative energy flux has divergence (2.1250a):*

$$\frac{\partial G^d}{\partial xi} = \frac{\partial G^q}{\partial x_i} - \frac{\partial}{\partial x_j}\left(v_i\,\tau_{ij}\right) + \bar{v}\,\frac{\partial I_i}{\partial x_i},\qquad(2.1250a)$$

and consists of three terms (2.1250b), namely: (i) the heat flux in the first term on the r.h.s. of (2.1223b); (ii) the energy flux (2.1250c) in the second term on the r.h.s. of (2.1223b) associated with viscosity; (iii) the energy flux associated with mass diffusion (2.1250d) ≡ (2.1244c) that involves the mass flux and relative affinity:

$$\vec{G}^d = \vec{G}^q + \vec{G}^v + \vec{G}^m:\quad G_i^v = v_i\,\tau_{ij}\quad \nabla.\vec{G}^m = \bar{vv}.\vec{I}.\qquad(2.1250\text{b--d})$$

In order to make explicit the fundamental equations of a viscous, thermally conducting two-phase flow (Note 2.11) it is necessary to specify three diffusion relations: (i) the viscous stresses in the momentum equations (2.1212a,b), for example, for a Newtonian fluid (2.1212c) leading to the Navier-Stokes equation (2.1213a–d); (ii/iii) the heat (mass) fluxes that appear in the equations of entropy (2.1215) [mass fraction (2.1241b)] and also of energy (2.1249a–c) ≡ (2.1251):

$$\frac{\partial}{\partial t}\left[\rho\left(u+\frac{v^2}{2}\right)\right] + \frac{\partial}{\partial x_i}\left[\left(u+\frac{p}{\rho}+\frac{v^2}{2}\right)\rho v_i + v_j\,\tau_{ij}\right] + \frac{\partial G^q}{\partial x_i} + \bar{v}\,\frac{\partial I_i}{\partial x_i} = 0.\qquad(2.1251)$$

Adding the mass fraction equation in the conservative form (2.1241b) to the system (2.1236) leads to *the fundamental equations of a viscous thermally conducting two-phase flow (2.1215):*

$$0 = \frac{\partial}{\partial t}\begin{bmatrix}\rho\\ \rho v_x\\ \rho v_y\\ \rho v_z\\ \rho u_0\\ \rho\xi\end{bmatrix} + \frac{\partial}{\partial x}\begin{bmatrix}\rho v_x\\ p+\rho v_x^2-\tau_{xx}\\ \rho v_x v_y-\tau_{xy}\\ \rho v_x v_z-\tau_{xz}\\ \rho h_0+G_x^q-v_i\tau_{ix}\\ \rho\xi v_x-I_x\end{bmatrix}$$

$$+\frac{\partial}{\partial y}\begin{bmatrix}\rho v_y\\ \rho v_y v_x-\tau_{yx}\\ p+\rho v_y^2-\tau_{yy}\\ \rho v_y v_z-\tau_{yz}\\ \rho h_0+G_y^q-v_i\tau_{iy}\\ \rho\xi v_y-I_y\end{bmatrix} + \frac{\partial}{\partial z}\begin{bmatrix}\rho v_z\\ \rho v_z v_x-\tau_{zx}\\ \rho v_z v_y-\tau_{zy}\\ p+\rho v_z^2-\tau_{zz}\\ \rho h_0+G_z^q-v_i\tau_{iz}\\ \rho\xi v_z-I_z\end{bmatrix}\qquad(2.1252)$$

involving three dissipative processes: (i) the viscous stresses τ_{ij}, for example, for a Newtonian fluid (2.1212c) ≡ (2.1238a–d); (ii,iii) the heat $\vec{G}^q$ and mass $\bar{I}$ fluxes, that are generally coupled (Note 2.13).

NOTE 2.13   ENTROPY PRODUCTION BY COUPLED HEAT AND MASS DIFFUSION

The entropy production in a domain D_3 of volume dV is the sum (2.1253a) of the contributions due to heat (2.346c) and mass (2.414g) diffusion leading to (2.1253b):

$$\dot{\bar{S}}_{qm} = \dot{\bar{S}}_q + \dot{\bar{S}}_m = \int_{D_3} \frac{-\nabla . \vec{G}^q + v \nabla . \vec{I}}{T} dV. \qquad (2.1253a, b)$$

An integration by parts leads to (b):

$$\dot{\bar{S}}_{qm} = -\int_{D_3} \nabla . \left(\frac{\vec{G}^q - \bar{v}\,\vec{I}}{T} \right) dV + \int_{D_3} \left[\vec{G}^q . \nabla \left(\frac{1}{T} \right) - \vec{I} . \nabla \left(\frac{\bar{v}}{T} \right) \right] dV. \qquad (2.1254)$$

Using the divergence theorem (III.5.163a–c) ≡ (2.1255c) the volume integral first on the r.h.s. of (2.1254) becomes a surface integral over the boundary, where (2.1255d) the heat (2.1255a) and mass (2.1255b) vanish:

$$\vec{G}^q . \vec{N} \big|_{\partial D_3} = 0 = \vec{I} . \vec{N} \big|_{\partial D_3} \quad \int_{D_3} \nabla . \left(\frac{\vec{G}^q}{T} - \frac{\bar{v}\,\vec{I}}{\rho T} \right) dV$$

$$= \int_{\partial D_3} \left(\frac{\vec{G}^q}{T} - \frac{\bar{v}\,\vec{I}}{\rho T} \right) d\vec{S} = 0. \qquad (2.1255a\text{–}d)$$

Thus the entropy production per unit time and volume (2.1256a) is given by (2.1256b, c) the integrand of the second term on the r.h.s. of (2.1254):

$$\dot{S}_{qm} = \frac{d\dot{\bar{S}}_{qm}}{dV} = \vec{G}^q . \nabla \left(\frac{1}{T} \right) - \vec{I} . \nabla \left(\frac{\bar{v}}{T} \right) = -\left(\vec{G}^q - \bar{v}\,\vec{I} \right) . \frac{\nabla T}{T^2} - \frac{\vec{I} . \nabla \bar{v}}{T}. \qquad (2.1256a\text{–}c)$$

In the presence of temperature (2.1257a, b) [mass fraction (2.1257d, e)] gradients alone, the heat (2.350a) ≡ (2.1224a) ≡ (2.1257c) [mass (2.415b) ≡ (2.1257f)] flux is specified by the Fourier (1818) [Fick (1855)] law as minus the temperature (mass fraction) gradient multiplied by the thermal conductivity k (by $\chi_m \rho$ where χ_m has the dimension length square per unit time as in the mass diffusivity):

$$\nabla T \neq 0 = \nabla \xi : \quad \vec{G}^q = -k\nabla T; \quad \nabla T = 0 \neq \nabla \xi : \quad \vec{I} = -\chi_m \rho \nabla \xi. \qquad (2.1257a\text{–}f)$$

In the presence of both temperature (2.1258a) and mass fraction (2.1258b) gradients there is cross-coupling in the heat (2.1258c) and mass (2.1258d) fluxes:

$$\nabla T \neq 0 \neq \nabla \xi; \quad \vec{G}^q = -k\nabla T - \alpha\nabla \xi, \quad \vec{I} = -\chi_m \rho \nabla \xi - \beta \nabla T, \qquad (2.1258a\text{–}d)$$

and the cross-coupling coefficients α, β are related by the Onsager (1922) reciprocity principle (subsection 2.4.6) to ensure entropy production (2.14).

NOTE 2.14   THERMAL CONDUCTIVITY, MASS DIFFUSIVITY AND CROSS-COUPLING COEFFICIENTS

Substituting the combined Fourier-Fick law (2.1258c, d) in the entropy production (2.1256c) by coupled thermal – mass diffusion leads to (2.1259a):

$$\dot{S}_{qm} = \left(k - \bar{v}\,\beta\right)\frac{\left|\nabla T\right|^{2}}{T^{2}} + \left(\alpha - \chi_{m}\,\rho\bar{v}\,\right)\frac{\nabla T.\nabla\xi}{T^{2}} - \frac{\bar{I}.\nabla\bar{v}}{T}. \qquad (2.1259a)$$

The relative affinity generally depends (2.1259b) on temperature, mass fraction and pressure, leading to (2.1259c):

$$\bar{v} = \bar{v}\left(T,\xi,p\right)\quad \nabla\bar{v} = \left(\frac{\partial\bar{v}}{\partial T}\right)_{\xi,p}\nabla T + \left(\frac{\partial\bar{v}}{\partial\xi}\right)_{T,p}\nabla\xi + \left(\frac{\partial\bar{v}}{\partial p}\right)_{T,\xi}\nabla p. \quad (2.1259b, c)$$

The last term in (2.1259c) would lead to products $\nabla T.\,\nabla p$ and $\nabla\xi.\,\nabla p$ without fixed sign, that would violate entropy production. Thus pressure changes are neglected (2.1260a) and the last term on the r.h.s. of (2.1259a) becomes (2.1260b), leading by substitution of (2.1259c) and (2.1258b, d) to (2.1260c):

$$p = const: \quad -\frac{\bar{I}.\nabla\bar{v}}{T}$$

$$= \left(\frac{\chi_{m}\,\rho}{T}\nabla\xi + \frac{\beta}{T}\nabla T\right).\left[\left(\frac{\partial\bar{v}}{\partial\xi}\right)_{T,p}\nabla\xi + \left(\frac{\partial\bar{v}}{\partial T}\right)_{\xi,p}\nabla T\right]$$

$$= \frac{\chi_{m}\,\rho}{T}\left(\frac{\partial\bar{v}}{\partial\xi}\right)_{T,p}\left|\nabla\xi\right|^{2} + \frac{\beta}{T}\left(\frac{\partial\bar{v}}{\partial T}\right)_{\xi,p}\left|\nabla T\right|^{2} \qquad (2.1260a\text{–}d)$$

$$+ \left[\frac{\chi_{m}\,\rho}{T}\left(\frac{\partial\bar{v}}{\partial T}\right)_{\xi,p} + \frac{\beta}{T}\left(\frac{\partial\bar{v}}{\partial\xi}\right)_{T,p}\right](\nabla T.\nabla\xi).$$

Substituting (2.1260d) in (2.1259a) specifies *the total entropy production per unit time and unit volume due to coupled thermal conduction and mass diffusion (2.1261):*

$$\dot{S}_{qm} = \left[k - \bar{v}\,\beta + \beta T\left(\frac{\partial\bar{v}}{\partial T}\right)_{\xi,p}\right]\frac{\left|\nabla T\right|^{2}}{T^{2}} + \frac{\chi_{m}\,\rho}{T}\left(\frac{\partial\bar{v}}{\partial\xi}\right)_{T,p}\left|\nabla\xi\right|^{2}$$

$$+ \left[\alpha - \chi_{m}\,\rho\bar{v} + \chi_{m}\,\rho T\left(\frac{\partial\bar{v}}{\partial T}\right)_{\xi,p} + \beta T\left(\frac{\partial\bar{v}}{\partial\xi}\right)_{T,p}\right]\frac{\nabla T.\nabla\xi}{T^{2}}, \qquad (2.1261)$$

that must be positive (2.1262a) implying that: (i–ii) the coefficients of the square terms must be positive (2.1262a, b); (iii) the coefficients of the cross-term must vanish (2.1262d) thus relating the coupling coefficients (α, β) in (2.1258c, d):

$$\dot{S}_{qm} > 0: \quad k > \beta \left[\bar{v} - T \left(\frac{\partial \bar{v}}{\partial T} \right)_{\xi, p} \right], \quad \chi_m \left(\frac{\partial \bar{v}}{\partial \xi} \right)_{T, p} > 0, \qquad (2.1262\text{a–c})$$

$$\alpha + \beta T \left(\frac{\partial \bar{v}}{\partial \xi} \right)_{T, p} = \chi_m \, \rho \left[\bar{v} - T \left(\frac{\partial \bar{v}}{\partial T} \right)_{\xi, p} \right]. \qquad (2.1262\text{d})$$

In the case of decoupled heat and mass diffusion (2.1263a, b) the term (2.1262d) does not appear, and (2.1262b) [(2.1262c)] lead to positive thermal conductivity (2.1263c) ≡ (2.350d) [positive thermal diffusivity (2.1263e) ≡ (2.415j) if (2.1263d) ≡ (2.415i) is met]:

$$\alpha = 0 = \beta: \quad k > 0, \quad \left(\frac{\partial \bar{v}}{\partial \xi} \right) > 0 \Rightarrow \chi_m > 0. \qquad (2.1263\text{a–e})$$

The fundamental equations of the viscous, thermally conducting two-phase flow (Note 2.12) can be written explicitly (Note 2.15) using the diffusion relation for: (i) the viscous stresses of a Newtonian fluid (2.1212c); (ii–iii) the coupled heat (2.1258c) and mass (2.1258d) fluxes.

NOTE 2.15   VELOCITY, TEMPERATURE, PRESSURE, MASS DENSITY AND MASS FRACTION

The fundamental equations for the viscous thermally conducting flow of a two-phase fluid involve one vector (velocity) and four scalar (temperature, pressure, density and mass fraction) variables, that appear explicitly in one vector and four scalar equations, namely: (i) continuity (2.1211a) ≡ (2.1264a) for the total mass density; (ii) mass fraction (2.1264b) of one phase (2.1241b) using the mass flux (2.1258d); (iii) state (2.1214a) ≡ (2.1264c):

$$\frac{\partial \rho}{\partial t} + \nabla \cdot \left(\rho \, \bar{v} \right) = 0, \quad \rho \frac{d \xi}{dt} = \nabla \cdot \left(\rho \, \chi_m \nabla \xi + \beta \nabla T \right), \quad p = p(\rho, T); \quad (2.1264\text{a–c})$$

(iv) momentum for a Newtonian viscous fluid (2.1213a) ≡ (2.1265):

$$\rho \frac{dv_i}{dt} + \frac{\partial p}{\partial x_i} = \frac{\partial}{\partial x_j} \left[\zeta \left(\frac{\partial v_i}{\partial x_j} + \frac{\partial v_j}{\partial x_i} \right) \right] + \frac{\partial}{\partial x_i} \left[\left(\eta - \frac{2}{3} \zeta \right) \frac{\partial v_j}{\partial x_j} \right]; \qquad (2.1265)$$

(v) entropy (2.1227b–d) ≡ (2.1266a–c):

$$\frac{1}{\gamma-1}\left(\frac{dp}{dt}-\gamma RT\frac{d\rho}{dt}\right)=\rho c_p\frac{dT}{dt}-\frac{dp}{dt}=\rho c_V\frac{dT}{dt}-RT\frac{d\rho}{dt}$$

$$=\nabla.\left(k\nabla T+\alpha\nabla\xi\right)+\eta\left(\nabla.\bar{v}\right)^2+\frac{\zeta}{2}\left[\frac{\partial v_i}{\partial x_j}+\frac{\partial v_j}{\partial x_i}-\frac{2}{3}\delta_{ij}\left(\nabla.\bar{v}\right)\right]^2,$$

(2.1266a–c)

for a perfect gas (2.1229a–h) with heat flux (2.1258c). The shear ζ and bulk η viscosities, the thermal conductivity, the mass diffusivity χ_m and thermal-mass coupling coefficients (α,β) need not be constant and are related by (2.1262d). The fundamental equations of a viscous thermally conducting two-phase flow of a perfect gas (Note 2.15) are considered next (Note 2.16) in one-dimensional cases, involving only one independent variable, either time or a spatial coordinate.

NOTE 2.16   STEADY ONE-DIMENSIONAL FLOW WITH TRIPLE DIFFUSION

The simplest solution with only time as independent variable (2.1267a, b) would lead to the trivial case (2.1267c–g) of all dependent variables constant (2.1267), namely density, mass fraction, velocity, temperature and pressure:

$$\frac{\partial}{\partial x_i}=0\neq\frac{\partial}{\partial t}:\quad 0=\frac{\partial\rho}{\partial t}=\frac{\partial\xi}{\partial t}=\frac{\partial\bar{v}}{\partial t}=\frac{\partial T}{\partial t}=\frac{\partial p}{\partial t}$$

(2.1267a–g)

$$\Rightarrow\left\{\rho,\xi,\bar{v},T,p\right\}=\text{constant}.$$

(2.1267h)

The simplest non-trivial solution of the fundamental equations for a viscous thermally conducting two-phase flow is steady (2.1268a) one-dimensional (2.1268b)

$$\frac{\partial}{\partial t}=0:\quad\left\{\rho,\xi,\bar{v},T,p\right\}=\left\{\rho(x),\xi(x),\bar{e}_x\,v(x),T(x),p(x)\right\},$$

(2.1268a, b)

in which case the material derivative (2.1212b) $\equiv$ (2.1269a) simplifies to (2.1269b, c) where prime denotes (2.1270d) derivative with regard to x:

$$\frac{d\rho}{dt}\equiv\frac{\partial\rho}{\partial t}+\left(\bar{v}.\nabla\right)\rho\rightarrow\frac{d\rho}{dt}=v\frac{d\rho}{dx}=v\rho',\quad\rho'\equiv\frac{d\rho}{dx}.$$

(2.1269a–d)

The case of constant diffusion parameters (2.1270a) leads to five scalar equations, namely: (i) continuity (2.1264a) $\equiv$ (2.1270b); (ii) mass fraction (2.1264b) $\equiv$ (2.1270c);

$$\left\{k,\alpha,\beta,\zeta,\eta\right\}=\text{const}:\quad\left(\rho v\right)'=0,\quad\rho v\xi'=\left(\rho\chi_m\xi'\right)'+\beta T'';$$

(2.1270a–c)

*(iii) viscous momentum (2.1265) ≡ (2.1271b) with **total viscosity** (2.1271a):*

$$\bar{\zeta} \equiv \eta + \frac{4}{3}\zeta : \quad \rho v v' + p' = \bar{\zeta} v'', \quad p = \rho \bar{\bar{R}} T; \tag{2.1271a–c}$$

(iv) state for a perfect gas (2.1214b) ≡ (2.1271c); (v) entropy (2.1266b) ≡ (2.1272b):

$$\bar{\bar{\zeta}} = \eta + \frac{8}{9}\zeta : \quad v\left(\rho c_V T' - RT \rho'\right) = kT'' + \alpha \xi'' + \bar{\bar{\zeta}}\left(v'\right)^2, \tag{2.1272a, b}$$

with **modified total viscosity** *(2.1272a).* This system of equations (Note 2.16) can be solved (Notes 2.18–2.19) in linearized form (Note 2.17).

NOTE 2.17   LINEARIZED FLOW OF PERFECT GAS WITH VISCOUS, THERMAL AND MASS DIFFUSION

The fundamental equations for the steady (2.1268a) one-dimensional (2.1268b) viscous, thermally conducting two-phase flow of a perfect gas (2.1271c) with constant diffusion coefficients (2.1270a) can be simplified as follows: (i) the equation of continuity (2.1270b) implies that the mass flux is constant (2.1273a):

$$\rho(x)v(x) \equiv j = const : \quad jv' + p' = \bar{\zeta} v''; \tag{2.1273a, b}$$

(ii) substitution of (2.1273a) linearizes the momentum equation (2.1271b) ≡ (2.1273b); (iii) in the mass fraction equation (2.1270c) is made the approximation (2.1274a) of constant **mass diffusion factor** leading to (2.1274b):

$$\chi_m \rho \equiv \psi = const : \quad j\xi' = \psi \xi'' + \beta T''; \tag{2.1274a, b}$$

(iv) the entropy equation (2.1272b) ≡ (2.1275):

$$j c_V T' - RT v \rho' = kT'' + \alpha \xi'' + \bar{\bar{\zeta}}\left(v'\right)^2, \tag{2.1275}$$

is linearized (2.1276c) with the approximations (2.1276a, b)

$$\bar{\bar{\zeta}}\left(v'\right)^2 \ll kT'' \gg RT v \rho' : \quad c_V jT' = kT'' + \alpha \xi''. \tag{2.1276a–c}$$

The equations of mass fraction (2.1274b) and linearized entropy (2.1276c) are coupled, but decoupled from the momentum equation (2.1273b) leading to the system (2.605):

$$0 = \begin{bmatrix} c_V \, j \dfrac{d}{dx} - k \dfrac{d^2}{dx^2} & -\alpha \dfrac{d^2}{dx^2} \\ -\beta \dfrac{d^2}{dx^2} & j \dfrac{d}{dx} - \psi \dfrac{d^2}{dx^2} \end{bmatrix} \begin{bmatrix} T(x) \\ \xi(x) \end{bmatrix}. \tag{2.1277}$$

The determinant of the matrix in (2.1277) specifies the differential operator (2.1278a,b):

$$0 = \left\{ \left(j c_V \dfrac{d}{dx} - k \dfrac{d^2}{dx^2} \right) \left(j \dfrac{d}{dx} - \psi \dfrac{d^2}{dx^2} \right) - \alpha \beta \dfrac{d^4}{dx^4} \right\} T, \xi(x)$$

$$= \left\{ (k\psi - \alpha\beta) \dfrac{d^4}{dx^4} - j(k + c_V \psi) \dfrac{d^3}{dx^3} + j^2 \, c_V \dfrac{d^2}{dx^2} \right\} T, \xi(x), \tag{2.1278a, b}$$

that applies both to the temperature and mass fraction. The solution of (2.1278b) specifies the temperature (Note 2.18) and substitution in (2.1276c) specifies the mass fraction (Note 2.19).

NOTE 2.18   TEMPERATURE DUE TO COUPLED THERMAL AND MASS DIFFUSION

The temperature satisfies the linear fourth-order ordinary differential equation (2.1278b) with constant coefficients, that is actually of the second-order (2.1279b) for the dependent variable (2.1279a):

$$\theta \equiv T'': \quad a\theta'' - j(k + c_V \psi)\theta' + j^2 \, c_V \theta = 0, \tag{2.1279a, b}$$

where the coefficient (2.1280a) of the highest order derivative is the determinant of (2.1258c, d) $\equiv$ (2.1280b) the heat and mass fluxes as linear functions of the temperature and mass fraction gradients:

$$a \equiv k\psi - \alpha\beta: \quad \begin{bmatrix} \vec{G}^q \\ \vec{I} \end{bmatrix} = - \begin{bmatrix} k & \alpha \\ \beta & \psi \end{bmatrix} \begin{bmatrix} \nabla T \\ \nabla \xi \end{bmatrix}, \tag{2.1280a, b}$$

where was used (2.1274a). The particular integrals of a linear unforced ordinary differential equation with constant coefficients (Section IV.1.3) are exponentials (IV.1.56a) $\equiv$ (2.1281a) and substitution in (2.1279b) leads to a characteristic polynomial (2.1281b) of the second-degree:

$$\theta(x) = e^{\lambda x}: \quad 0 = a\lambda^2 - j(k + c_V \psi)\lambda + j^2 \, c_V = a(\lambda - \lambda_+)(\lambda - \lambda_-), \tag{2.1281a–c}$$

with roots (2.1281c) = (2.1282a):

$$2a\lambda_{\pm} = j\left(k+c_V\psi\right) \pm j\left|\left(k+c_V\psi\right)^2 - 4c_V a\right|^{1/2}. \qquad (2.1282\text{a})$$

Substitution of (2.1280a) in (2.1282a) specifies the roots (2.1282b,e)

$$2\frac{k\psi - \alpha\beta}{j}\lambda_{\pm} = k+c_V\psi \pm \left|\left(k+c_V\psi\right)^2 - 4c_V\left(k\psi - \alpha\beta\right)\right|^{1/2}$$

$$= k+c_V\psi \pm \left|\left(k-c_V\psi\right)^2 + 4c_V\alpha\beta\right|^{1/2}. \qquad (2.1282\text{b, c})$$

From (2.1282c) follows that the roots $\lambda_{\pm}$ are real for positive product of the cross-coupling coefficients (2.1283a). Also the latter product is less than the product of the diagonal diffusion coefficients in (2.1280b), namely the thermal conductivity and mass diffusivity if the determinant (2.1280a) is positive $a > 0$ implying (2.1283b). From (2.1282b) follows that the roots are positive (2.1283c) with the ordering (2.1283d)

$$0 < \alpha\beta < k\psi: \quad 0 < \lambda_- < \lambda_+, \quad \lambda_+ + \lambda_- = j\frac{k+c_V\psi}{k\psi - \alpha\beta}, \qquad (2.1283\text{a–e})$$

and with sum (2.1283e). The solutions (2.1281a) are finite in the range (2.1284a) and their linear combination specifies the general integral (2.1284b):

$$-\infty < x \le 0: \quad \theta(x) = A_+\left(\lambda_+\right)^{-2}\exp\left(\lambda_+ x\right) + A_-\left(\lambda_-\right)^{-2}\exp\left(\lambda_- x\right) \equiv T''(x), \quad (2.1284\text{a–c})$$

where $A_{\pm}$ are arbitrary constants.

The constants in (2.1284b) were chosen so that a double integration (2.1279a) ≡ (2.1284c) leads to the temperature (2.1285b) in the range (2.1285a) ≡ (2.1284a):

$$-\infty < x \le 0 \quad T(x) = B_- + B_+ x + A_+\exp\left(\lambda_+ x\right) + A_-\exp\left(\lambda_- x\right), \quad (2.1285\text{a, b})$$

where $B_{\pm}$ are arbitrary constants. From (2.1283c, d) follows that the last two terms on the r.h.s. of (2.1285b) vanish as $\to -\infty$, and thus the temperature at $-\infty$ is finite (2.1286a) if the constant (2.1286b) vanishes; in this case the constant (2.1286c, d) coincides with the asymptotic temperature

$$T(-\infty) < \infty \Rightarrow B_+ = 0, \quad T_\infty \equiv \lim_{x\to-\infty} T(x) = B_-. \qquad (2.1286\text{a–d})$$

This determines two arbitrary constants (2.1286b, d) in the temperature profile (2.1285a,b)≡ (2.1287a,b):

$$-\infty < x \le 0: \quad T(x) = T_\infty + A_+\exp\left(\lambda_+ x\right) + A_-\exp\left(\lambda_- x\right). \qquad (2.1287\text{a, b})$$

The remaining two constants $A_\pm$ are determined from the temperature (2.1288a, b) and its gradient (2.1288c, d) at the boundary:

$$T_0 \equiv T(0) = T_\infty + A_+ + A_-, \quad T_0' = \lim_{x \to 0} \frac{dT}{dx} = \lambda_+ A_+ + \lambda_- A_-. \quad \text{(2.1288a–d)}$$

The system (2.1288b, d) may be solved for the constants $A_\pm$ leading to (2.1289a)

$$\{(T_0 - T_\infty)\lambda_+ - T_0', T_0' - \lambda_-(T_0 - T_\infty)\} = (\lambda_+ - \lambda_-)\{A_-, A_+\}$$

$$= j \frac{\left|(k - c_V \psi)^2 + 4 c_V \alpha \beta\right|^{1/2}}{k\psi - \alpha\beta} \{A_-, A_+\}, \quad \text{(2.1289a, b)}$$

where (2.1282c) was used in (2.1289b). Assuming (2.1290a) from (2.1283d) follows (2.1290b) and hence the constants $A_\pm$ have opposite signs (2.1290c, d) with larger modulus (2.1290e) for the positive constant:

$$(T_0 - T_\infty)\lambda_- > T_0' > 0: \quad (T_0 - T_\infty)\lambda_+ > T_0', \quad A_- > 0 > A_+, \quad \text{(2.1290a–d)}$$

$$|A_-| > |A_+|, \quad T_0 > T_\infty, \quad \text{(2.1290e, f)}$$

using (2.1290c–e) in (2.1288a) follows (2.1290f) that the boundary temperature is higher than the asymptotic temperature (2.1286c).

From (2.1287b) the temperature gradient (2.621a–c):

$$T'(x) = A_+ \lambda_+ \exp(\lambda_+ x) + A_- \lambda_- \exp(\lambda_- x), \quad \text{(2.1291)}$$

vanishes (2.1292a) at the point (2.1292b), or (2.1292c) using (2.1290c, d):

$$T'(x_1) = 0: \quad (\lambda_+ - \lambda_-)x_1 = \log\left(-\frac{\lambda_- A_-}{\lambda_+ A_+}\right) = \log\left(\frac{\lambda_- A_-}{\lambda_+ |A_+|}\right) > 0, \quad \text{(2.1292a–e)}$$

since from (2.1283c) and (2.1290e) there is no conclusion on the sign of (2.1292d). If (2.1292d) is positive (2.1292e) the temperature (2.1287b) has no stationary point in the range (2.1287a) and decays (2.1290f) monotonically from the wall to infinity (Figure 2.53). Thus *for the steady (2.1268a) one-dimensional (2.1268b) linearized (2.1276a, b) two-phase flow of a viscous, thermally conducting fluid, with the approximation (2.1274a), the temperature satisfies the linear unforced fourth-order ordinary differential equation with constant coefficients (2.1278a,b), whose solution, finite in the range (2.1287a), is (2.1287b) where: (i) the exponents are given by (2.1282b,c); (ii) the coefficients (2.1289a,b) are determined by the asymptotic temperature (2.1286c) and the temperature (2.1288a) and temperature gradient (2.1288c) at the*

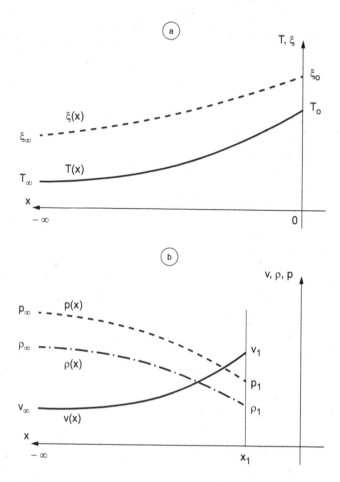

FIGURE 2.53 The steady one-dimensional linearized thermally conducting two-phase flow of a Newtonian viscous flow with coupled heat and mass diffusion leads to: (a) temperature and mass fraction decaying from the wall to constant asymptotic values; (b) in the far-field of temperature close to the asymptotic value to velocity decaying and pressure and mass density increasing to constant asymptotic values.

boundary. If the condition (2.1290a) is met the wall temperature is higher than the asymptotic temperature (2.1290f), and if the condition (2.1292e) is met the decay is monotonic (Figure 2.53). The temperature profile (Note 2.18) can be used to determine the mass fraction profile (Note 2.19).

NOTE 2.19   COUPLING OF TEMPERATURE AND MASS FRACTION DUE TO DOUBLE DIFFUSION

The mass fraction is specified by (2.1276c) ≡ (2.1293a) in terms of the temperature (2.1287b) leading to (2.1293b):

$$\alpha \, \xi'' = j \, c_V \, T' - k \, T'' = \lambda_+ \left(c_V \, j - \lambda_+ \, k \right) A_+ \exp\left(\lambda_+ x \right)$$
$$+ \lambda_- \left(c_V \, j - \lambda_- \, k \right) A_- \exp\left(\lambda_- x \right). \qquad (2.1293a, b)$$

A double integration leads to (2.624a–b):

$$\xi(x) = \left(\frac{c_V \, j}{\lambda_+ \alpha} - \frac{k}{\alpha} \right) A_+ \exp\left(\lambda_+ x \right)$$
$$+ \left(\frac{c_V \, j}{\lambda_- \alpha} - \frac{k}{\alpha} \right) A_- \exp\left(\lambda_- x \right) + C_+ \, x + C_-, \qquad (2.1294)$$

where appear two new constants of integration $C_\pm$. From (2.1283c, d) follows that the first two terms on the r.h.s. of (2.1294) are finite in the range (2.1287a) and thus a finite asymptotic mass fraction (2.1295a) implies the vanishing of one constant (2.1295b); the other constant (2.1295d) is specified by the asymptotic mass fraction (2.1295c):

$$\xi(-\infty) < \infty : \quad C_+ = 0, \quad \xi_\infty \equiv \lim_{x \to -\infty} \xi(x) = C_-. \qquad (2.1295a\text{–}d)$$

Substituting (2.1295b, d) in (2.624a–b) specifies the mass fraction (2.1296b) with coefficients (2.1296a):

$$E_\pm = \left(\frac{c_V \, j}{\lambda_\pm \alpha} - \frac{k}{\alpha} \right) A_\pm : \quad \xi(x) = \xi_\infty + E_+ \exp\left(\lambda_+ x \right) + E_- \exp\left(\lambda_- x \right). \qquad (2.1296a, b)$$

If the condition (2.1297a) is met, from (2.1283d) it implies (2.1297c) and together with (2.1290a) $\equiv$ (2.1297b) implies (2.1297d, e):

$$c_V \, j > \lambda_+ \, k, \quad \left(T_0 - T_\infty \right) \lambda_- > T_0' : \quad c_V \, j > \lambda_- \, k, \quad E_- > 0 > E_+. \qquad (2.1297a\text{–}e)$$

From (2.1283d) follows (2.1297f), and (2.1290e; 2.1296a) imply (2.1297g); substitution of (2.1297g) in (2.1296b) with $x = 0$ shows that the mass fraction is larger at the wall than asymptotically at infinity (2.1297h):

$$c_V \, j - \lambda_- \, k > c_V \, j - \lambda_+ \, k :$$

$$|E_-| > |E_+|, \quad \xi_0 > \xi_\infty, \quad \log\left(\frac{\lambda_-}{\lambda_+} \frac{E_-}{|E_+|} \right) > 0. \qquad (2.1297f\text{–}i)$$

From (2.1283d; 2.1297g) follows that the sign of (2.1297i) is not fixed; if the sign is positive, as in (2.1292e), then $\xi(x)$ does not have a stationary point for $x < 0$ and thus the decay of the mass fraction from the wall to infinity is monotonic (Figure 2.53). Thus *in the same conditions as the temperature (2.1287a,b) the mass fraction is given by (2.1296b) were: (i) the exponents are given by (2.1282b,c)); (ii) the coefficients (2.1296a; 2.1289a,b) involve the asymptotic temperature (2.1286c) and the temperature (2.1288a) and its gradient (2.1288c) at the boundary; (iii) the mass fraction can be chosen only at one point, for example, asymptotically (2.1295c); (iv) if the conditions (2.1297a, b) are met the mass fraction is higher at the wall than asymptotically at infinity (2.1297h) and if the condition (2.1297c) is met the decay is monotonic (Figure 2.53).* Having determined the temperature (Note 2.18) and mass fraction (Note 2.19) the solution for the viscous thermally conducting two-phase flow is completed (Note 2.20) by the velocity, mass density and pressure.

NOTE 2.20   VELOCITY, MASS DENSITY AND PRESSURE IN A FLOW WITH TRIPLE DIFFUSION

The velocity is specified by the viscous momentum equation (2.1273b) that has a first integral (2.1298a) with constant (2.1298b) specified by the pressure (2.1298c), velocity (2.1298d) and dilatation (2.1298e) at the boundary:

$$j v(x) + p(x) - \bar{\zeta} v'(x) \equiv b j = v_0 j + p_0 - \bar{\zeta} v_0',$$

$$\{ p_0, v_0, v_0' \} \equiv \left\{ p(0), v(0), \lim_{x \to 0} \frac{dv}{dx} \right\}.$$
(2.1298a, b)
(2.1298c–e)

The pressure is related to: (i) the density by the equation of state (2.1271c) ≡ (2.1299a) for an ideal gas; (ii) to the velocity (2.1299b) by the conservation of the mass flux (2.1273a):

$$p(x) = \bar{R} T(x) \rho(x) = \frac{\bar{R} j T(x)}{v(x)} \qquad v + \frac{\bar{R} T}{v} - \frac{\bar{\zeta}}{j} v' = b,$$
(2.1299a–c)

leading to the first-order non-linear differential equation (2.1299c) for the velocity with the temperature (2.1287b) in a coefficient, corresponding to a Riccati equation (Section IV.3.5). A simpler solution is obtained asymptotically for large distance (2.1300a, b) when the temperature (2.1287b) is approximately constant and equal to the asymptotic temperature (2.1300b) and the differential equation for the velocity (2.1299c) simplifies to (2.1300c):

$$x < x_1 \equiv - \max \left[(\lambda_{\pm})^{-1} \right]: \quad T(x) = T_\infty; \quad \frac{\bar{\zeta}}{j} \frac{dv}{dx} = \frac{v^2 - b v + \bar{R} T_\infty}{v}.$$
(2.1300a–c)

All terms in the numerator of the r.h.s. of (2.1300c) have the dimensions of velocity squared, because: (i) in the second term b has the dimensions of velocity (2.1298b); (ii) in the last terms appears (2.1233a) $\equiv$ (2.1301a) the square of the asymptotic isothermal sound speed leading to (2.1301b):

$$\overline{RT_\infty} = \left(c_{t\infty}\right)^2 \quad : \frac{\overline{\zeta}}{j} \frac{dv}{dx} = \frac{v^2 - bv + \left(c_{t\infty}\right)^2}{v}. \qquad (2.1301\text{a, b})$$

The numerator in (2.1301b) is quadratic in the velocity (2.633a–d) with roots (2.634a–d):

$$v^2 - bv + \left(c_{t\infty}\right)^2 = \left(v - v_+\right)\left(v - v_-\right), \qquad (2.1302\text{a})$$

$$2v_\pm = b \pm \left|b^2 - 4\left(c_{t\infty}\right)^2\right|^{1/2}. \qquad (2.1302\text{b})$$

If the conditions (2.1303a, b) are met:

$$b \equiv v_0 + \left(p_0 - \overline{\zeta}\, v_0'\right)/j > 0,$$

$$b^2 = \left[v_0 + \left(p_0 - \overline{\zeta}\, v_0'\right)/j\right]^2 > 4\left(c_{t\infty}\right)^2: \quad v_+ > v_- > 0, \qquad (2.1303\text{a–d})$$

the roots (2.12302b) are real and positive and ordered by (2.1303c, d). The differential equation for the velocity (2.1301b) can be written in dimensionless form (2.1304a–c):

$$\frac{j}{\overline{\zeta}}\, dx = \frac{v\, dv}{v^2 - bv + \left(c_{t\infty}\right)^2} = \frac{v\, dv}{\left(v - v_+\right)\left(v - v_-\right)}$$

$$= \left(\frac{v_+}{v - v_+} - \frac{v_-}{v - v_-}\right)\frac{dv}{v_+ - v_-}. \qquad (2.1304\text{a–c})$$

The separable non-linear first-order differential equation (2.1304c) is integrated in the far-field from the position (2.1300a) with velocity (2.1305a) to an arbitrary point x with velocity v leading to (2.1305b):

$$v_1 \equiv v\left(x_1\right); \quad \frac{j}{\overline{\zeta}}\left(x - x_1\right)$$

$$= \frac{v_+}{v_+ - v_-} \log\left(\frac{v - v_+}{v_1 - v_+}\right) - \frac{v_-}{v_+ - v_-} \log\left(\frac{v - v_-}{v_1 - v_-}\right), \qquad (2.1305\text{a, b})$$

that can be rearranged:

$$\exp\left[\frac{j}{\varsigma}(x-x_1)\right]=\left(\frac{v-v_+}{v_1-v_+}\right)^{v_+/(v_+-v_-)}\times\left(\frac{v_1-v_-}{v-v_-}\right)^{v_-/(v_+-v_-)}. \tag{2.1306}$$

If $x \to x_1$ the l.h.s. of (2.1306) and both factors on the r.h.s. of (2.638a–d) are unity, confirming (2.1305a). In the asymptotic limit $x \to -\infty$ the l.h.s. of (2.1306) is zero and the r.h.s. vanishes if the first factor is zero showing that v_+ is the asymptotic velocity (2.1307a, b):

$$\lim_{x\to-\infty} v(x)=v_+\equiv v_\infty; \quad \lim_{x\to\infty} v(x)=v_-<v_+ \tag{2.1307a–d}$$

if $v = v_-$ then the r.h.s. of (2.638a–d) is $+\infty$ implying $x \to \infty$ that is (2.1307c), satisfying (2.1303c) $\equiv$ (2.1307d). The limit (2.1307c) lies outside the ranges (2.1300a) and (2.1287a) and would lead by (2.1283c, d) to infinite temperature (2.1287b) and mass fraction (2.1296a,b). It is shown next that the velocity is larger than the asymptotic velocity (2.1308a) implying (2.1308b) from (2.1304b), so that moving away from the boundary (2.1308c) the velocity decreases (2.1308d), in agreement with (2.1308a):

$$v(x)>v_+ \Rightarrow \frac{dv}{dx}>0: \quad dx<0 \Rightarrow dv<0. \tag{2.1308a–d}$$

It can be shown that other assumptions would lead to contradictions between (A.a) and (A.d):

$$v(x)<v_- \Rightarrow \frac{dv}{dx}>0: \quad dx<0, \quad dv<0, \tag{A.a–d}$$

$$v_-<v(x)<v_+ \Rightarrow \frac{dv}{dx}<0: \quad dx<0, \quad dv>0, \tag{B.a–d}$$

and also (B.a) and (B.d). Thus *in the same conditions as for the temperature (2.1287a,b; 2.1282b, 2.1289a,b) and mass fraction (2.1296a,b) the velocity is given by (2.1306) in the far-field (2.1300a) and decays monotonically (2.1308d) to the asymptotic value (2.1307a, b). As the velocity decreases (Figure 2.53) the mass density (2.1299a) increases as well as (2.1270c) the pressure (2.1309b):*

$$\rho(x)=\frac{j}{v(x)}, \quad p(x)=\frac{\bar{\bar{R}}T_\infty j}{v(x)}. \tag{2.1309a, b}$$

As a final remark (Note 2.21) the six physical diffusion mechanisms and their interactions (chapter) are listed (Table 2.27).

TABLE 2.27

Distinct Physical Diffusion and Dissipation Mechanisms and their Interactions

Letter	Mechanism	Section	Interaction
A	heat conduction	2.4.1–2.4.2	B,D
B	mass diffusion	2.4.18	A
AB	heat & mass	N2.13–N2.14	-
C	viscosity	2.4.9–2.4.13	-
AC	viscous & thermal	N2.4–N2.9	-
ABC	viscous & thermal & mass	N2.10–N2.20	-
D	electric (Ohm)	2.4.3–2.4.4	A
F	electromagnetic (Hall)	2.4.14–2.4.15	-
AD	thermoelectric (Ohm)	2.4.5–2.4.8	-
AD	thermoelectric (Thomson)	2.4.16	-
AD	thermoelectric (Peltier)	2.4.17	
ADF	piezoelectromagnetism	2.4.19–2.4.22	

G – thermal radiation

Note: Five main diffusion mechanisms have been considered in Chapter 2, namely: (A) heat conduction; (B) mass diffusion; (C) viscous stresses; (D/E) Ohm/Hall electric currents. In addition, five diffusion couplings: (AB/AC) heat conduction with mass diffusion (viscous stresses); (AD) thermoelectric Ohm/Thomson/Peltier effect. Finally, two triple diffusions: (ABC) thermal and mass diffusion in a viscous fluid; (ADF) piezoelectromagnetism with heat conduction in an elastic solid.

NOTE 2.21   SIX PHYSICAL DIFFUSION MECHANISMS

Table 2.27 lists five basic diffusion mechanisms: (A) heat conduction, specified by Fourier's law (subsections 2.4.1–2.4.2); (B) mass diffusion specified by Fick's law (subsection 2.4.18); (C) viscous stresses for a Newtonian fluid (subsections 2.4.9–2.4.13); (D) electric current specified by Ohm's law in an electric field (subsections 2.4.3–2.4.4); (E) Hall current in an electromagnetic field (subsections 2.4.14–2.4.15). This leads to five couplings: (AB) coupled heat and mass diffusion (Notes 2.13); (AC) simultaneous but not coupled viscosity and heat conduction (Notes 2.5–2.9); (AD) three thermoelectric couplings, associated with Ohm/Thomson/Volta-Seebeck-Peltier effects (subsections 2.4.5–2.4.8/2.4.16/2.4.17). As triple interactions are considered: (ABC) the flow of a Newtonian viscous fluid with coupled heat and mass diffusion (Notes 2.10–2.20); (ADF) piezoelectromagnetism for an elastic solid slab with electric currents and thermal conduction (subections 2.4.19–2.4.22). A seventh dissipation mechanism is (G) thermal radiation at high temperatures, that may affect another irreversible process, namely shock waves (Sections 2.7–2.9). The basic physical diffusion mechanisms and some of their combinations in dissipative processes are illustrated in the Diagram 2.5 culminating in (a) solids [(b) fluids] with anisothermal piezoelectromagnetism (triple diffusion) combining elastic (viscous) stresses with thermal and electric (and mass) diffusion in the presence of electromagnetic (pressure) fields.

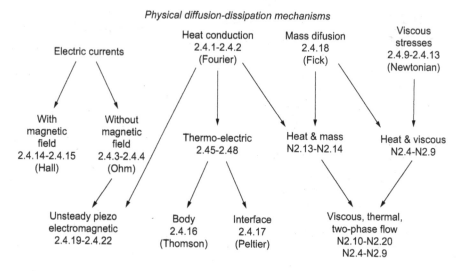

Physical diffusion-dissipation mechanisms

DIAGRAM 2.5 Four basic diffusion mechanisms, namely (i) electric currents, (ii) thermal conduction, (iii) mass diffusion and (iv) viscous stresses and several of their combinations culminating with (a) anisothermal piezoelectromagnetism in solids and (b) viscous thermally-conducting two-phase flow.

2.10 CONCLUSION

The first principle of thermodynamics (Diagram 2.1) states that the sum of the heat and the work equals the internal energy, that is a function of state, that is in an evolution depends only on the initial and final states, and not on the path between then. The work of the inertia force (gravity field) can be incorporated as the kinetic (gravity potential) energy in an augmented internal energy. The other forms of work are the product of intensive (extensive) parameters, namely: (i) electric displacement (field) for the electrical work; (ii) magnetic induction (field) for the magnetic work; (iii) stress (strain) tensor for the elastic work; (iv) chemical potential (mole number) for the chemical work. A particular isotropic case of the (iii) elastic work is the mechanical work (Figure 2.1) associated with the pressure (volume) as intensive (extensive) parameter. The heat is associated with the temperature (entropy) as the intensive (extensive) parameter. The extensive parameters are additive for two systems, and the intensive parameters are equal in equilibrium (Figures 2.2 and 2.3). The constitutive relations express each of the four intensive parameters as functions of all four extensive parameters with the constitutive tensors as the coefficients. The linear constitutive relations are symmetric, implying that there are ten linear constitutive tensors, namely four (six) relating each constitutive tensor to the corresponding (other) extensive parameter(s) [Table 2.1 (2.2)]. This leads to physical analogies (Table 2.3) among mechanics, electricity, magnetism and elasticity.

Whereas the first principle of thermodynamics concerns the equilibrium states of a system, the second principle of thermodynamics applies to the evolution between two states, and in the more (less) general form (Figure 2.5) requires the entropy (internal energy) to be maximum (minimum); this requires the thermodynamic surface of internal energy as a function of entropy and volume to be convex (Figure 2.5), and excludes saddle points on a concave surface (Figure 2.6). The second principle of thermodynamics leads to inequalities for the thermodynamic derivatives (Figure 2.4); there are twelve non-inverse thermodynamic derivatives (Table 2.5) which can be expressed in terms of only three of them (Table 2.4). The second principle of thermodynamics states that the entropy of an isolated system cannot decrease, and is constant (must increase) in a reversible (irreversible) process. The transfer of a gas between two reservoirs (Figure 2.7) with fixed (variable) volume separated by a wall with an orifice (a porous wall) are examples (a) [(b)] of irreversible thermodynamic processes. The first (second) principle of thermodynamics concerning equilibrium states (evolution between states) leads to constitutive (Tables 2.1–2.2) [diffusive (Tables 2.6–2.7)] relations between extensive and intensive thermodynamic parameters (fluxes and gradients), both with symmetries. The physical mechanisms of diffusion leading to energy dissipation include electric currents, heat conduction, viscous stresses and mass diffusion and their combinations (Table 2.27, Diagram 2.5, Figures 2.9–2.11 and 2.53).

The third principle of thermodynamics is unrelated to the first two and states (Figure 2.8) that in an isochoric (isobaric) process, that is at constant volume (pressure), as temperature tends to zero: (i) the entropy and specific heat at constant volume (pressure) scale linearly; (ii) the internal energy (enthalpy) scale quadratically. The three principles of thermodynamics apply to all substances (Diagram 2.2) such as solids, liquids and gases, and different substances are distinguished by their equations of state. A simple equation of state applies (Tables 2.9–2.11) to perfect (ideal) gases for which the specific heats at constant volume and pressure depend only on temperature (are constant); in both cases differ by a gas constant. A basic thermodynamic system involves only heat (work) with temperature (pressure) as the intensive parameter and entropy (volume) as the extensive parameter. Thus four particular thermodynamic processes are isothermal/isobaric/adiabatic/isochoric if, respectively, temperature/pressure/entropy/volume are kept constant. These thermodynamic processes can be combined in succession in a thermodynamic cycle in which the system returns to the initial state. The work (Figure 2.12) and heat are not functions of state, but their sum, the internal energy, is. Thus in a thermodynamic cycle heat and work can be exchanged keeping the internal energy constant. This is the basis for several engineering devices.

The Carnot cycle (Figure 2.13) consists of two adiabatics and two isothermals, and its efficiency (Table 2.12) can be calculated for: (i) the piston engine (Figure 2.13) that burns fuel to produce heat and extract work (Figure 2.15), for example, to power a car; (ii/iii) to a refrigerator (heat pump) using work [Figure 2.15] to extract (supply) heat from a cold (to a hot) body. A variation of the Carnot cycle (Figure 2.13) applied to the piston engine (Figure 2.14) that has one crank and one rod, is the Atkinson engine (Figure 2.16) with two cranks and rods; this is equivalent to variable valve timing in the piston engine, so that Atkinson cycle (Figure 2.17) shortens the compression (II) and extends the expansion (IV) relative (Figure

2.14) to the Carnot cycle, increasing the efficiency. Both the Carnot (Figure 2.13) and Atkinson (Figure 2.17) cycles apply air breathing engines, that cannot operate in a closed space, and in the case of a submarine would require a schnorkel for the inlet air and exhaust gases, limiting the operations to periscopic depth. For submarine operations in deep sea, air independent propulsion is needed, for example, the Stirling engine (Figure 2.19) and cycle (Figure 2.20). In the case of the Carnot/Atkinson/Stirling cycles (Figures 2.13/2.17/2.20) the gas velocities are small and can be neglected. This is not the case for jet propulsion, requiring the consideration of flows.

The simplest compressible flow (Table 2.13) is: (i) irrotational, that is without vortices; (ii) homentropic, that is with constant entropy; (iii) steady, that is independent of time. In this case all flow variables, namely the velocity, pressure, mass density, temperature and sound speed can be expressed in terms of any one of them. The piston engine (Figure 2.14) driving a propeller (Figure 2.21) is limited in the maximum speed by the velocity at the tip of the blades approaching the sound speed. This limitation is overcome by the turbojet engine (Figure 2.22) that can be used to reach supersonic speeds. The variations of the turbojet engine (Figure 2.22) include (Figure 2.23) the turboprop (turboshaft) used to power the propeller of an aircraft (rotor of an helicopter), and the turbofan (Figure 2.23) that uses two concentric shafts and a by-pass flow to decrease fuel consumption (Table 2.14) and increase thrust (Table 2.15). There are other types of engines (Figure 2.24) like the (a) ramjet and (b) pulse jet with fewer moving parts operating in the atmosphere, and (c) rocket and (d) ionic propulsion that can be used in the vacuum of space. A variety of other engines (Figure 2.25) use mechanical and electrical propulsion and hybrid combinations (Diagram 2.3). The turbojet (Figure 2.22) uses the Brayton cycle (Figure 2.26) that can be compared (Figure 2.27) with the Carnot (Figure 2.13) and Stirling (Figure 2.20). cycles.

The flow of a fluid (Diagram 2.4) can be continuous or have discontinuities (Table 2.18 and Figure 2.29) of three types: (a) with tangential velocity, that may be discontinuous, as well as other flow variables, except for the pressure that must be continuous; (b) normal for a normal shock, across which the velocity decreases, and pressure, and other flow variables increase; (c) oblique shock with continuous tangential velocity and decreasing normal velocity. Across a normal shock the flow changes discontinuously from supersonic to subsonic, with decreasing velocity, and increasing pressure, mass density, temperature, sound speed, internal energy and enthalpy (Figure 2.28). A normal shock is an irreversible process (Figure 2.30), implying an increase in entropy (Table 2.19) and a pressure ratio larger than for an adiabatic flow (Figure 2.33). The speed of advance of the normal shock front is the arithmetic mean of upstream and downstream velocities (Figure 2.31) and exceeds the geometric mean (Figure 2.32), that equals the critical sound speed, and is conserved across the shock. A sound source (Figure 2.34) generates spherical waves at rest (a), and as its velocity increases (b) forms a normal shock at the sound speed (c) and a Mach cone at supersonic speed (d). Locally the case (c) [(d)] is a normal (oblique) shock wave.

The distinction should be made (Figure 2.35) between (a) an oblique shock wave [(b) and adiabatic expansion turn] in a concave (convex) corner, across which the normal velocity decreases (increases), so that the process is thermodynamically irreversible (reversible) with entropy growth (constant entropy). The locus of the downstream velocity for an oblique shock wave (Figure 2.36) is a folio or strophoid (a), that becomes a circle (b), in the case of a strong shock (Figure 2.38) with a large upstream Mach number. An oblique shock (Figure 2.35) turns the velocity away from the normal, with the deflection angle (Table 2.20) being maximum for an intermediate angle of incidence (Figure 2.37). For a given angle of deflection there are two oblique shocks (Figure 2.36), the strong (weak) shock tending to a normal shock (adiabatic acoustic wave) when the deflection angle tends to zero; the strong and weak shock coincide at the maximum deflection angle. Conversely a shock wave with the same upstream Mach number can be compared with different incidence angles (Table 2.22) leading to distinct deflection angles. A shock wave can turn gradually along a smooth concave wall (Figure 2.39) and at a sharp convex corner an expansion front (Table 2.21) can form an expansion fan (Figure 2.39). A flow can be accelerated or decelerated in a nozzle continuously or form shock waves.

A nozzle is a duct of varying cross-section (Figure 2.41) whose walls can be replaced by streamlines (Figure 2.42). A subsonic flow, for example, incompressible that conserves the volume flux, is (a) accelerated in a converging nozzle with converging streamlines, and is (b) decelerated in a diverging nozzle with diverging streamlines. For a supersonic flow the variations of mass density reverse the result, with (c) acceleration in a diverging nozzle with diverging streamlines and (d) deceleration in a converging nozzle with converging streamlines. Thus the acceleration of a flow from subsonic to supersonic requires a throated nozzle (Figure 2.40) with subsonic flow and converging streamlines (Figure 2.43) before the throat or smallest cross-section of the streamlines, where the flow is sonic, accelerating further to supersonic in the divergent section with diverging streamlines. The flow rate in a convergent-divergent nozzle is determined by the sonic condition at the throat, and can be varied changing the throat area or using perforated walls (Figure 2.46). Changing the exhaust (Figure 2.46) affects the exit pressure leading (Figures 2.44 and 2.45) to: (a) perfect expansion for pressure equal to ambient, and tangential shear layers issued from the lips; (b) underexpansion for pressure higher than ambient, and a diamond pattern of oblique shock waves in the exhaust jet; (c) overexpansion for pressure lower than ambient, implying that there must be a normal shock inside the duct at a location that ensures pressure match at the exit or a normal or oblique shock forms at the exit. Thus the type of nozzle exit or exhaust flow depends on ambient or atmospheric conditions.

In order to compare aircraft and engine performance an International Standard Atmosphere (I.S.A.) has been defined that assumes hydrostatic equilibrium (Figure 2.47) of a perfect gas under a uniform gravity field, with (Figure 2.48) two layers: (i) troposphere with temperature decreasing linearly from 15 C at sea level to –56.5 C at the tropopause at 11 km altitude; (ii) isothermal stratosphere at –56.5 C from the tropopause up to 25 km. The temperature profile allows the calculation of pressure, mass density and sound speed at all altitudes from the values at sea level (Table 2.23). The I.S.A. corresponds to the atmosphere of the Earth at mid latitude and mid season, and can be modified for higher or lower temperature or altitude. The use of I.S.A.

allows calculation of the various flow regimes in a convergent-divergent nozzle, including cases: (a) unchoked (choked) with subsonic (sonic) flow at the throat; (b) unshocked (shocked) with adiabatic flow throughout (normal shock in the divergent section). The nozzles can be used for turbojet (Table 2.24) and rocket (Table 2.25) engines, and can lead to the formation of an oblique shock (Table 2.26) at the exit. The shock waves form not only in nozzles but also as a bow wave in front of sharp or blunt bodies in supersonic flow (Figure 2.52).

Bibliography

The bibliography of the series "Mathematics and Physics for Science and Technology" is quite extensive since it covers a variety of subjects. In order to avoid overlaps, each volume contains only a part of the bibliography on the subjects most closely related to its content. The bibliography of earlier volumes is generally not repeated, and some of the bibliography may be relevant to earlier and future volumes. The bibliography covered in the four published volumes is:

A. General
   a. Overviews
      1. General mathematics: book 1
      2. Theoretical physics: book 2
      3. Engineering technology: book 3
   b. Reference
      4. Collected works: book 6
      5. Generic Encyclopaedias: book 10

B. Mathematics
   c. Theory of functions
      6. Real functions: book 1 and 2
      7. Complex analysis: book 1 and 2
      8. Generalized functions: book 3
   d. Differential and integral equations
      9. Ordinary differential equations: book 4
      10. Partial differential equations: book 10
      11. Non-linear differential and integral equations: book 5
   e. Geometry
      12. Tensor calculus – book 3
   f. Higher analysis
      13. Special functions: book 8

C. Physics
   g. Classical mechanics
      14. Material particles: book 7
   h. Thermodynamics
      15. Thermostatics: book 2 and 11
      16. Heat: books 2 and 11
   i. Fluid mechanics
      17. Hydrodynamics: book 1
      18. Aerodynamics: book 1
   j. Solid mechanics
      19. Elasticity: book 2
      20. Structures: book 2

This choice of subjects is explained next.

The general bibliography consists of overviews and reference works. The overviews have been completed with mathematics, physics, and engineering, respectively, in volumes I, II and III corresponding to books 1, 2 and 3. The reference bibliography starts with the collected works of notable authors in book 6 of volume IV and generic encyclopedias in book 10 of Volume V, that is the first book of this set. Concerning mathematics, the bibliography on the theory of functions has appeared in volumes I, II and III corresponding to books 1, 2 and 3. The bibliography on ordinary differential equations has appeared in book 4 and on non-linear differential equations in book 5, both in volume IV and partial differential equations in book 10 of volume V. Volume IV also contains in book 8 the bibliography on special functions. The bibliography on tensor calculus appears in book 3 that coincides with volume III. Concerning physics, the bibliography on material particles appears in book 7 of volume IV. The bibliography on thermodynamics and heat appears in book 2 that coincides with volume II. The bibliography on solid mechanics, including elasticity and structures, appears in book 2 that coincides with volume II. The bibliography on fluid mechanics, including hydrodynamics and aerodynamics appears in book 1 that coincides with volume I. The present book 11 in volume V includes next the bibliography on:

1. Thermodynamics
2. Heat

1. THERMODYNAMICS

Azevedo, E. G. *Termodinâmica aplicada*. Escolar Editora, Lisbon 1995, 2ª edição 2000.

Barrère, M. and Prud'Homme, R. *Aerothermochimie des ecoulement homogénes*. Gauthier-Villars, Paris 1970.

Candel, S. M. *Mécanique des fluides*. Dunod, Paris 1995.

Davies, J. T. and Rideal, E. K. *Interfacial phenomena*. Academic Press, New York 1961.

Denbigh, K. *The principles of chemical equilibrium*. Cambridge University Press, Cambridge, UK 1957.

Fourier, J. *Theorie analytique de le chaleur*. Translation inst. English 1978, Cambridge University Press, Cambridge, UK 1822.

Frank-Kamenetskii, D. A. *Diffusion and heat exchange in chemical kinetics*. Princeton University Press, Princeton, NJ 1955.

Hahn, H. G. *Methode der finiten Elemente in der Festigkeitsleher*. Akademische Verlag Gesellschaft, Frankfurt 1975.

Kestin, J. *Thermodynamics*. Blaisdell, Waltham, MA 1966.

Kuo, K. K. *Principles of combustion*. Wiley, New York 1986.

Leal, L. G. *Advanced transport phenomena*. Cambridge University Press, Cambridge, UK 2007.

Meyer, E. and Schiffner, E. *Technische Termodynamic*. VC17 Verlaggessellschaft, Weinheim 1986.

Pauling, L. *The nature of the chemical bond*. Cornell University Press, Ithaca, NY 1939, 3rd ed. 1960.

Philips, N. V. *Problems of low temperature physics and thermodynamics*. Pergamon, Oxford 1962.

Poisont, T. and Veynante, D. *Theoretical and numerical combustion*. R. T. Edwards, Philadelphia 2001.

Roberts, J. K. and Miller, A.R. *Heat and thermodynamics*. Blackie, London 1928, 3rd ed. 1940.

Zemansky, M. S. and Dittman, R. H. *Heat and thermodynamics*. McGraw-Hill, New York 1937, 6th ed. 1979.

2. HEAT

Becquerel, J. *Thermodynamique*. Hermann, Paris 1924.

Bird, R. B., Stewart, W. E. and Lightfoot, E. N. *Transport phenomena*. John Wiley, New York 1960.

Burgers, J. M. *The non-linear diffusion equation*. Reidel, Dordrecht 1974.

Carslaw, H. S. and Jaeger, H. S. *Heat conduction in solids*. Clarendon, Oxford, 1ST edition 1946, 2nd ed. 1959.

Chandrasekhar, S. *Radiative transfer*. Oxford University Press, London; New York 1950, reprinted Dover 1960.

Damaskin, B. and Petri, O. *Foundations of theoretical electrochemistry*. Mir Publishers, Moscow 1978.

Eckert, E. R. G. and Drake, R. M. *Heat and mass transfer*. McGraw-Hill, New York 1959.

Leontiev, A. *Theorie des échanges de chaleur et masse*. Nauka, Moscow 1979, Editions Mir 1985.

MacAdams, W. H. *Heat transmission*. McGraw-Hill, New York 1954.

Ozisik, M. N. *Boundary-value problems in heat conduction*. International Textbook Company, New York; Scranton, PA 1968, reprinted Dover 1989.

Planck, M. *The theory of heat radiation*. Berlin 1912, 2nd ed. Blakiston, London 1914, reprinted Dover 1959.

Shah, R. K. and Sekulic, D. P. *Fundamentals of heat exchanger design*. Wiley, New York 2003.

Slater, J. C. *Introduction to chemical physics*. McGraw-Hill, New York 1939.

References

Bernoulli, J. 1638. *Hydrodynamics*, Johannis Reinholdi Dulseckeri, Argentorati, Basel.

Boyle, R. 1660. *New Experiments Physico-Mechanical: Touching the Spring of the Air and their Effects Printed by H. Hall*, Printer to the University for T. Robinson.

Boyle, R. 1662. *Boyle papers*.

Mariotte, E. Lettres écrites par MM. Mariotte, Pecquet, et Perrault, sur le sujet d'une nouvelle découverte touchant la veüe faite par M. Mariotte, 1676.

Gay-Lussac, J. L. 1802. "Recherches sur la dilatation des gas et des vapeurs", *Chimie* **43**, 137–175.

Dalton, J. 1802a. "Essay II: On the force of steam or vapour from water and various other liquids, both in vacuum and in air", *Memoirs of the Literary and Philosophical Society of Manchester* **8**, 550–574.

Dalton, J. 1802b. "Essay IV: On the expansion of electric fluids by heat", *Memoirs of the Literary and Philosophical Society of Manchester* **8**, 595–602.

Stirling, R. 1816. "Improvements for diminishing the consumption of fuel and in particular an engine capable of being applied to the moving machinery on a principle entirely new", *English Patent 4081*.

Fourier, J. B. J. 1818. *Theorie analytique de le chaleur*, repr. Dover, Paris, 1955.

Navier, C. L. M. H. 1822. "Mémoires sur les Lois du Movement des Fluides", *Mémoires de l'Académie des Sciences de Paris* **6**, 389.

Clapeyron, B. P. W. 1834. "Memoire sur la puissance motrice de la chaleur", *Journal de l'École Royale Polytechnique* 14, fasc. 23, Paris.

Stokes, G. G. 1845. "On the Theories of Internal Friction of Fluids in Moton", *Transactions of the Cambridge Philosophical Society* **7**, 287 (Papers 1, 75).

Clausius, R. 1850. "Ueber die bewegende Kraft der Wärme und die Gesetze, welche sich darus für die Wärmelehre selbst ableiten lassen", *Annalen der Physik* **79**, 368–397.

Clausius, R. 1854. "Ueber eine veränderte Form der mechanischen Wärmetheorie", *Annalen der Physik* **93**, 481–506.

Carnot, S. 1878. *Reflexions sur la puissance du feu et sur les machines propres à developer cette puissance*, Gauthier-Villars, Paris.

Atkinson, J. 1882. "Gas engine", *U.S. Patent 367496*, issued 1887-08-02.

Thomson, J. J. 1882. "On an absolute thermometric scale founded on Carnot's theory", *Mathematical Papers* **1**, 302, Cambridge University Press.

Fick, A. 1885a. "Ueber diffusion", *Annalen der Physik* **94**, 59–86.

Fick, A. 1885b. "On liquid diffusion", *Philosophical Magazine* **10**, 30–39.

Tait, P. G. 1888. "Report on some of the physical properties of fresh water and sea water", *Physics and Chemistry* **2**, 1–76.

Lie, S., Engel, F. 1888. *Theorie der Transformation Gruppen*, 3 vols, Teubner, Leipzig 1888–1893

Diesel, R. 1913. *Einstellung der Dieselmotors*, Springer, Berlin.

Index

Pages in italics refer figures and pages in **bold** refer tables.